UNDERSTANDING ELEMENTARY CALCULUS

UNDERSTANDING ELEMENTARY CALCULUS

PRINCIPLES, PROBLEMS, AND SOLUTIONS

Charles Godino

Brooklyn College

W. A. BENJAMIN, INC.

New York   *Amsterdam*

1969

UNDERSTANDING ELEMENTARY CALCULUS

Principles, Problems, and Solutions

*This Manuscript was put into production on December 12, 1968;
this volume was published on August 30, 1969*

W. A. BENJAMIN, INC.

New York, New York 10016

To Emily, Lucy, and Frank

PREFACE

This volume covers material from analytic geometry, differential calculus, and integral calculus. It is designed to supplement textbooks that are commonly used in a first course in calculus. The terminology, notation, and treatment are modern, and they compare favorably with the material contained in more recent calculus texts. A student should have some background in plane geometry and trigonometry before attempting to use this supplement.

In compiling this book, we have tried to find a middle ground between a regular calculus textbook and a full-fledged problems book. Consequently, this supplement is more than a problems book but less than a regular textbook. Most problem books limit themselves to bare statements of definitions and theorems and concentrate on problems and their solutions. We have attempted to develop and discuss many of the more difficult topics covered in a first course in calculus. This approach should be an advantage to the average student, especially when the explanations complement or offer an alternative to a textbook discussion.

We have included a judicious sampling of worked-out examples with a generous number of diagrams. Often, we have tried to express a method of attack or a point of view rather than a simple recipe. An effort has been made to explain *why* a problem is done in a certain way and not just *how* it is done. Some of the examples are routine and will enhance the student's background. However, there are a great many that do not usually appear in calculus texts either because of their complexity or because of space limitations. Even some of the more simple examples will give a student an opportunity to learn a technique without becoming involved in a long, tedious calculation.

We have dispensed with many of the usual proofs and derivations found in calculus texts. However, we have included proofs and derivations that are not universally used, and in many instances we have tried to exhibit methods that augment the ones found in standard texts. In fact, we often give an alternate definition or proof when one exists and then we make a comparison between the two definitions or proofs.

Another important feature of this book is the collection of problems that follow each section. A mathematician is essentially a problem solver, and one learns to solve problems only by gaining experience in actually working them out on his own. There are simple problems intended to drill the student in fundamentals. There are more difficult and challenging ones designed to extend the student's knowledge and increase his skill in methematical reasoning. There are also some worthwhile theoretical problems that require the student to go beyond the level of the usual course material. The most difficult ones have been starred as a warning to the student. The answers to all problems are given in the back of the book since we feel that the student should have a quick check on his work.

Since there are many good calculus texts on the market, the student should never limit himself to merely the textbook used in class. The student should read other texts and thus broaden his background. In the bibliography, we have compiled a list of some of the books most commonly used as class texts and reference texts for a first course in calculus. We have also included a number of books that may help the student become more adept at problem solving.

Charles Godino

Brooklyn, New York
March 1969

ACKNOWLEDGMENTS

The author is indebted to the various consultants for their patience in reading the first draft; their suggestions and constructive criticism were instrumental in shaping the final form of this book. Thanks are also due to the staff of W. A. Benjamin, Inc., for the extra time they have spent on this project and to Jerry Kovacic, Rose Ranieri, Carl Cirillo, and Professor Frank Rush for their assistance in compiling the answers at the back of the book. Finally, the author wishes to express his gratitude to his wife, Emily, for her patient and tireless efforts in typing the first draft, the revisions, and the final copy.

NOTE TO THE STUDENT

We suggest that you use this book in the following way. Either just before or just after you cover a topic in class, refer to the section or sections in this book that cover similar material. You should then make a thorough reading of the section or sections, and pay particular attention to the worked-out examples. There will also be suggested exercises that appear periodically throughout each section, and these should be done before you go on since they are usually an integral part of the material being presented. Finally, the problems at the end of each section should be attempted—as many as time permits These problems are designed to help you to review the topics covered in the section and to test your ability to apply what you have learned to many related problems. Answers to *all* the problems are given in the back of the book.

CONTENTS

Part III **Integral Calculus**

Part I

ANALYTIC GEOMETRY

RECTANGULAR COORDINATES

We assume that you are familiar with the rectangular coordinate system in the plane using the horizontal x axis and the vertical y axis that intersect at a point called the origin and that divide the plane into four quadrants (see Figure 1.1).

Also we assume that you know that there is a one-to-one correspondence between points in the plane and ordered pairs of real numbers; that is, each point P has a unique pair of coordinates (a, b) associated with it, while each pair of real numbers (a, b) corresponds to a unique point P in the plane. The number a is called the x coordinate of P, and b is called the y coordinate of P (see Figure 1.2).

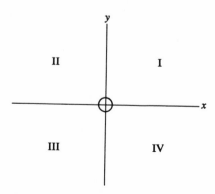

Figure 1.1

3

Finally, all points on the *x* axis have *y* coordinate zero, all points on the *y* axis have *x* coordinate zero, while the origin has coordinates (0, 0).

PROBLEMS

1.1   If a rectangle has endpoints of one diagonal given by (5, 2) and (9, 4), find the other two vertices.   Is there more than one answer?

1.2   If a rhombus (a parallelogram with all four sides having equal length) has one side determined by the segment between (1, 2) and (4, 6), find the other two vertices.   Is there more than one answer?

1.3   If a parallelogram has three vertices given by the points $A(2, 3)$, $B(3, 6)$, $C(6, 3)$, find the coordinates of the fourth vertex if (a) AC is a diagonal, (b) BC is a diagonal, (c) AB is a diagonal.

1.4   Find the exact points on the *x* axis corresponding to the fractions 7/9, 11/13, 17/7, and 3/14.   (*Hint:* For 7/9, draw a slant line from the origin (see Figure 1.3) through the first quadrant.   Measure nine equal segments from the origin along the slant line, connecting the endpoint of the last segment to the point (1, 0) by the line segment *L*.   Draw a line segment parallel to *L* from each of the equally spaced points on the slant line to the *x* axis.)

1.5   Find the exact points on the *x* axis corresponding to the irrational numbers $\sqrt{2}$, $\sqrt{3}$, and π.   (*Hint:* For $\sqrt{2}$, construct an isosceles right triangle with both legs 1 unit in length; for $\sqrt{3}$, construct an equilateral triangle with sides 1 unit in length, then draw an altitude; for π, construct a circle of radius 1 and take the length of the semicircle.)

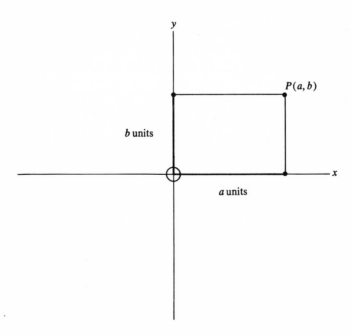

Figure 1.2

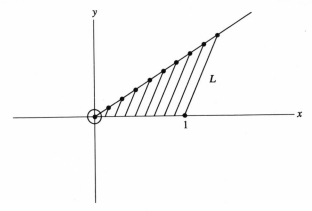

Figure 1.3

DISTANCE BETWEEN POINTS AND DIVISION OF A LINE SEGMENT

We assume that you know each of the following facts.

1. "The absolute value of a real number a," denoted by $|a|$, is a if a is nonnegative and $-a$ if a is negative. For example, $|2| = 2$, $|-5| = 5$, $|\sqrt{2}| = \sqrt{2}$, and $|-\pi| = \pi$. Observe also that $|a| = |-a|$ for every real number a.

2. If A and B are two points on the x axis (or y axis) with x coordinates (or y coordinates) a and b, respectively, then the distance between A and B is given by $|a - b|$ or $|b - a|$ (since $|a - b| = |-(a - b)| = |b - a|$). For example, the distance between the two points on the x axis with coordinates 2 and 5, respectively, is $|2 - 5| = |-3| = 3$, while the distance between the two points on the y axis with coordinates -2 and -3, respectively, is $|(-2) - (-3)| = |1| = 1$.

3. The distance between two points A and B in the plane having coordinates (a_1, a_2) and (b_1, b_2) respectively, is given by $(|a_1 - b_1|^2 + |a_2 - b_2|^2)^{1/2}$ or simply by $[(a_1 - b_1)^2 + (a_2 - b_2)^2]^{1/2}$, since $|a - b|^2 = (a - b)^2$ for any two real numbers a and b (see Figure 2.1).

4. The line segment joining A to B is divided into a ratio of m to n by the point C having coordinates (c_1, c_2), where $c_1 = a_1 + [m/(m + n)](b_1 - a_1)$ and $c_2 = a_2 + [m/(m + n)](b_2 - a_2)$ (see Figure 2.2).

5. From the preceding formula, the midpoint of the segment joining A to B is

$$\left(\frac{b_1 + a_1}{2}, \ \frac{b_2 + a_2}{2} \right)$$

6

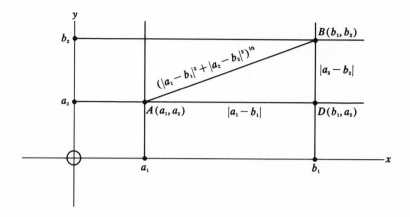

Figure 2.1

while the points of trisection are

$$\left(\frac{b_1 + 2a_1}{3}, \ \frac{b_2 + 2a_2}{3}\right)$$

and

$$\left(\frac{2b_1 + a_1}{3}, \ \frac{2b_2 + a_2}{3}\right)$$

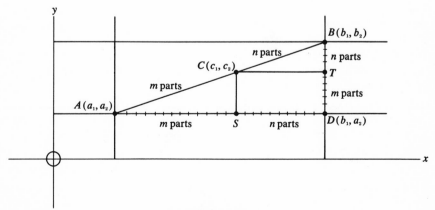

Figure 2.2

PROBLEMS

2.1 Determine which of the following triangles are isosceles, equilateral, or right triangles and find the area of each triangle.   The vertices of each triangle are the three given points.
  (a) $(1, 4)$ $(10, 6)$ $(2, 2)$
  (b) $(1, -1)$ $(1, 7)$ $'(8, 3)$
  (c) $(-1, 0)$ $(3, 0)$ $(1, 2\sqrt{3})$
  (d) $(1, -1)$ $(3, 0)$ $(2, 2)$
  (e) $(2, 3)$ $(4, -7)$ $(6, 2)$

2.2 Prove that for any real number $x > 1$, any triangle having sides of length $x^2 - 1$, $x^2 + 1$, and $2x$ is a right triangle.

2.3 Find the midpoint of the line segment having endpoints
  (a) $(3, -2)$ and $(5, 4)$
  (b) $(-1, -2)$ and $(7, 3)$

2.4 Find the points of trisection of the line segment having endpoints
  (a) $(4, -1)$ and $(7, 5)$
  (b) $(-5, 2)$ and $(8, -4)$

2.5 In each of the following, find the point that divides the given line segment into the given ratio.
  (a) the segment from $(-6, 2)$ to $(-1, -4)$, ratio 1 to 5
  (b) the segment from $(3, 7)$ to $(-2, 4)$, ratio 3 to 1
  (c) the segment from $(3, 6)$ to $(8, 11)$, ratio 2 to 3
  (d) the segment from (x_1, x_2) to (y_1, y_2), ratio r to s

THE LINE

Definition 3.1   Let α be the angle between a line L and the positive x axis, where α has its initial side on the positive x axis and its terminal side on L. Then α is called "the angle of inclination of L." For example, horizontal lines have angle of inclination $\alpha = 0°$; vertical lines have $\alpha = 90°$. In any event, $0° \leq \alpha < 180°$.

Definition 3.2   Let L be any nonvertical line having its angle of inclination equal to α. Then, the slope of L is defined as $\tan \alpha$. Note that all horizontal lines have slope zero while vertical lines have no defined slope (even though they have angles of inclination).

Theorem 3.1   If $A(a_1, a_2)$ and $B(b_1, b_2)$ are two points on a nonvertical line L, the slope of L is equal to $(b_2 - a_2)/(b_1 - a_1)$.

The proof follows from the fact that if α is the angle of inclination, then $\tan \alpha = (b_2 - a_2)/(b_1 - a_1)$ (see Figure 3.1).

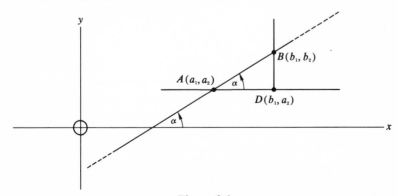

Figure 3.1

9

Corresponding to every nonvertical line L, there is an equation of the form $y = mx + b$. where m is the slope of the line L and b is the y coordinate of the point of intersection of L with the y axis [the point $(0, b)$ is called the y intercept of L]. This equation is called the *slope–intercept form* of the equation of a line. Since every point on a vertical line L has the same x coordinate, we can write the equation of a vertical line that intersects the x axis at the point $(k, 0)$ as $x = k$. There are many forms of the equation of a nonvertical line, but they can all be reduced to the $y = mx + b$ form. Some of these other forms are as follows.

1. The line through the point (x_1, y_1) and having slope m has equation $y - y_1 = m(x - x_1)$, which is called the *point–slope form*.

2. The line through the points (x_1, y_1) and (x_2, y_2) has equation $y - y_1 = [(y_1 - y_2)/(x_1 - x_2)](x - x_1)$, which is called the *two-point form*.

3. The line crossing the x axis at $(a, 0)$ and the y axis at $(0, b)$ has equation $(x/a) + (y/b) = 1$, which is simply called the *intercept form*.

4. The general linear equation $Ax + By + C = 0$ [see Problem 3.3(e)].

To find the equation of a slant line, we must know either two points on the line or one point and the slope. Depending on the information given, we use one of the preceding forms and ultimately reduce the equation to the slope–intercept form.

EXAMPLE 3.1   Find the equation of the line with slope 7 that intersects the y axis at $(0, 2)$.

SOLUTION $m = 7$, $b = 2$, using the slope–intercept form, we arrive at the equation $y = 7x + 2$.

EXAMPLE 3.2   Find the equation of the line with slope 2 that passes through the point $(3, -1)$.

SOLUTION $(x_1, y_1) = (3, -1)$, $m = 2$, using the point-slope form, we arrive at the equation $y - (-1) = 2(x - 3)$, which reduces to $y = 2x - 7$.

EXAMPLE 3.3   Find the equation of the line that passes through the points $(1, 7)$ and $2, 5)$.

SOLUTION $(x_1, y_1) = (1, 7)$, $(x_2, y_2) = (2, 5)$; using the two-point form, we arrive at the equation $y - 7 = [(7 - 5)/(1 - 2)](x - 1)$, which reduces to $y = -2x + 9$.

It is clear that *parallel* lines have equal angles of inclination and hence *equal* slopes. On the other hand, *perpendicular* lines have slopes that are *negative reciprocals* of each other. This fact is not so obvious; therefore we include a short proof.

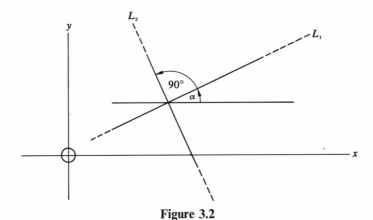

Figure 3.2

Theorem 3.2 Let L_1 with slope m_1 and L_2 with slope m_2 be two perpendicular slant lines. Then, $m_2 = -(1/m_1)$ or $m_1 m_2 = -1$.

PROOF Let α be the angle of inclination of L_1 (see Figure 3.2), then $90° + \alpha$ is the angle of inclination of L_2 and we have $m_1 = \tan \alpha$, $m_2 = \tan(90° + \alpha)$. But $\tan(90° + \alpha) = -\cot \alpha = -(1/\tan \alpha) = -(1/m_1)$ (see the Appendix). Therefore, $m_2 = -(1/m_1)$.

EXAMPLE 3.4 Let the line L have equation $y = 2x + 1$. Find the equation of the line passing through the point $(2, 1)$ that is parallel to L and of the line through $(2, 1)$ that is perpendicular to L.

SOLUTION The line parallel to L has slope 2 and passes through $(2, 1)$; therefore, the equation is $y - 1 = 2(x - 2)$, which reduces to $y = 2x - 3$.
The line perpendicular to L has slope $-\frac{1}{2}$ and passes through $(2, 1)$. Therefore, the equation is $y - 1 = -\frac{1}{2}(x - 2)$, which reduces to $y = -\frac{1}{2}x + 2$.

Thus far we have dealt with the problem of finding equations of lines; now we would like to look at an equation and be able to tell something about the graph of the line. For example, whenever we have an equation in the form $y = mx + b$, we know that the line has slope m, y intercept $(0, b)$, and x intercept $(-b/m, 0)$.

EXAMPLE 3.5 Find the slope and the intercepts of the line having equation $2x + y = 3$.

SOLUTION First we put the equation in the form $y = -2x + 3$. Hence, the slope is -2, the y intercept is $(0, 3)$, and the x intercept is $(\frac{3}{2}, 0)$. Sketch the graph as an exercise.

Figure 3.3   Figure 3.4

Finally, we discuss the problem of determining whether a given set of points are all on the same line. Clearly, any two distinct points will be on the line determined by them. For three distinct points A, B, and C, simply look at the slopes of the segments AB and AC (see Figures 3.3 and 3.4). If they are equal, the segments are parallel; since they have the point A in common, they must lie on the same line.

The problem of four or more distinct points is best solved by finding the equation of the line determined by the first two points and then substituting the remaining points into this equation. If they all satisfy the equation, all the points are collinear; if one point fails to satisfy the equation, the points are not collinear.

PROBLEMS

3.1   Determine which of the following sets of points are collinear.
   (a) $(-2, 0)$ $(2, -1)$ $(10, -3)$
   (b) $(0, 4)$ $(-3, 2)$, $(6, 7)$
   (c) $(-2, 0)$, $(2, -1)$, $(-6, 1)$, $(10, -4)$
   (d) $(0, 1)$, $(1, 4)$, $(-1, -2)$, $(2, 7)$, $(3, 10)$
3.2   Find the equation of each of the following lines.
   (a)   through $(2, -1)$ and $(-3, 2)$
   (b)   through $(0, 4)$ and $(-2, 0)$
   (c)   through $(0, b)$ and $(a, 0)$
   (d)   through $(2, 1)$ with slope 3
   (e)   through $(-1, 2)$ with slope -5
   (f)   through $(0, -2)$ with slope $\frac{1}{2}$
   (g)   through $(1, 3)$ and parallel to the line having equation $y = -3x + 2$
   (h)   through $(3, -2)$ and perpendicular to the line having equation
          $y = \frac{1}{2}x + 1$
3.3   In each of the following, find the slope and the intercepts, and sketch the graph of the given equation.
   (a) $x - 2y = -3$
   (b) $2x + y + 6 = 0$
   (c) $y = -5x + 2$
   (d) $5x + y = 0$
   (e) $Ax + By + C = 0$ Does this represent the equation of a line for all real values of A, B, and C? Discuss the various cases in detail.

THE INTERSECTION OF LINES AND THE ANGLE BETWEEN LINES

The general problem of finding the intersection point of two slant lines involves the simultaneous solution of two linear equations, since the point of intersection has coordinates that satisfy both of the equations. You should be familiar with the techniques for solving such equations; consequently, we proceed directly to the examples.

EXAMPLE 4.1 Find the intersection point of the lines with equations $y = 2x - 4$ and $y = -3x + 1$.

SOLUTION $2x - 4 = y = -3x + 1$ implies $2x - 4 = -3x + 1$. Hence $5x = 5$, $x = 1$, and $y = -2$. Therefore $(1, -2)$ is the point of intersection (see Figure 4.4).

EXAMPLE 4.2 Find the intersection point of the lines with equations $2x - y = -3$ and $x + 2y + 1 = 0$.

SOLUTION Putting the equations in the slope–intercept form, we obtain $y = 2x + 3$ and $y = -(\frac{1}{2})x - (\frac{1}{2})$. Therefore, $2x + 3 = -(\frac{1}{2})x - (\frac{1}{2})$, so $(\frac{5}{2})x = (-\frac{7}{2})$. Hence $x = -\frac{7}{5}$, $y = \frac{1}{5}$, and the point of intersection is $(-\frac{7}{5}, \frac{1}{5})$.

Finally, as an exercise, verify that if L_1 is a line with equation $y = m_1 x + b_1$ and L_2 is another line, not parallel to L_1, with equation $y = m_2 x + b_2$, the x coordinate of the point of intersection of L_1 and L_2 is given by $-(b_2 - b_1)/(m_2 - m_1)$. Use this formula to obtain the answers given in the preceding two examples.

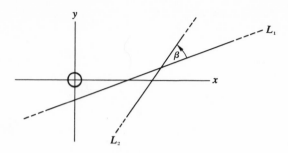

Figure 4.1

We now come to the discussion of the angle made by two intersecting lines. We state the main theorem without proof (the proof may easily be obtained from the formula for the tangent of the difference of two angles).

Theorem 4.1   Let L_1 and L_2 be two nonparallel, nonperpendicular lines with slopes m_1 and m_2, respectively. Let β be the positive angle with initial side on L_1 and terminal side on L_2 (see Figure 4.1). Then,

$$\tan \beta = (m_2 - m_1)/(1 + m_2 m_1), \quad 0° < \beta < 180°$$

It is clear that L_1 parallel to L_2 means $\beta = 0°$ and L_1 perpendicular to L_2 means $\beta = 90°$. If L_1 is a vertical line while L_2 is a slant line with slope m_2 (see Figure 4.2), we have $\tan \beta = -(1/m_2)$. If L_2 is vertical while L_1 is a slant line with slope m_1 (see Figure 4.3), we have $\tan \beta = (1/m_1)$.

EXAMPLE 4.3   Find the tangent of the angle between the lines with equations $y = 2x - 4$ and $y = -3x + 1$.

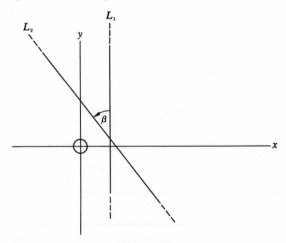

Figure 4.2

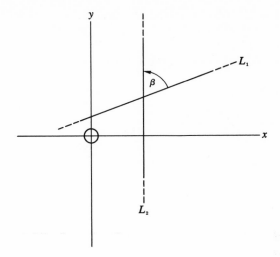

Figure 4.3

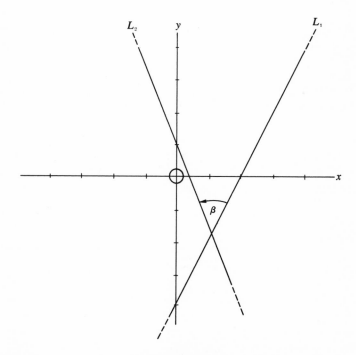

Figure 4.4

SOLUTION   The intersection point of these lines was found to be $(1, -2)$ in Example 4.1.   A sketch of the graphs appears in Figure 4.4, where

$$L_1: y = 2x - 4 \qquad m_1 = 2$$
$$L_2: y = -3x + 1 \qquad m_2 = -3$$
$$\tan \beta = \frac{m_2 - m_1}{1 + m_2 m_1} = \frac{-3 - 2}{1 + (-3)(2)} = \frac{-5}{-5} = 1$$

Therefore, in this case, we can actually find the angle between the lines easily since $\tan \beta = 1$ means $\beta = 45°$.   Notice also that the angle from L_2 to L_1 is simply $180° - 45° = 135°$.

EXAMPLE 4.4   Find the tangent of the angle between the lines with equations $y = 2x + 3$ and $y = -\frac{1}{2}x - \frac{1}{2}$.

SOLUTION   The intersection point of these lines was found to be $(-\frac{7}{5}, \frac{1}{5})$ in Example 4.2.   In this case, however, we observe that the lines are perpendicular since $m_2 = -(1/m_1)$.   Consequently, the *tangent* of the angle does not exist, but the angle itself is $90°$.   As an exercise, sketch the graphs.

EXAMPLE 4.5   Find the tangent of the angle between the lines with equations $y = 2x + 1$ and $y = -x + 2$.

SOLUTION

$$L_1: y = 2x + 1 \qquad m_1 = 2$$
$$L_2: y = -x + 2 \qquad m_2 = -1$$
$$\tan \beta = \frac{-1 - 2}{1 + (-2)} = \frac{-3}{-1} = 3.$$

As an exercise, sketch the graphs.

PROBLEMS

4.1   Find the point of intersection of the following pairs of lines.
   (a) $x + y = -3$ $x - y = -1$
   (b) $y = -2x + 7$ $4x + 2y = 1$
   (c) $y = x + 7$ $y = 10x - 2$
   (d) $3x - 2y = 5$ $x + 5y = 1$
   (e) $ax + by = c$ and $dx + ey = f$
4.2   Find the tangent of the angle between each pair of lines in the preceding problem.

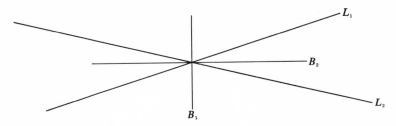

Figure 4.5

4.3* Let B_1 and B_2 be the lines bisecting the angles formed by two intersecting straight lines L_1 and L_2. Show $B_1 \perp B_2$ (see Figure 4.5).

4.4 Find the vertices of the triangle whose sides lie along the lines with equations $3x - 2y = 4$, $4x + y = 20$, and $5x + 4y = 14$.

DISTANCE FROM A POINT TO A LINE

The distance from a point (p, q) to a horizontal line with equation $y = b$ is clearly $|q - b|$ and the distance from a point (p, q) to a vertical line with equation $x = a$ is simply $|p - a|$ (see Figure 5.1).

Consequently, we need only consider the case of the distance from a point (p, q) to a slant line L with equation $y = mx + b$ (see Figure 5.2). The best way to obtain such a distance formula involves the use of the angle β between L and the line $x = p$. We know from Section 4 that $\tan \beta = 1/m$, and this implies that $\sin \beta = 1/(1 + m)^{1/2}$. However, $\sin \beta = D/H$ (check Figure 5.2 again). Therefore, $D/H = 1/(1 + m^2)^{1/2}$, which implies $D = H/(1 + m^2)^{1/2} = |mp + b - q|/(1 + m^2)^{1/2}$. Notice that $mp + b$ is just the y coordinate of

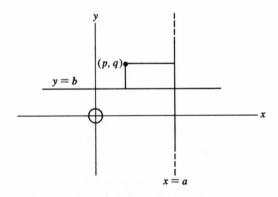

Figure 5.1

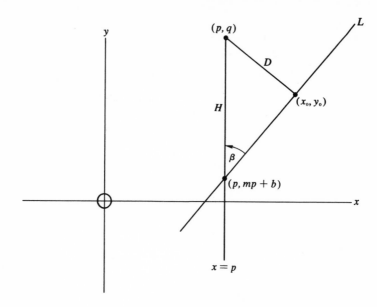

Figure 5.2

the intersection of the vertical line $x = p$ with L, while q is simply the y coordinate of (p, q). Hence, the numerator of the fraction that represents D is merely the length H of the vertical line segment from (p, q) to L.

EXAMPLE 5.1 Find the distance from the point $(-3, 7)$ to the line with equation $4x + 3y = 5$.

SOLUTION Putting the equation into the slope–intercept form yields $y = -\frac{4}{3}x + \frac{5}{3}$. Therefore, $m = -\frac{4}{3}$, $b = \frac{5}{3}$, $p = -3$, $q = 7$, and

$$D = \frac{|mp + b - q|}{(1 + m^2)^{1/2}} = \frac{|(-4/3)(-3) + (5/3) - 7|}{[1 + (16/9)]^{1/2}} =$$

$$\frac{|-3 + (5/3)|}{(25/9)^{1/2}} = \frac{|-4/3|}{(5/3)} = \frac{(4/3)}{(5/3)} = \frac{4}{5}$$

We now discuss the problem of finding the distance between two parallel lines L_1 and L_2. The solution to this type of problem involves the selection of any point (p, q) on L_1 and the computation of the distance from (p, q) to L_2. Consequently, we have reduced the problem to that of finding the distance from a point to a line.

EXAMPLE 5.2 Find the distance between the parallel lines $y = -\frac{4}{3}x + \frac{7}{3}$ and $y = -\frac{4}{3}x + \frac{5}{6}$.

SOLUTION   An obvious point on the first line is the y intercept $(0, \frac{7}{3})$. Therefore, we simply find the distance from $(0, \frac{7}{3})$ to $y = -\frac{4}{3}x + \frac{5}{6}$. Using the distance formula, we obtain

$$D = \frac{|(-4/3)(0) + (5/6) - (7/3)|}{[1 + (16/9)]^{1/2}} = \frac{|5/6 - 14/6|}{(25/9)^{1/2}} = \frac{(9/6)}{(5/3)} = \frac{9}{10}$$

PROBLEMS

5.1   Find the distance from the point (2, 4) to each of the following lines.
(a) $y = 2x + 1$
(b) $y = -2x + 8$
(c) $y = -\frac{1}{4}x$
5.2   Repeat the preceding problem using the point $(-1, 2)$.
5.3   Find the distance between the following pairs of parallel lines.
(a) $y = 3x + 7$ $y = 3x - 2$
(b) $y = -x + 2$ $y = -x + 5$
(c) $y = 6x - 2$ $y = 6x + 4$
(d) $2x + y - 3 = 0$ $2x + y + 1 = 0$
(e) $3x + 2y + 1 = 0$ $6x + 4y - 2 = 0$
5.4   Find the lengths of the three altitudes in each of the triangles with the following vertices.
(a) $(1, 4)$ $(10, 6)$ $(2, 2)$
(b) $(1, -1)$ $(3, 0)$ $(2, 2)$
5.5   Find a point on the line $y = -2x + 1$ that is at a distance 2 from the line $y = x + 1$.
5.6* Develop the formula given in the text for the distance from a point to a line (see Figure 5.2) by first finding the line L' through (p, q) and perpendicular to L. Then determine the intersection point of L and L', namely (x_0, y_0). Finally, find the distance between (x_0, y_0) and (p, q).
5.7* Prove that if we are given the equation of a line L in the general form $Ax + By + C = 0$, the distance from any point (p, q) to L is given by

$$\frac{|Ap + Bq + C|}{(A^2 + B^2)^{1/2}}$$

Section 6

PROBLEMS FROM PLANE GEOMETRY

We now apply techniques of analytic geometry to some familiar problems from plane geometry. These problems are concerned basically with triangles, parallelograms, and arbitrary quadrilaterals. The key to the easy solution of such problems is the judicious choice of coordinates for the vertices. We can make such a choice because the coordinate axes may be introduced after the figure is given in the plane. For example, whether dealing with a triangle or a quadrilateral, we can always choose one vertex at the origin and another on the x axis. In the special cases of equilateral, isosceles, and right triangles, the third vertex can be written in terms of the first two vertices. In the special case of a parallelogram, the fourth vertex can always be written in terms of the first three vertices. We must be careful, however, not to use a diagram that tacitly assumes a property that is not actually given. For example, in an *arbitrary* triangle we cannot choose the third vertex on the y axis, and in an *arbitrary* quadrilateral we cannot have a pair of opposite parallel sides. Some sample diagrams are shown in Figures 6.1, 6.2, and 6.3.

EXAMPLE 6.1 Prove that the line segments that join the midpoints of the sides of any quadrilateral, if taken in order, form a parallelogram.

SOLUTION We use a diagram like Figure 6.2, clearly labeling the midpoints of the sides and the line segments that join them. Since this problem involves midpoints, our arithmetic will be simplified if we place a factor of 2 with the coordinates of each vertex (see Figure 6.4). There are two ways of solving this problem. First, we can show that *one pair* of opposite sides is parallel *and* equal. Another way would be to show that *each pair* of opposite sides is parallel.

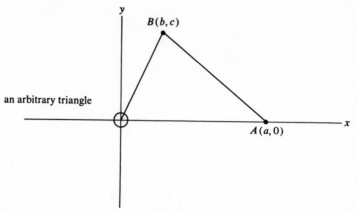

Figure 6.1

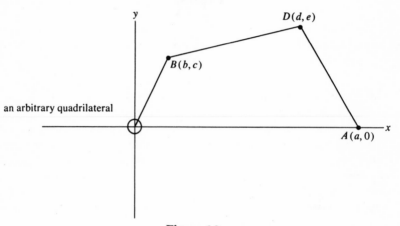

Figure 6.2

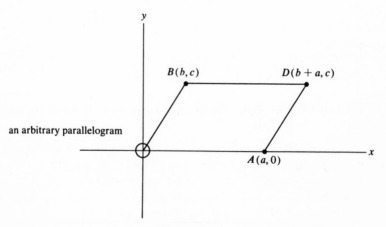

Figure 6.3

22

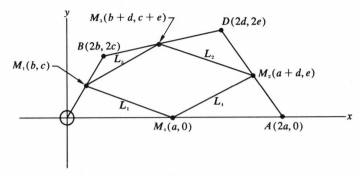

Figure 6.4

Method 1   The slope of L_1 is $(c - 0)/(b - a) = c/(b - a)$. The slope of L_2 is $[(c + e) - e]/[(b + d) - (a + d)] = c/(b - a)$. Therefore L_1 is parallel to L_2.

The length of L_1 is $[(b - a)^2 + c^2]^{1/2}$, while the length of L_2 is $\{[(b + d) - (a + d)]^2 + [(c + e) - e]^2\}^{1/2} = [(b - a)^2 + c^2]^{1/2}$. Therefore, the length of L_1 equals the length of L_2.

Method 2   After showing that L_1 is parallel to L_2 as we did in Method 1, we then show that L_3 is parallel to L_4.

The slope of L_3 is $[(c + e) - c]/[(b + d) - b] = e/d$. The slope of L_4 is $(e - 0)/[(a + d) - a] = e/d$. Therefore, L_3 is parallel to L_4.

EXAMPLE 6.2   Prove that the diagonals of a parallelogram are perpendicular if and only if the parallelogram is a rhombus.

SOLUTION   This is a twofold statement. First it states that if the diagonals of a parallelogram are perpendicular, it is a rhombus. The second statement is that the diagonals of a rhombus are perpendicular. We prove the first statement by using a diagram like Figure 6.3, clearly labeling the perpendicular diagonals L_1 and L_2 (see Figure 6.5). We must

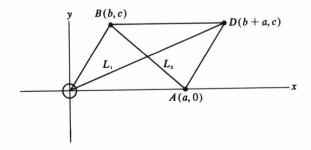

Figure 6.5

show that the adjacent sides OA and OB have equal length.   Now, L_1 perpendicular to L_2 means that the product of their slopes is -1.   The slope of L_1 is $c/(b+a)$, while the slope of L_2 is $c/(b-a)$.   Therefore, $[c/(b+a)][c/(b-a)] = -1$.   This   implies   that $c^2/(b^2 - a^2) = -1$; thus, $c^2 = a^2 - b^2$.   Therefore, $b^2 + c^2 = a^2$ and $a = (b^2 + c^2)^{1/2}$.   But the length of OA is a and the length of OB is $(b^2 + c^2)^{1/2}$; therefore, we have completed the proof of the first statement.

To prove that the diagonals of a rhombus are perpendicular, we start with the fact that $a = (b^2 + c^2)^{1/2}$.   We then repeat the steps in the proof of the first statement in *reverse order*, arriving at $[c/(b+a)]$ $\times [c/(b-a)] = -1$, which implies that L_1 is perpendicular to L_2.

EXAMPLE 6.3   Prove that in any triangle the perpendicular bisectors of the sides intersect at a single point.

SOLUTION   We use a diagram like Figure 6.1, clearly labeling the three perpendicular bisectors L_1, L_2, and L_3.   Once again, our arithmetic can be simplified if we multiply the coordinates of each vertex by 2 (see Figure 6.6).   First, we find the equations of the lines L_1, L_2, and L_3; then we locate the point of intersection P of L_1 and L_2 and show that this point is also on the line L_3.

Since L_1 is the vertical line through $(a, 0)$, its equation is $x = a$. L_2 is the line through (b, c) with slope equal to the negative reciprocal of the slope of OB.   Therefore, the slope of L_2 is $-b/c$ and the equation is $y - c = (-b/c)(x - b)$, which reduces to $y = (-bx/c) + (b^2/c) + c$.   Substituting $x = a$ into this equation yields the y coordinate of the intersection point P of L_1 and L_2, namely, $y = (-ba/c) + (b^2/c) + c = (b^2 + c^2 - ab)/c$.   Therefore, $(a, [b^2 + c^2 - ab]/c)$ is the point of intersection P of L_1 and L_2.

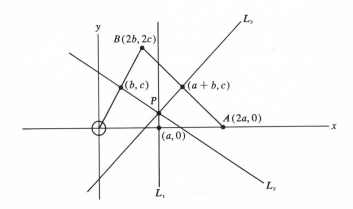

Figure 6.6

L_3 is the equation of the line through $(a + b, c)$ with slope equal to the negative reciprocal of the slope of AB. Therefore, the slope of L_3 is $(a - b)/c$ and the equation is $y - c = [(a - b)/c][x - (a + b)]$, which reduces to $y = [(a - b)/c]x + (b^2 - a^2 + c^2)/c$.

To prove P is on L_3, we must show that the coordinates of P satisfy the equation of L_3. Substituting $x = a$ in the equation for L_3, we obtain

$$y = \frac{(a - b)}{c} a + \frac{b^2 - a^2 + c^2}{c} = \frac{a^2 - ab + b^2 - a^2 + c^2}{c} = \frac{b^2 + c^2 - ab}{c}$$

which is the y coordinate of P. Therefore, the coordinates of P do satisfy the equation of L_3 and P is thus on L_3. This means that P is on all three lines; therefore, P is the single point of intersection of L_1, L_2, and L_3.

PROBLEMS

6.1 Prove that the line segment joining the midpoints of two sides of a triangle is parallel to the third side and is half its length.

6.2 Prove that the diagonals of a quadrilateral bisect each other if and only if the figure is a parallelogram.

6.3 Prove that the lines drawn from a vertex of a parallelogram to the midpoints of the opposite sides trisect a diagonal.

6.4 Prove that the line segment joining the midpoints of the nonparallel sides of a trapezoid bisects the diagonals.

6.5 Prove that in any triangle. if the sum of the squares of two of the sides equals the square of the third side, the triangle is a right triangle. Is this the Pythagorean theorem?

6.6 Prove that in any right triangle, the midpoint of the hypotenuse is equidistant from the three vertices.

6.7 Prove each of the following statements.
   (a) The medians of a triangle meet at a single point that lies two thirds of the way from each vertex to the midpoint of the opposite side.
   (b) The perpendicular bisectors of the sides in a triangle intersect at a single point (see Example 6.3).
   (c) The altitudes in a triangle intersect at a single point.
   (d) The three intersection points from parts (a), (b), and (c) are collinear.

6.8* Is it true that the bisectors of the angles in a triangle intersect at a single point? If yes, prove it. If no, exhibit an appropriate diagram.

6.9* Through any point P within a triangle, draw line segments parallel to the three sides and terminating at the sides. Show that the sum of the ratios of these segments to their corresponding sides is equal to 2. [*Hint:* Use Figure 6.7 and show that $(R'/R) + (S'/S) + (T'/T) = 2$. Try to make a judicious choice of coordinates for the three vertices.]

6.10* Through a point P within a parallelogram with vertices A, B, C, and D, draw two lines EF and GH parallel to the sides. Show that the diagonals EG and HF of the smaller parallelograms intersect on the diagonal AC of the given parallelogram. (*Hint:* Use Figure 6.8 and make a judicious choice of coordinates for the points A, B, C, and D.)

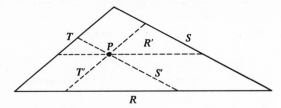

Figure 6.7

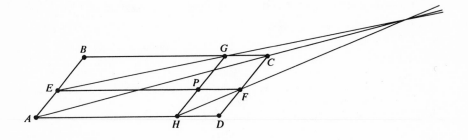

Figure 6.8

THE CIRCLE

Definition 7.1 A circle is the graph of all points in the plane at a given distance (the radius) from a fixed point (the center).

It follows from the definition that a point (x, y) is on a circle of radius r and center (h, k) if and only if $[(x - h)^2 + (y - k)^2]^{1/2} = r$ (see Figure 7.1). From this equation we obtain the standard form of the equation, namely, $(x - h)^2 + (y - k)^2 = r^2$.

If we are given the equation of a circle in the form $x^2 + y^2 + Cx + Dy + E = 0$, we can "complete the square" and obtain the equivalent equation $[x + (C/2)]^2 + [y + (D/2)]^2 = \{[(C^2 + D^2)/4] - E\}$. This is the equation of the circle with radius $\{[(C^2 + D^2)/4] - E\}^{1/2}$ and center $(-C/2, -D/2)$. The expression for the radius may be simplified to $\frac{1}{2}(C^2 + D^2 - 4E)^{1/2}$. Consequently, a circle is determined by its center and radius, and we can, in practical cases, work with the standard form $(x - h)^2 + (y - k)^2 = r^2$. Observe also that whenever a circle has its center at the origin, its equation will reduce to $x^2 + y^2 = r^2$.

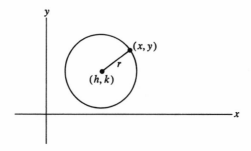

Figure 7.1

EXAMPLE 7.1   Find the equation of the circle with radius 5 and center $(2, -3)$.

SOLUTION $r = 5$, $h = 2$, $k = -3$.   Therefore, the equation is $(x - 2)^2 + (y + 3)^2 = 25$.

EXAMPLE 7.2   Find the center and radius of the circle having equation $x^2 + y^2 + 2x - 4y - 11 = 0$.

SOLUTION   Completing the square, we obtain $(x^2 + 2x + 1) + (y^2 - 4y + 4) - 11 = 1 + 4$, which leads to $(x + 1)^2 + (y - 2)^2 = 16$.   Therefore, this circle has center $(-1, 2)$ and radius 4.

There are three other basic types of problems that deal with finding equations of circles.   We list these with a brief word on each.

1. Find the equation of a circle with center (h, k) and passing through the point (a, b).   [See Problem 7.1(c) and (d).]

This problem gives us the center, so we need only find the radius by computing the distance between (h, k) and (a, b).

2. Find the equation of a circle with endpoints of a diameter given by (a, b) and (c, d).   [See Problem 7.1(e) and (f).]

This problem gives us neither the center nor the radius of the circle directly. The midpoint of the segment from (a, b) to (c, d) is certainly the center.   Once we have the center, we find the radius by computing the distance from the center to either (a, b) or (c, d) (see Figure 7.2).

3. Find the equation of a circle that passes through three points $A(a_1, a_2)$, $B(b_1, b_2)$, and $C(c_1, c_2)$.

This problem can be solved in two ways, one basically *algebraic* and the other basically *geometrical*.   The algebraic method depends on the fact that if A, B, and C are on a circle with equation $(x - h)^2 + (y - k)^2 = r^2$, their coordinates must satisfy the equation of the circle.   Substituting these coordinates for x and y, we obtain three equations in the unknowns h, k, and r.   We then solve these three equations in three unknowns, obtaining the desired values of h, k, and r.

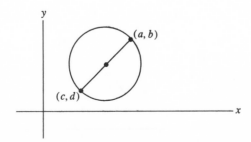

Figure 7.2

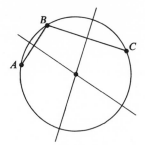

Figure 7.3

The geometrical method depends on the fact that in any circle the perpendicular bisector of a chord passes through the center of the circle. Consequently, if we know *two* chords, we can find the two corresponding perpendicular bisectors and their intersection point will be the center of the circle. Therefore, knowing three points on the circle, we know the two chords AB and BC and we use their perpendicular bisectors to find the center (see Figure 7.3). After finding the center, we obtain the radius by computing the distance from the center to any one of the three points A, B, and C on the circumference.

EXAMPLE 7.3 Find the equation of the circle that passes through the points $(0, 0)$, $(\sqrt{3}/2, \frac{1}{2})$, and $(1, 1)$.

SOLUTION *First Method* Substituting $(0, 0)$ into $(x - h)^2 + (y - k)^2 = r^2$, we obtain $h^2 + k^2 = r^2$. Substituting $(1, 1)$, we obtain $(1 - h)^2 + (1 - k)^2 = r^2$. Substituting $(\sqrt{3}/2, \frac{1}{2})$, we obtain $(\sqrt{3}/2 - h)^2 + (\frac{1}{2} - k)^2 = r^2$. Solving simultaneously, we arrive at $h = 0$, $k = 1$, and $r = 1$. Therefore, the equation is $x^2 + (y - 1)^2 = 1$. The graph appears in Figure 7.4.

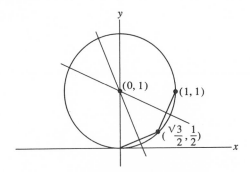

Figure 7.4

Second Method   The slope of the chord from $(0, 0)$ to $(\sqrt{3}/2, \frac{1}{2})$ is $\frac{1}{2}/\sqrt{3}/2 = 1/\sqrt{3}$.   Therefore, the slope of the perpendicular bisector is $-\sqrt{3}$ and the midpoint of the chord is $(\sqrt{3}/4, \frac{1}{4})$.   The equation of the perpendicular bisector is $y - \frac{1}{4} = -\sqrt{3}[x - (\sqrt{3}/4)]$, which reduces to $y = -\sqrt{3}x + 1$.

In a similar way, we obtain $y = (\sqrt{3} - 2)x + 1$ as the equation of the perpendicular bisector of the chord from $(\sqrt{3}/2, \frac{1}{2})$ to $(1, 1)$.

The intersection of these two lines is found to be the point $(0, 1)$; therefore $(0, 1)$ is the center of the circle.   We find the radius to be 1 in the usual way; therefore, the equation of the circle is $x^2 + (y - 1)^2 = 1$.

We now come to the problem of finding tangent lines to circles at specified points.   We use the fact that any tangent line must be perpendicular to the radius line drawn to the point of contact.   Therefore, if we know the slope of the radius line, the slope of the tangent line is the negative reciprocal of it.

EXAMPLE 7.4   Find the equation of the tangent line to the circle $(x - 2)^2 + (y + 1)^2 = 18$ at the point $(5, 2)$.

SOLUTION   The radius line from $(2, -1)$ to $(5, 2)$ has slope $[2 - (-1)]/(5 - 2) = 3/3 = 1$.   Therefore, the tangent line is the line through $(5, 2)$ with slope -1 and has equation $y - 2 = -1(x - 5)$, which reduces to $y = -x + 7$.

PROBLEMS

7.1   Find the standard form of the equation of each of the following circles.
   (a)   center $(-1, 3)$, radius 5
   (b)   center $(2, -3)$, radius 2
   (c)   center $(1, -1)$ passing through $(3, 5)$
   (d)   center $(2, 3)$ passing through $(5, 7)$
   (e)   with endpoints of a diameter given by $(1, -2)$ and $(4, 7)$
   (f)   with endpoints of a diameter given by $(-2, 3)$ and $(6, -7)$
   (g)   circumscribed about the right triangle with vertices $(1, 4)$, $(10, 6)$, and $(2, 2)$.   (See Problem 6.6.)
   (h)   through $(5, 3)$ and $(-2, 2)$ with radius 5.   (Is there only one answer to this problem?)
7.2   Graph the following equations.
   (a) $x^2 + y^2 = 5$
   (b) $(x - 1)^2 + (y + 2)^2 = 4$
   (c) $(x - 2)^2 + (y - 3)^2 = 12$
   (d) $x^2 + 10x + y^2 - 16y - 11 = 0$
   (e) $x^2 - 4x + y^2 = 12$
   (f) $x^2 + y^2 + Cx + Dy + E = 0$.   Does this represent a circle for all values of C, D, and E?

7.3 Find the equation of the tangent line to each of the following circles at the indicated point.
- (a) $x^2 + y^2 + 2x - 4y - 5 = 0$ at (2, 1)
- (b) $(x - 1)^2 + (y + 1)^2 = 5$ at (2, −3)
- (c) $x^2 + y^2 = 25$ at (3, 4)
- (d) $x^2 + y^2 = r^2$ at (a, b)

7.4 Prove that in any circle, the line segment from the midpoint of a chord to the center is perpendicular to the chord.

7.5 Find the standard form of the equation of each of the following circles.
- (a) through (4, 2), (0, 0), and (1, 1)
- (b) through (3, 0), (5, −2), and (3, −4)
- (c) with center on the line $y = x + 1$ and tangent to the line $y = -\frac{1}{3}x + 5$ at the point (6, 3)
- (d) with center (1, −1) and a tangent line given by $y = 2x + 1$
- (e) through (−1, −2) and (6, −1) that is tangent to the x axis

INTERSECTIONS OF LINES AND CIRCLES

In a plane there are three possibilities for the intersection of a line L with a circle C: two points of intersection; one point of intersection (which means L is tangent to C); or no point of intersection (see Figure 8.1). In any event, we simply solve simultaneously the equation of the line and the equation of the circle. There are three possibilities with regard to the simultaneous solutions, each possibility corresponding to one of those mentioned in the first sentence, namely, two solutions, one solution, or no real solutions.

If L is horizontal ($y = b$) or vertical ($x = a$), we merely substitute for x or y in the circle equation $(x - h)^2 + (y - k)^2 = r^2$. We then solve the resulting quadratic equation (see Problems 8.1 and 8.2). If L is a slant line with equation $y = mx + b$, we can substitute $mx + b$ for y in the circle equation and then solve the resulting quadratic equation in x (see Problems 8.3 and 8.4).

The next problem is that of the intersection of two circles. Once again, there are three possibilities: two, one, or no points of intersection (see Figure 8.2).

Suppose we have two circles C_1 and C_2 with equations $E_1 = 0$ and $E_2 = 0$ where E_1 and E_2 are in the form $(x - h)^2 + (y - k)^2 - r^2$. If we subtract

Figure 8.1

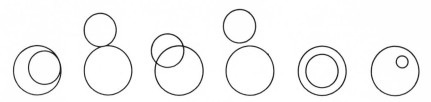

Figure 8.2

the two equations, we will get a *linear* equation $E_3 = E_1 - E_2 = 0$. Then E_3 has the property that if (p, q) is a point of intersection of C_1 and C_2, (p, q) satisfies $E_3 = 0$ [since (p, q) satisfies $E_1 = 0$ and $E_2 = 0$].

Now, if (r, s) is a second point of intersection of C_1 and C_2, (r, s) must also satisfy $E_3 = 0$. But $E_3 = 0$ is the equation of a line and hence is determined by the two intersection points. In this case, the line with equation $E_3 = 0$ is called "the common chord" of the two intersecting circles. If the circles are tangent to each other, the line with equation $E_3 = 0$ is just the common tangent line at the point of intersection. Finally, if C_1 and C_2 fail to intersect we can sometimes still compute the linear equation $E_3 = 0$. In this case, the line with equation $E_3 = 0$ is a line that is perpendicular to the line through the centers of C_1 and C_2 and that fails to intersect either circle.

Question Under what conditions will the line $E_3 = 0$ be the perpendicular bisector of the line joining the centers of C_1 and C_2? [See Problem 8.15(*b*) and (c).]

In the cases where $E_3 = 0$ is the equation of the common chord or common tangent, we find the intersection points of the line $E_3 = 0$ with either circle C_1 or circle C_2. These intersection points will be the intersection points of the circles C_1 and C_2. It is also very worthwhile to sketch a graph *before* you attempt a problem on the intersection of two circles. In most instances where the circles fail to intersect, a rough sketch will tell you immediately and save you the trouble of going through all the algebraic manipulations only to find there is no real solution.

In Figure 8.3, we give a few diagrams that represent some of the situations just described. In each case, the circles represent C_1 and C_2, the line represents the line with equation $E_3 = 0$. (Why is there a line missing in one of the diagrams?)

Figure 8.3

EXAMPLE 8.1   Find the intersection of the line $y = x + 4$ with the circle $x^2 + y^2 = 16$.

SOLUTION   Substituting, we obtain the quadratic equation

$$x^2 + (x + 4)^2 = 16.$$

This leads to $2x^2 + 8x + 16 = 16$, which reduces to $2x(x + 4) = 0$. Therefore, we obtain two roots, $x = 0$ and $x = -4$. When $x = 0$, we have $y = 4$; when $x = -4$, $y = 0$. Therefore, the two points of intersection are $(0, 4)$ and $(-4, 0)$.

EXAMPLE 8.2   Find the intersection points of the two circles $x^2 + y^2 = 16$ and $(x + 2)^2 + (y - 2)^2 = 8$.

SOLUTION   Let $E_1 = (x + 2)^2 + (y - 2)^2 - 8$; let $E_2 = x^2 + y^2 - 16$. Then $E_3 = E_1 - E_2 = x^2 + 4x + 4 + y^2 - 4y + 4 - 8 - x^2 - y^2 + 16 = 0$. Therefore, $E_3 = x - y + 4 = 0$ and the common chord has equation $y = x + 4$. We now look for the intersection points of the common chord with either one of the two given circles. From Example 8.1, we see that the intersection points of the line $y = x + 4$ with the circle $x^2 + y^2 = 16$ are $(0, 4)$ and $(-4, 0)$. Therefore, these are the points of intersection of the two circles. A sketch appears in Figure 8.4.

EXAMPLE 8.3   Find the intersection points of the circles $(x - 2)^2 + (y - 1)^2 = 1$ and $(x - 5)^2 + (y - 1)^2 = 4$.

SOLUTION   Let $E_1 = (x - 2)^2 + (y - 1)^2 - 1$ and $E_2 = (x - 5)^2 + (y - 1)^2 - 4$. Subtracting, we obtain

$$x^2 - 4x + 4 + y^2 - 2y + 1 - 1 - (x^2 - 10x + 25) - (y^2 - 2y + 1) + 4 = 0$$

which leads to $E_3 = 6x - 21 + 3 = 0$. Therefore $6x = 18$, $x = 3$, and the equation of the common chord is $x = 3$. Substituting $x = 3$ into

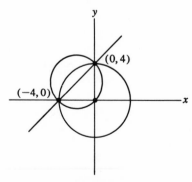

Figure 8.4

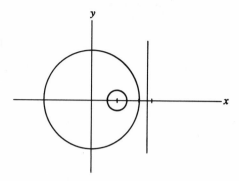

Figure 8.5

the first equation, we obtain $1 + (y - 1)^2 = 1$, which reduces to the quadratic $(y - 1)^2 = 0$.   Therefore, there is one distinct root $y = 1$ and the two circles are tangent at (3, 1).

We now give a few examples in which we have two circles that fail to intersect.   In one obvious situation, namely, that of concentric circles, the subtraction of the two equations does *not* yield a linear equation; hence, $E_3 = 0$ does not exist in this case.

EXAMPLE 8.4   Find the intersection of the circles $x^2 + y^2 = 16$ and $(x - 2)^2 + y^2 = 1$.

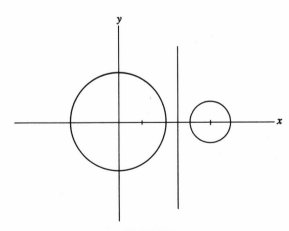

Figure 8.6

SOLUTION A sketch of the graphs (see Figure 8.5) reveals that the circles do not intersect, however, subtracting yields $E_3 = 4x - 19 = 0$. Therefore, $E_3 = 0$ is the equation of the vertical line through (19/4, 0).

EXAMPLE 8.5 Find the intersection of the circles $x^2 + y^2 = 4$ and $(x - 4)^2 + y^2 = 1$.

SOLUTION A sketch of the graphs (see Figure 8.6) shows that the circles do not intersect. Subtracting the equations yields $E_3 = 8x - 19 = 0$. Therefore, $E_3 = 0$ is the equation of the vertical line through (19/8, 0).

PROBLEMS

Find the intersection points of each of the following pairs of graphs.

8.1 $x^2 + y^2 = 25$ $x = 4$
8.2 $x^2 + y^2 = 13$ $y = 2$
8.3 $x^2 + y^2 = 8$ $y = x$
8.4 $x^2 + y^2 = 2$ $y = x + 4$
8.5 $(x - 2)^2 + (y - 1)^2 = 5$ $2x + y = 13$
8.6 $(x - 2)^2 + (y - 1)^2 = 5$ $2x + y = 10$
8.7 $x^2 + y^2 = 4$ $(x - 2)^2 + y^2 = 4$
8.8 $(x - 6)^2 + (y - 3)^2 = 16$ $(x - 7)^2 + (y - 4)^2 = 1$
8.9 $(x - 4)^2 + y^2 = 4$ $(x - 1)^2 + y^2 = 1$
8.10 $x^2 + y^2 = 1$ $(x - 4)^2 + (y - 4)^2 = 4$
8.11 $x^2 + y^2 + 2x - 4y - 11 = 0$ $x^2 + y^2 - 2x - 3 = 0$
8.12 $(x + 1)^2 + (y + 1)^2 = 5$ $(x + 3)^2 + y^2 = 10$
8.13 $(x + 3)^2 + (y - 2)^2 = 4$ $(x + 1)^2 + (y - 1)^2 = 9$
8.14 $(x + 3)^2 + (y - 2)^2 = 4$ $(x + 3)^2 + y^2 = 16$
8.15* Let $(x - h)^2 + (y - k)^2 = r^2$ and $x^2 + y^2 = s^2$ be the equations of two circles. Let $E_1 = (x - h)^2 + (y - k)^2 - r^2$ and $E_2 = x^2 + y^2 - s^2$. Compute the linear equation $E_3 = E_1 - E_2 = 0$. Answer each of the following.

(a) Under what conditions will $E_3 = 0$ be meaningless?
(b) Prove that the line with equation $E_3 = 0$ is perpendicular to the line through the centers of the circles.
(c) Find a condition that forces $E_3 = 0$ to be the perpendicular bisector of the line segment joining the centers of the circles.
(d) Does it make any difference in parts (b) and (c) if the two circles fail to intersect?
(e) Prove that if the two circles fail to intersect the line $E_3 = 0$ does not intersect either circle.
(f) Let (p, q) be the intersection point of the line $E_3 = 0$ with the line through the centers of the circles. Try to develop a formula for the ratio of the distances from (0, 0) to (p, q) and from (p, q) to (h, k). When will this ratio be 1?

8.16 (a) Find the equation of a line passing through the point $(3, -1)$ and tangent to the circle $x^2 + y^2 = 2$. Is there more than one answer to this problem? (b) Repeat part (a) using the point (1, 3).

Section 9

THE PARABOLA

Definition 9.1 A parabola is the graph of all points in the plane that are equidistant from a fixed point (the focus) and a fixed line (the directrix) not passing through the fixed point.

Instead of giving separate definitions of terms connected with the graph of a parabola, we exhibit Figures 9.1 and 9.2 with these terms properly labeled. There are, of course, parabolas that possess slant lines as their axis of symmetry. By a rotation of axes, however, these can be reduced to either of the former two cases. Consequently, we deal exclusively with parabolas having either vertical or horizontal axes of symmetry.

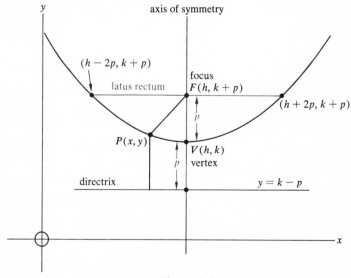

Figure 9.1

37

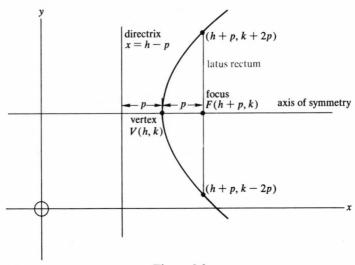

Figure 9.2

Using the definition of a parabola along with the distance formula, we find that the standard form of the equation of the parabola in Figure 9.1 is given by $(x - h)^2 = 4p(y - k)$. For the parabola in Figure 9.2, the corresponding equation is $(y - k)^2 = 4p(x - h)$.

Note that if the vertex is at the origin, the preceding equations reduce to $x^2 = 4py$ and $y^2 = 4px$, respectively. Observe, too, that in all cases where the parabola opens either upward or to the right, p is the distance from the vertex to the focus. In cases where the parabola opens downward or to the left, p is the negative of the distance from the vertex to the focus.

If we are given the equation of a parabola in the form $x^2 + Cx + Dy + E = 0$ or in the form $y^2 + Cx + Dy + E = 0$, we complete the square for the terms that involve the quadratic variable and then put the equation in the standard form. Thus we can deal exclusively with equations in their standard form.

EXAMPLE 9.1 Find the equation of the parabola with focus $(0, 5)$ and directrix $y = 1$.

SOLUTION   The focus at $(0, 5)$ and the directrix line $y = 1$ indicate that the vertex is on the y axis, halfway between the points $(0, 5)$ and $(0, 1)$. Thus, the vertex is $(0, 3)$ and $h = 0$, $k = 3$. Since the vertex is below the focus, the parabola opens upward; hence, $p = +2$. Therefore, the equation is $x^2 = 4(2)(y - 3) = 8(y - 3)$. A sketch of the graph appears in Figure 9.3.

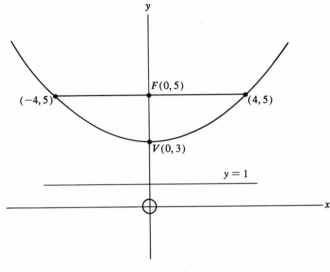

Figure 9.3

EXAMPLE 9.2 Find the equation of the parabola with vertex at $(0, 1)$ and focus $(0, -6)$.

SOLUTION   Since the focus is below the vertex, the parabola opens downward; hence, $p = -7$. Now $h = 0$ and $k = 1$; therefore, the equation is $x^2 = 4(-7)(y - 1)$, which reduces to $x^2 = -28(y - 1)$.

EXAMPLE 9.3 Graph the parabola $(y - 2)^2 = 8(x - 3)$, carefully labeling all pertinent points and lines.

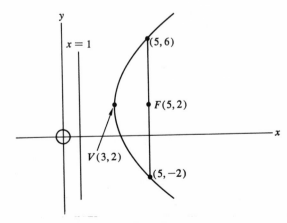

Figure 9.4

SOLUTION   Comparing the given equation with the standard form $(y - k)^2 = 4p(x - h)$, we arrive at the values: $h = 3$; $k = 2$; $p = 2$. Therefore, the vertex is (3, 2) and the directrix is $x = 1$.   The graph appears in Figure 9.4.

EXAMPLE 9.4   Graph the parabola $y^2 + 8y + 12x - 44 = 0$, carefully labeling all pertinent points and lines.

SOLUTION   Completing the square, we obtain $(y^2 + 8y + 16) = -12x + 60$.   Therefore, $(y + 4)^2 = -12(x - 5)$ and we have a parabola with vertex (5, −4) opening to the left.   The graph is shown in Figure 9.5.

EXAMPLE 9.5   Find the equation of the parabola having a vertical axis of symmetry and passing through the points (0, 0), (8, 0), and (12, −6).

SOLUTION   Since the parabola has a vertical axis of symmetry, its equation must have the form $(x - h)^2 = 4p(y - k)$.   Substituting the three given points, we obtain these three equations in the three unknowns h, k, and p: $(-h)^2 = 4p(-k)$; $(8 - h)^2 = 4p(-k)$; and $(12 - h)^2 = 4p(-6 - k)$.   The first two equations yield the result $h^2 = (8 - h)^2$, which implies $h^2 = 64 - 16h + h^2$.   Therefore, $16h = 64$ and $h = 4$. Substituting $h = 4$ into the first and third equations, we obtain

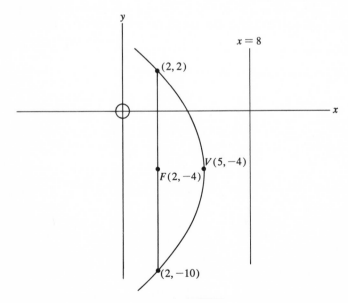

Figure 9.5

$-4pk = 16$ and $64 = -4pk - 24p$. The last equation reduces to $64 = 16 - 24p$, which leads to $24p = -48$. Therefore, $p = -2$, and since $-4pk = 16$ we must have $k = 2$. Hence, the required equation is $(x - 4)^2 = -8(y - 2)$.

Finally, as an exercise, find the equation of the parabola having a *horizontal* axis of symmetry and passing through the points $(16, -9)$, $(0, -1)$, and $(0, 7)$.

PROBLEMS

9.1 Find the equation of the graph of all points that are equidistant from the point $(8, 2)$ and the line $x = 2$.

9.2 Find the standard form of the equation of each of the following parabolas.
  (a) with vertex $(0, 0)$ and focus $(0, 8)$
  (b) with vertex $(0, 0)$ and directrix $y = 4$
  (c) with vertex $(3, 2)$ and focus $(-3, 2)$
  (d) with vertex $(0, 3)$ and focus $(4, 3)$
  (e) with focus $(0, 3)$ and directrix $x = -4$

9.3 Graph the following equations, carefully labeling all pertinent points and lines.
  (a) $y = 4x^2$
  (b) $x = -8y^2$
  (c) $y = x^2 + 4x + 5$
  (d) $x = 2y^2 + y - 3$
  (e) $y = -2x^2 - x + 3$
  (f) $4x^2 - y + 24x + 24 = 0$
  (g) $x + 2y^2 - 12y + 19 = 0$

9.4 Find the equation of a parabola with axis of symmetry $x = 4$ and tangent to the line $y = x + 1$. Is there more than one answer to this problem?

9.5 Find the equation of at least one parabola with vertex $(2, 0)$ and passing through the points $(0, 2)$ and $(4, 2)$.

THE ELLIPSE

Definition 10.1   An ellipse is the graph of all points in the plane the sum of whose distances from two fixed points (the foci) is a constant.

The simplest method of constructing an ellipse from the definition is the following.

1. Choose two fixed points F and F'.   Let the distance between F and F' be designated by d.

2. Take a piece of string having length $D > d$.   Attach one end of the string at the point F, the other end at the point F'.

3. Using the tip of a pen or pencil, move the string away from the segment FF' until it is pulled tight.   Keeping the string taut, move the tip of the pencil along the string around the segment FF', thus tracing the path of an ellipse with foci at F and F' with D equal to the constant mentioned in the definition (see Figure 10.1).

Observe that if F and F' are the same point, we attach the two ends of the string at only one point and our pen or pencil traces the curve of a circle

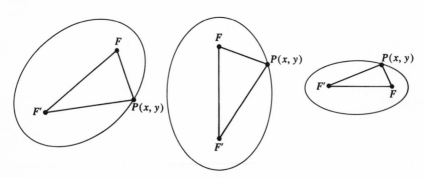

Figure 10.1

with center at F and radius $D/2$.   Consequently, a circle is a special case of an ellipse with $F = F'$.

We deal exclusively with ellipses in which the line segment FF' is either vertical or horizontal (since a suitable rotation of axes reduces any other ellipse to one of these).   Figures 10.2 and 10.3 are labeled diagrams that give the definitions of the terms connected with the ellipse.

Using the definition of the ellipse along with the distance formula, we find that the standard form of the equation of the ellipse in Figure 10.2 is given by $[(x - h)^2/a^2] + [(y - k)^2/b^2] = 1$.   For the ellipse in Figure 10.3, the corresponding equation is $[(y - k)^2/a^2] + [(x - h)^2/b^2] = 1$.   Notice that if the center is at the origin, the preceding equations reduce to $(x^2/a^2) + (y^2/b^2) = 1$ and $(y^2/a^2) + (x^2/b^2) = 1$, respectively.

If we are given the equation of an ellipse in the form $Ax^2 + By^2 + Cx + Dy + E = 0$, we complete the square on both the x and y terms.   This leads to an equivalent equation in the standard form.   Hence, we may deal only with equations in their standard form.

EXAMPLE 10.1   Find the equation of the ellipse with center at the origin, foci at $(4, 0)$ and $(-4, 0)$, and a major axis 12 units in length.

SOLUTION   If the foci are at $(4, 0)$ and $(-4, 0)$, then $c = 4$.   The center $(0, 0)$ is the midpoint of the major axis; hence, $a = 6$ and it follows that $b^2 = 6^2 - 4^2 = 20$.   Therefore, the equation is $(x^2/36) + (y^2/20) = 1$ (see Figure 10.4).

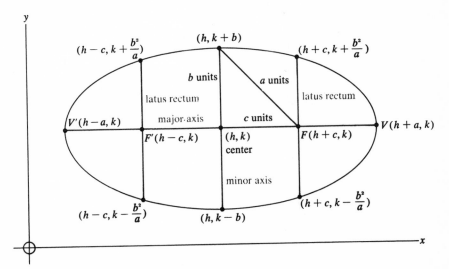

Figure 10.2

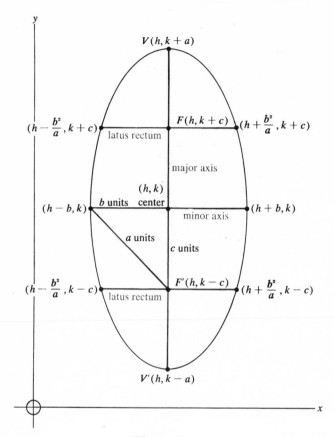

Figure 10.3

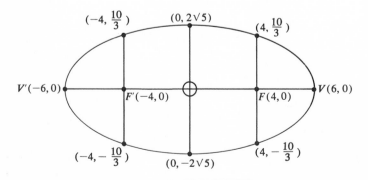

Figure 10.4

EXAMPLE 10.2   Find the equation of the ellipse having endpoints of its major axis at (4, 2) and (4, 10) and endpoints of its minor axis at (2, 6) and (6, 6).

SOLUTION   Plotting the given points and lines, we obtain the graph in Figure 10.5.   Therefore, $a = 4$, $b = 2$, $h = 4$, and $k = 6$.   The equation is $[(y - 6)^2/16] + [(x - 4)^2/4] = 1$.

EXAMPLE 10.3   Find the equation of the ellipse with one vertex at (1, 3), one focus at (3, 3), and center at (6, 3).

SOLUTION   Plotting the given points, we obtain the graph in Figure 10.6. Therefore, $a = 5$, $c = 3$, and since $a^2 - c^2 = b^2$, we have $b^2 = 16$ and $b = 4$.   The equation is $[(x - 6)^2/25] + [(y - 3)^2/16] = 1$.

EXAMPLE 10.4   Graph the ellipse with equation $2x^2 + 3y^2 + 12x - 24y + 60 = 0$.

SOLUTION   Completing the square, we obtain

$$2(x^2 + 6x + 9) + 3(y^2 - 8y + 16) = 18 + 48 - 60.$$

Therefore, $2(x + 3)^2 + 3(y - 4)^2 = 6$ and the equation in standard form is $[(x + 3)^2/3] + [(y - 4)^2/2] = 1$.   Consequently, the center is at $(-3, 4)$, $a = \sqrt{3}$, and $b = \sqrt{2}$.   The graph appears in Figure 10.7.

EXAMPLE 10.5   Graph the ellipse with equation $4x^2 + 25y^2 = 100$.

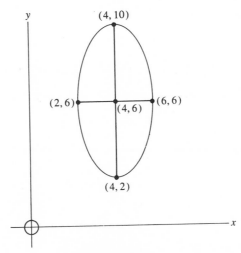

Figure 10.5

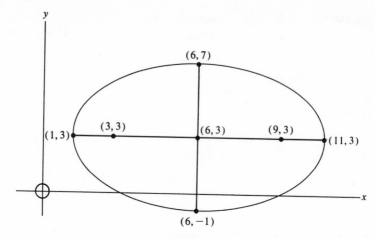

Figure 10.6

SOLUTION We divide by 100 and obtain the standard form of the equation $(x^2/25) + (y^2/4) = 1$. This implies that $a = 5$, $b = 2$, and hence $c = \sqrt{21}$. The graph is shown in Figure 10.8.

EXAMPLE 10.6 Find the equation of the ellipse with center at the origin, major axis on the x axis, and passing through the points $(3, \sqrt{2})$ and $(\sqrt{6}, 2)$.

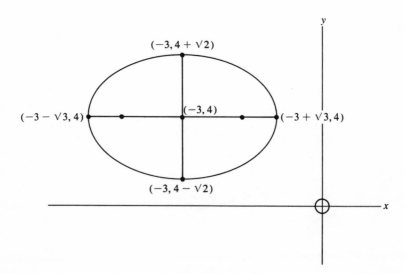

Figure 10.7

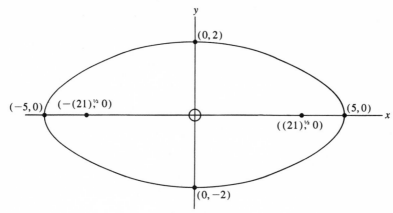

Figure 10.8

SOLUTION The equation has the form $(x^2/a^2) + (y^2/b^2) = 1$. Substituting the given points, we obtain the two equations in two unknowns $(9/a^2) + (2/b^2) = 1$ and $(6/a^2) + (4/b^2) = 1$. Solving these equations simultaneously, we arrive at $a^2 = 12$ and $b^2 = 8$. Therefore, the equation is $(x^2/12) + (y^2/8) = 1$.

PROBLEMS

10.1 Find the equation of the graph of all points the sum of whose distances from the two points (0, 3) and (4, 3) is equal to 8.

10.2 Find the standard form of the equation of each of the following ellipses.
  (a) with one vertex (3, 0) and endpoints of the major axis at (0, 4) and (0, −4)
  (b) with one focus (2, 0), one vertex (5, 0), and center at (0, 0)
  (c) with one vertex (6, 3) and foci at (4, 3) and (−4, 3)
  (d) with center (5, 1), one vertex (5, 4), and one end of the minor axis at (3, 1)

10.3 Graph each of the following equations, carefully labeling all pertinent points and lines.
  (a) $25x^2 + 4y^2 = 100$
  (b) $9x^2 + 4y^2 = 36$
  (c) $4(x + 2)^2 + 3(y - 1)^2 = 12$
  (d) $[(x - 2)^2/16] + [(y + 1)^2/9] = 1$
  (e) $x^2 + 8y^2 = 1$
  (f) $4x^2 + 9y^2 + 8x - 36y + 4 = 0$
  (g) $4x^2 + y^2 + 8x - 4y - 8 = 0$

10.4 Find the equation of at least one ellipse that passes through the points (0, 0), (8, 0), and (4, 2).

THE HYPERBOLA

Definition 11.1 A hyperbola is the graph of all points in the plane the difference between whose distances from two fixed points (the foci) is a constant (see Figure 11.1).

We deal first with hyperbolas in which the line segment through the foci is either vertical or horizontal. Later we touch on the case where this segment is a slant line. Once again we give labeled diagrams (Figures 11.2 and 11.3) showing the terms used to describe the various points and lines connected with a hyperbola.

Using the definition of the hyperbola along with the distance formula, we find that the standard form of the equation of the hyperbola in Figure 11.2 is given by $[(x - h)^2/a^2] - [(y - k)^2/b^2] = 1$. For the hyperbola in Figure 11.3, the corresponding equation is $[(y - k)^2/a^2] - [(x - h)^2/b^2] = 1$.

If we are given the equation of a hyperbola in the form $Ax^2 + By^2 + Cx + Dy + E = 0$, we complete the square on both the x and y terms. This leads to an equivalent equation in the standard form. Hence, we deal only with equations in their standard form.

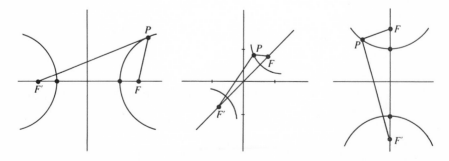

Figure 11.1

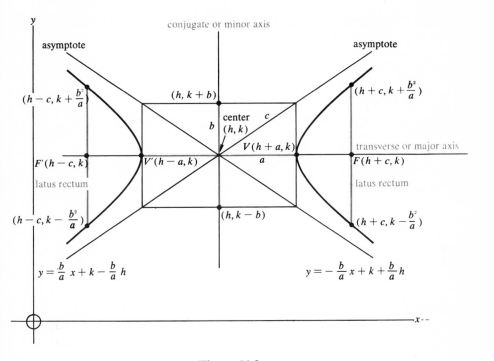

Figure 11.2

EXAMPLE 11.1 Find the equation of the hyperbola with vertices at the points $(-5, 2)$ and $(-1, 2)$ and foci at $(-7, 2)$ and $(1, 2)$.

SOLUTION   Plotting the given points, we obtain the sketch in Figure 11.4. The center is halfway between the vertices, so its coordinates are $(-3, 2)$. Since a is the distance from the center to a vertex, $a = 2$. Since c is the distance from the center to a focus, $c = 4$. Now $b^2 = c^2 - a^2 = 16 - 4 = 12$; therefore, $b = \sqrt{12}$. The equation is $[(x + 3)^2/4] - [(y - 2)^2/12] = 1$.

EXAMPLE 11.2 Find the equation of the hyperbola with foci at $(0, 10)$ and $(0, -10)$ with asymptotes $y = \frac{4}{3}x$ and $y = -\frac{4}{3}x$.

SOLUTION   Since the asymptotes have the form $y = (a/b)x$ and $y = (-a/b)x$, we know that $a/b = \frac{4}{3}$. Since the foci are at $(0, 10)$ and $(0, -10)$, we know that $h = 0, k = 0$, and $c = 10$. But $a^2 + b^2 = c^2 = 100$. Therefore, we must solve $(a/b) = \frac{4}{3}$ and $a^2 + b^2 = 100$ simultaneously. The first equation reduces to $a = (4b/3)$. Substituting into the second equation yields $(16b^2/9) + b^2 = 100$. This reduces to $(25b^2/9) = 100$;

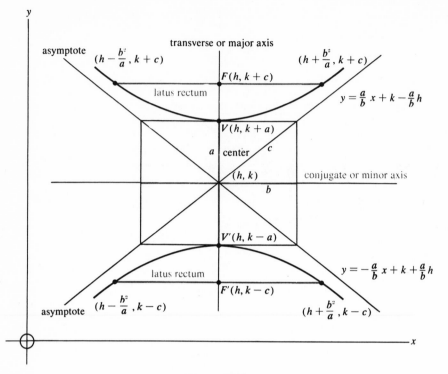

Figure 11.3

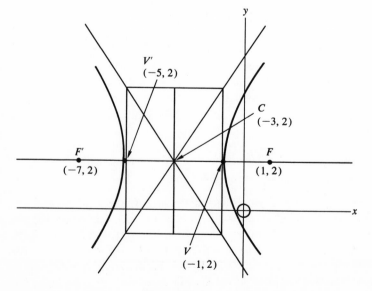

Figure 11.4

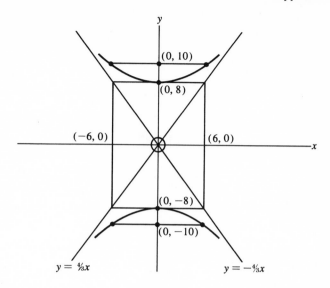

Figure 11.5

hence, $b^2 = 36$. It follows that $a^2 = 64$ and the equation is $(y^2/64) - (x^2/36) = 1$. A sketch of the graph appears in Figure 11.5.

EXAMPLE 11.3 Sketch the graph of the hyperbola with equation $25x^2 - y^2 = 100$, carefully labeling all pertinent points and lines.

SOLUTION Dividing by 100, we obtain the standard form of the equation $(x^2/4) - (y^2/100) = 1$. Therefore, the center is the origin, $a = 2$, $b = 10$, and since $a^2 + b^2 = c^2$, we have $c = (104)^{1/2}$. The asymptotes are $y = \pm 5x$. In sketching the graph, it is best to draw the asymptotes first, mark the vertices, and then draw the curve (see Figure 11.6).

EXAMPLE 11.4 Sketch the graph of the hyperbola with equation $9y^2 - x^2 - 2x - 5 = 0$, carefully labeling all pertinent points and lines.

SOLUTION Completing the square on the x terms, we obtain the equation $9y^2 - (x^2 + 2x + 1) = 4$. Dividing by 4, we arrive at the standard form $[y^2/(4/9)] - [(x + 1)^2/4] = 1$. Therefore, the center is $(-1, 0)$, $a = \frac{2}{3}$, $b = 2$, and since $a^2 + b^2 = c^2$, we have $c = (40/9)^{1/2} = \frac{2}{3}(10)^{1/2}$. The asymptotes are $y = \pm (a/b)x = \pm (\frac{2}{3}/2)x = \pm \frac{1}{3}x$. We sketch the asymptotes first, label the vertices, and then draw the graph (see Figure 11.7).

We now complete our discussion of the hyperbola by considering the class of hyperbolas that have asymptotes equal or parallel to the x and y axes. In this situation, the transverse axis is on or parallel to the line $y = x$ or the line

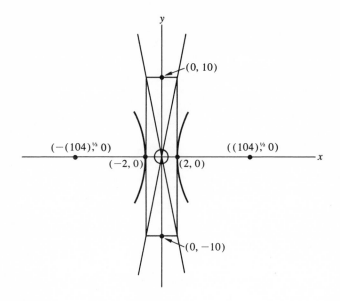

Figure 11.6

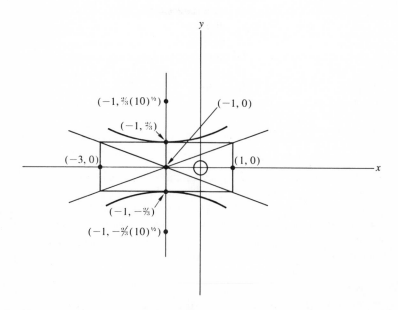

Figure 11.7

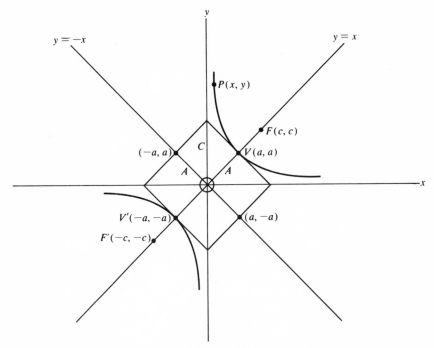

Figure 11.8

$y = -x$. Since the asymptotes are perpendicular, these hyperbolas are all rectangular hyperbolas. This means that $C^2 - A^2 = A^2$, where C is the distance from the center to the foci and A is the distance from the center to the vertices. A sketch of one of these types is shown in Figure 11.8; note that it resembles the sketch in Figure 11.2 after the rotation of the x and y axes through an angle of 45°. In most analytic geometry textbooks, this situation is treated after a discussion of translations and rotations of axes. In some calculus texts, the properties of these hyperbolas are simply stated and examples are given. We attempt a straightforward derivation of the equation (leaving out some of the messy details), using the definition of a hyperbola and the known properties of hyperbolas that have their transverse axis parallel to the x or y axis.

Using the definition of a hyperbola, we know that the distance from a point $P(x, y)$ on this hyperbola to F' minus the distance from P to F must equal $2A$ (check Figure 11.8). Now A clearly equals $\sqrt{2}a$; therefore, we have the equation

$$[(x + c)^2 + (y + c)^2]^{1/2} - [(x - c)^2 + (y - c)^2]^{1/2} = 2\sqrt{2}a.$$

If we rewrite the equation in the form

$$[(x + c)^2 + (y + c)^2]^{1/2} = [(x - c)^2 + (y - c)^2]^{1/2} + 2\sqrt{2}a$$

we can simplify it by first squaring both sides and then, after isolating the one remaining radical on the right side, squaring both sides again.   This is the same technique used in most textbooks when they *derive* the equation of a hyperbola.   This simplification results in the equation

$$(c^2 - 2a^2)x^2 + (c^2 - 2a^2)y^2 + 2c^2xy = 4a^2(c^2 - a^2)$$

But $C = \sqrt{2}c$, $A = \sqrt{2}a$, and since the hyperbola is rectangular, $C^2 - A^2 = A^2$.   Therefore, $C^2 = 2A^2$, $2c^2 = 2(2a^2)$, and we obtain $c^2 = 2a^2$.   Consequently, the coefficient of x^2 and y^2 in the preceding equation is zero and the equation reduces to $2(2a^2)xy = 4a^2(c^2 - a^2)$.   But $c^2 = 2a^2$ means $c^2 - a^2 = a^2$, so we may further reduce the equation to $4a^2xy = 4a^4$.   Hence, the final form of the equation is $xy = a^2$.

If we had computed the equation of the hyperbola with branches in the second and fourth quadrants, we would have obtained the equation $xy = -a^2$. The vertices in this case would be at the points $(-a, a)$ and $(a, -a)$.

Finally, if we had computed the equation of the hyperbola with center (h, k) and asymptotes $x = h$ and $y = k$, we would have obtained the equation $(x - h)(y - k) = a^2$ if the vertices were $(h + a, k + a)$ and $(h - a, k - a)$ (see Figure 11.9) or the equation $(x - h)(y - k) = -a^2$ if the vertices were $(h - a, k + a)$ and $(h + a, k - a)$.

EXAMPLE 11.5   Find the equation of the hyperbola with foci $(2, 2)$ and $(-2, -2)$ having the x and y axes as asymptotes.

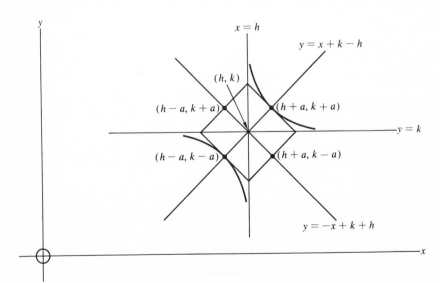

Figure 11.9

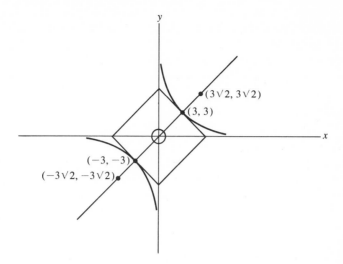

Figure 11.10

SOLUTION   From the preceding, we know that the equation has the form $xy = a^2$ where $c^2 = 2a^2$. Now $c = 2$, $c^2 = 4$, and therefore $a^2 = 2$. The equation is $xy = 2$.

EXAMPLE 11.6 Sketch the graph of the hyperbola with equation $xy = 9$, carefully labeling all pertinent points and lines.

SOLUTION   Once again, from the preceding, $xy = a^2$ is the form of the equation; therefore, $a^2 = 9$ and $a = 3$. Thus, this hyperbola has center $(0, 0)$; its asymptotes are the lines $x = 0$ and $y = 0$; and its vertices are $(3, 3)$ and $(-3, -3)$. The foci are (c, c) and $(-c, -c)$ where $c^2 = 2a^2 = 18$. Therefore, $c = 3\sqrt{2}$ and the foci are $(3\sqrt{2}, 3\sqrt{2})$ and $(-3\sqrt{2}, -3\sqrt{2})$. A sketch of the graph is shown in Figure 11.10.

EXAMPLE 11.7 Sketch the graph of the hyperbola with equation $(x - 2)(y + 3) = -16$, carefully labeling all pertinent points and lines.

SOLUTION   Comparing this equation with the form $(x - h)(y - k) = -a^2$, we see that the center is $(2, -3)$ and the asymptotes are the lines $x = 2$ and $y = -3$. Now $a^2 = 16$; hence, $a = 4$ and since the vertices are $(h - a, k + a)$ and $(h + a, k - a)$, we obtain the points $(6, -7)$ and $(-2, 1)$. Since $c^2 = 2a^2$, we have $c = 4\sqrt{2}$; therefore the foci are $(2 - 4\sqrt{2}, -3 + 4\sqrt{2})$ and $(2 + 4\sqrt{2}, -3 - 4\sqrt{2})$. A sketch of the graph appears in Figure 11.11.

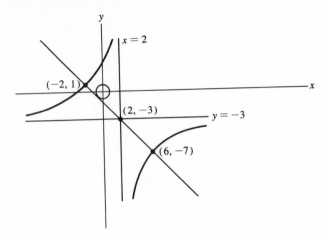

Figure 11.11

PROBLEMS

11.1   Find the equation of the graph of all points the difference between whose distances from the two points (10, 0) and (−10, 0) is equal to 16.

11.2   Find the standard form of the equation of each of the following hyperbolas.
  (a)   with one focus (6, 0) and vertices at (4, 0) and (−4, 0)
  (b)   with one vertex (0, −4) and foci at (0, −5) and (0, 1)
  (c)   with foci at (10, 0) and (−10, 0) having asymptotes $y = \frac{4}{3}x$ and $y = -\frac{4}{3}x$
  (d)   with vertices (7, 9) and (−3, −1) having asymptotes $x = 2$ and $y = 4$
  (e)   with foci (−2, 2) and (2, −2) having the x and y axes as asymptotes

11.3   Graph each of the following equations, carefully labeling all pertinent points and lines.
  (a) $4x^2 - y^2 = 16$
  (b) $16y^2 - x^2 = 1$
  (c) $[(x+1)^2/4] - [(y-2)^2/2] = 1$
  (d) $x^2 - y^2 - 2y = 2$
  (e) $xy = 64$
  (f) $xy = -9$
  (g) $(x+1)y = 2$
  (h) $(x-1)(y+1) = 1$
  (i) $x(y-1) = 3$
  (j) $(x+2)(y-1) = -1$

11.4   Find the equation of at least one hyperbola passing through the points (3, 3), (5, 0), and (5, 6).

INTERSECTIONS OF GRAPHS OF CONICS

We have already dealt with the intersection of two lines, two circles, and a line with a circle. We now discuss the intersection of the graphs of lines with parabolas, ellipses, and hyperbolas. This involves the simultaneous solution of a linear equation with a quadratic. The method is simply to solve the linear equation for either x or y and then substitute into the quadratic equation. Upon solving the resulting quadratic in x or y, we obtain two solutions, one solution, or no real solutions, corresponding to the three geometrical possibilities of two points, one point, or no points of intersection. This is essentially the same method used for finding the intersection points of lines with circles.

EXAMPLE 12.1 Find the intersection points of the parabola $y^2 = 4x$ with the line $y = \frac{1}{2}x + \frac{3}{2}$.

SOLUTION We may substitute $y = \frac{1}{2}x + \frac{3}{2}$ into the parabola equation and solve the resulting quadratic in x. In this case, however, it is simpler to solve the linear equation for x and then substitute into the equation of the parabola. This yields a quadratic in y.

Now, solving the linear equation for x, we arrive at $x = 2y - 3$. Substituting, we obtain $y^2 = 4(2y - 3) = 8y - 12$. The quadratic is therefore $y^2 - 8y + 12 = 0$, which factors into $(y - 6)(y - 2) = 0$. Hence, the roots are $y = 6$ and $y = 2$. The corresponding values of x are $x = 9$ and $x = 1$, respectively. Consequently, the points of intersection are $(9, 6)$ and $(1, 2)$. A sketch appears in Figure 12.1.

EXAMPLE 12.2 Find the intersection points of the parabola $y^2 = 4x$ and the line $y = x + 2$.

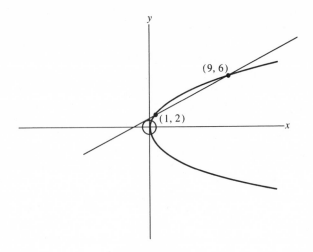

Figure 12.1

SOLUTION Solving the linear equation for x yields $x = y - 2$. Substituting for x in the parabola equation yields the equation $y^2 = 4(y - 2)$, which results in the quadratic $y^2 - 4y + 8 = 0$. But the discriminant of this quadratic is $b^2 - 4ac = (-4)^2 - 4(1)(8) = -16$. Therefore, there are no real roots and the two graphs fail to intersect. Of course, if we had sketched the graphs first, it would have been clear that there was no intersection. Consequently, it is always advisable to draw a rough sketch of the graphs *before* attempting an algebraic solution (see Figure 12.2).

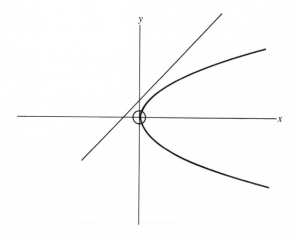

Figure 12.2

EXAMPLE 12.3   Find the intersection points of the hyperbola $xy = -6$ and the line $y = -\frac{3}{2}x + \frac{5}{2}$.

SOLUTION   Substituting $y = -\frac{3}{2}x + \frac{5}{2}$ into $xy = -6$, we obtain the equation $x(-\frac{3}{2}x + \frac{5}{2}) = -6$.   This results in the quadratic $-\frac{3}{2}x^2 + \frac{5}{2}x + 6 = 0$, which simplifies to $3x^2 - 5x - 12 = 0$.   Factoring, we obtain $(3x + 4)(x - 3) = 0$, therefore the roots are $x = -\frac{4}{3}$ and $x = 3$.   The corresponding y values are $y = 9/2$ and $y = -2$ and the points of intersection are $(-\frac{4}{3}, \frac{9}{2})$ and $(3, -2)$.   As an exercise, plot the graphs and indicate the points of intersection.

We now come to a discussion of the intersection of the graphs of circles, parabolas, ellipses, and hyperbolas with each other.   In the case of two circles, we already have a technique, but in the other situations, there are no standard techniques.   Sometimes subtracting the given equations yields a simple equation in one variable whose roots lead to the points of intersection of the graphs.   For example, in finding the points of intersection of the parabola $y^2 = 4x$ and the circle $x^2 + y^2 = 4$, subtraction yields the quadratic $x^2 + 4x - 4 = 0$.   Clearly, the x coordinate of any intersection point of the two graphs must also satisfy this quadratic equation; therefore, we find the roots of this quadratic equation in order to obtain the points of intersection.   It turns out that the root $x = -2 + 2\sqrt{2}$ gives rise to the two points of intersection $(-2 + 2\sqrt{2}, [-8 + 8\sqrt{2}]^{1/2})$ and $(-2 + 2\sqrt{2}, -[-8 + 8\sqrt{2}]^{1/2})$. If we try the same method on the equations $x^2 + y^2 = 4$ and $4x^2 + 3y^2 - x + 2y = 7$, however, we obtain another equation of the second degree in two variables.   Even multiplying the first equation by 4, thus eliminating the x^2 term during the subtraction, would still leave us with an equation in two variables of the second degree in y.   This means that the subtraction method does not work quite as well in general for arbitrary conics as it did in the case of circles.

The only general technique is the one that involves the solving of one of the two given equations for one of the variables and then substituting into the other one, which reduces the problem to the solution of an equation in one variable.

EXAMPLE 12.4   Find the intersection points of the parabola $y = x^2$ and the circle $x^2 + y^2 = 1$.

SOLUTION   The parabola equation is already written in a form that can be substituted into the circle equation to yield the quadratic $y^2 + y - 1 = 0$, which has roots $y = (-1 \pm \sqrt{5})/2$.   The value $y = (-1 + \sqrt{5})/2$ gives rise to two values of x (since $x = \pm\sqrt{y}$), namely, $x = [(-1 + \sqrt{5})/2]^{1/2}$

and $x = -[(-1 + \sqrt{5})/2]^{1/2}$. Therefore, two points of intersection are $([(-1 + \sqrt{5})/2]^{1/2}, \; (-1 + \sqrt{5})/2)$ and $(-[(-1 + \sqrt{5})/2]^{1/2}, (-1 + \sqrt{5})/2)$.

Now, the value $y = (-1 - \sqrt{5})/2$ yields two imaginary values of x, namely, $x = [(-1 - \sqrt{5})/2]^{1/2}$ and $x = -[(-1 - \sqrt{5})/2]^{1/2}$. Therefore, there are only two points of intersection. As an exercise, plot the graphs and indicate the points of intersection.

EXAMPLE 12.5 Find the intersection points of the hyperbola $xy = 1$ and the ellipse $x^2 + 4y^2 = 4$.

SOLUTION Solving the hyperbola equation for y yields $y = 1/x$. Substituting into the ellipse equation results in $x^2 + (4/x^2) = 4$, which reduces to $x^4 - 4x^2 + 4 = 0$. This is a quadratic in x^2 that gives $x^2 = [4 \pm (16 - 16)^{1/2}]/2 = 2$. Therefore, $x = \pm\sqrt{2}$ and the corresponding y values are $y = \pm 1/\sqrt{2}$. Therefore, the points of intersection are $(\sqrt{2}, 1/\sqrt{2})$ and $(-\sqrt{2}, -1/\sqrt{2})$. A sketch appears in Figure 12.3.

In the case of the intersection of two circles, there were at most two points of intersection. In the preceding examples involving general conics, we also dealt with problems where there were at most two intersection points. We now discuss some problems where the intersection of two conics involves more than two points. Some sample graphs depicting this situation are shown in Figure 12.4.

EXAMPLE 12.6 Find the intersection points of the circle $x^2 + y^2 = 3$ and the hyperbola $x^2 - y^2 = 1$.

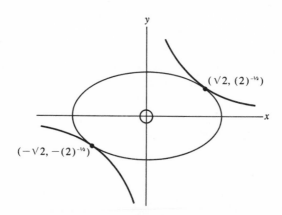

Figure 12.3

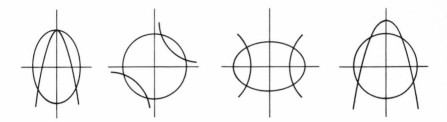

Figure 12.4

SOLUTION   Adding the two equations results in the equation $2x^2 = 4$ or $x^2 = 2$. Therefore, $x = \pm \sqrt{2}$ and since $y = \pm (x^2 - 1)^{1/2}$, we have the four points of intersection $(\sqrt{2}, 1)$, $(\sqrt{2}, -1)$, $(-\sqrt{2}, 1)$, and $(-\sqrt{2}, -1)$. As an exercise, plot the graphs and indicate the points of intersection.

EXAMPLE 12.7   Find the intersection points of the parabola $y = x^2 - 1$ and the hyperbola $y^2 - x^2 = 1$.

SOLUTION   Solving the first equation for x^2 yields $x^2 = y + 1$. Substituting for x^2 in the second equation results in the equation $y^2 - (y + 1) = 1$, which reduces to $y^2 - y - 2 = 0$. Factoring, we obtain $(y - 2)(y + 1) = 0$ and the roots are $y = 2$ and $y = -1$. Now $x^2 = y + 1$; therefore, $x = \pm (y + 1)^{1/2}$ and the values $y = 2$ and $y = -1$ give $x = \pm \sqrt{3}$ and $x = 0$, respectively. Hence, the three points of intersection are $(\sqrt{3}, 2)$, $(-\sqrt{3}, 2)$, and $(0, -1)$. A sketch of the graphs appears in Figure 12.5.

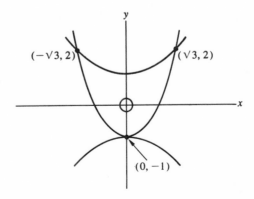

Figure 12.5

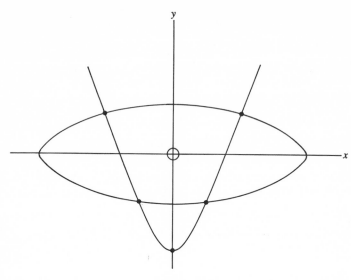

Figure 12.6

EXAMPLE 12.8   For each of the equations $y = x^2 - 2$ and $x^2 + 16y^2 = 16$, sketch the graph and determine how many points of intersection exist.

SOLUTION $y = x^2 - 2$ in standard form is $x^2 = (y + 2)$.   Therefore, the vertex of this parabola is $(0, -2)$ and it opens upward.   The equation $x^2 + 16y^2 = 16$ in standard form is $(x^2/16) + y^2 = 1$.   Therefore, the graph is an ellipse with center at the origin and vertices $(4, 0)$, $(-4, 0)$, $(0, 1)$, and $(0, -1)$.   A sketch of the graphs is shown in Figure 12.6. Therefore, there are *four* intersection points.

Finally, as an exercise, solve the two equations simultaneously and find the coordinates of the points of intersection.

PROBLEMS

12.1   Find the intersection points of the graphs of each of the following pairs of equations and plot the graphs.
(a) $y = \frac{1}{2}x + 2$ $y = 2x^2$
(b) $y = 2/x$, $y = x + 1$
(c) $y = x$ $x^2 + 3y^2 = 1$
(d) $y = x^2$ $x^2 + y^2 = 2$
(e) $y = 2x^2$ $2x^2 + y^2 = 2$
(f) $x^2 - y^2 = 1$ $3x^2 + y^2 = 1$
(g) $x^2 - y^2 = 1$ $xy = 2$
(h) $x^2 + 2y^2 = 1$ $2x^2 + y^2 = 1$
(i) $xy = 12$ $x + y = 7$

(j) $x^2 - y^2 = 5$ $3x + y = 11$

(k) $y = x^2 - 2$ $x^2 + 16y^2 = 16$

(l) $y^2 - x = 10$ $x^2 - y^2 = 1$

(m) $(x-4)^2 + 4(y-2)^2 = 16$ $(x-4)^2 - 4(y-2)^2 = 4$

12.2 Graph each of the following equations and label all points of intersection.

$(x-1)^2 + y^2 = 1$ $y = x^2$ $y = x + 7$

12.3* Graph each of the following equations and label all points of intersection.

$(x-2)^2 + y^2 = 4$ $y = (\sqrt{3}/3)x + 2\sqrt{3}$ $(x+2)^2 + y^2 = 4$

$y = -(\sqrt{3}/3(x + 2\sqrt{3}$ $x^2 - y^2 = 36$

Section 13

MISCELLANEOUS GRAPHS
AND EQUATIONS

The general problem of finding the graph of a given equation is not as simple as the problems associated with conics. In many cases, however, we can plot points with a reasonable amount of accuracy and then sketch the curve. For example, the graph of $y = x^3$ can be found by simply plotting the points $(0, 0)$, $(1, 1)$, $(-1, -1)$, $(2, 8)$, $(-2, -8)$, $(3, 27)$, $(-3, -27)$, and so on. A sketch is shown in Figure 13.1.

In most cases, we can analyze the equation and determine such things as intercepts, asymptotes, and the behavior of the graph for large and small

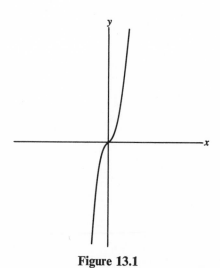

Figure 13.1

64

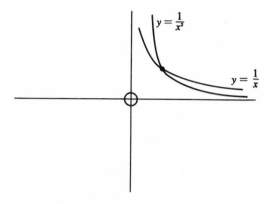

Figure 13.2

values of x. For example, the graph of $y = 1/x^2$ in the first quadrant closely resembles the graph of the hyperbola $y = 1/x$, since for small $x > 0$, y is very large, whereas for large $x > 0$, y is very close to zero. Both graphs have the positive x axis and the positive y axis as asymptotes (see Figure 13.2). As an exercise, plot the graph of $y = 1/x^2$ in the second quadrant. Which hyperbola does the graph resemble?

Other examples are the graphs of $y = x^4$, $y = x^6$, and $y = x^{2n}$ for n equal to any positive integer. All these graphs, although not parabolic, are similar to the graph of $y = x^2$ since they all pass through the points (0, 0) and (1, 1) and they all take on large positive values of y for large positive or negative values of x (see Figure 13.3).

Consequently, when we see an equation that does not fit into a known category, we can either plot points or look for similarities among the graphs we know. We now discuss a few special types of graphs.

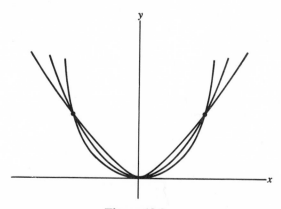

Figure 13.3

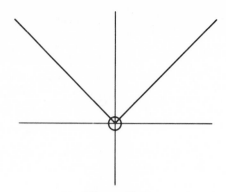

Figure 13.4

EXAMPLE 13.1 Find the graph of each of the following absolute value equations: (a) $y = |x|$, (b) $y = |x + 2|$, (c) $y = |x^2 - 1|$.

(a) In dealing with equations involving absolute values, we must allow for the situations when the expression inside the absolute value symbol is positive and when it is negative. For example, consider $y = |x|$. If $x \geq 0$, then $y = |x| = x$ and the graph is simply the set of points (x, x) that lie on the line $y = x$ in the first quadrant. For $x \leq 0$, we see that $y = |x| = -x$ and the graph is the set of points $(x, -x)$ that lie on the line $y = -x$ in the second quadrant. The complete graph is shown in Figure 13.4.

(b) A problem like $y = |x + 2|$ is handled in the same way. We first take $x + 2 \geq 0$. This means that $|x + 2| = x + 2$. But $x + 2 \geq 0$ only if $x \geq -2$. Therefore, the graph of $y = |x + 2|$ for $x \geq -2$

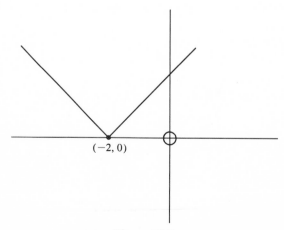

$(-2, 0)$

Figure 13.5

consists of the points on the line $y = x + 2$. On the other hand, if we take $x + 2 \leq 0$, then $|x + 2| = -x - 2$. But $x + 2 \leq 0$ only if $x \leq -2$. Therefore, the graph of $y = |x + 2|$ for $x \leq -2$ consists of the points on the line $y = -x - 2$. A sketch of the entire graph appears in Figure 13.5; note that it is identical to the one in Figure 13.4 except that the vertex is $(-2, 0)$ instead of $(0, 0)$. We can actually generalize to the statement: the graph of $y = |x - h|$ is the same as $y = |x|$ with the vertex moved to the point $(h, 0)$.

QUESTION Can you state that the graph of $y - k = |x - h|$ is the same as $y = |x|$ with vertex (h, k) instead of $(0, 0)$?

(c) When dealing with an equation like $y = |x^2 - 1|$, we continue the process of first taking $x^2 - 1 \geq 0$ and then taking $x^2 - 1 \leq 0$. In this case, $x^2 - 1 \geq 0$ if $x^2 \geq 1$, which means $x \geq 1$ or $x \leq -1$. Therefore, when $x \geq 1$ or $x \leq -1$, the graph of $y = |x^2 - 1|$ coincides with the parabola $y = x^2 - 1$ or $y + 1 = x^2$. However, when $x^2 - 1 \leq 0$, we have $x^2 \leq 1$; hence, x lies between -1 and $+1$ in this case. Therefore, the graph of $y = |x^2 - 1|$ between $x = -1$ and $x = +1$ coincides with the parabola $y = -x^2 + 1$ or $x^2 = -(y - 1)$. The solid curve in Figure 13.6 represents the graph of $y = |x^2 - 1|$. Note that this curve is just the reflection or mirror image of the dotted curve in the x axis. Consequently, in dealing with the preceding three equations, a simple method for obtaining the graph is to plot the graph of each equation without the absolute value signs and then reflect in the x axis that part of the graph which lies below the x axis. Use this method on the preceding equations and then determine why it works.

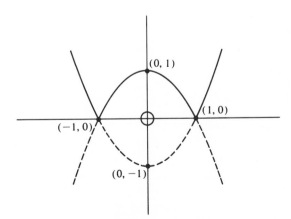

Figure 13.6

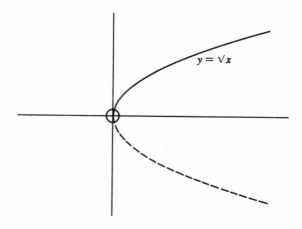

$y = \sqrt{x}$

Figure 13.7

EXAMPLE 13.2   Find the graph of each of the following square root equations:

(a) $y = \sqrt{x}$;   (b) $y = \sqrt{1 - x^2}$;   (c) $y = \sqrt{x^2 - 1}$.

(a) We observe that for all $x < 0$, y is imaginary. Therefore, there are no points on this graph to the left of the y axis. We also notice that, if we observe the convention that $\sqrt{a}$ always equals the positive square root of a, y can never be negative. Hence there exist no points of this graph below the x axis. Finally, squaring both sides of the equation, we obtain the relation $y^2 = x$. Therefore, the graph coincides with the graph of the parabola $y^2 = x$ in the *first* quadrant. The *solid* curve in Figure 13.7 represents the graph of $y = \sqrt{x}$. Note that if we were given the equation $y = -\sqrt{x}$, we would have the lower half of the parabola $y^2 = x$, that is, the dotted curve in Figure 13.7.

(b) The graph of $y = (1 - x^2)^{1/2}$ cannot possess any points with x coordinates greater than 1 or less than -1 (since y would then be imaginary). Also, y is always positive or zero since $(1 - x^2)^{1/2}$ is always positive or zero. We square both sides and obtain $y^2 = 1 - x^2$, which reduces to $x^2 + y^2 = 1$. Therefore, the graph of $y = (1 - x^2)^{1/2}$ coincides with the upper half of the circle $x^2 + y^2 = 1$. If we were given $y = -(1 - x^2)^{1/2}$, the corresponding graph would be the lower half of the circle $x^2 + y^2 = 1$. A sketch appears in Figure 13.8.

(c) Finally, we observe that the graph of $y = (x^2 - 1)^{1/2}$ has no points with x coordinate between -1 and $+1$. Since $x^2 - 1 \geq 0$, there are no points below the x axis. Squaring both sides, we obtain the equation

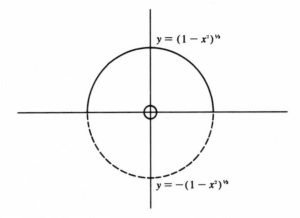

Figure 13.8

$y^2 = x^2 - 1$, which simplifies to $x^2 - y^2 = 1$. Therefore, the graph of $y = (x^2 - 1)^{1/2}$ coincides with the section of the graph of the hyperbola $x^2 - y^2 = 1$ that appears above the x axis. The solid curve in Figure 13.9 represents the graph.

QUESTION   What is the equation corresponding to the section of the graph of the hyperbola $x^2 - y^2 = 1$ that appears below the x axis (the dotted curve in Figure 13.9)?

EXAMPLE 13.3   Find the graph of each of the following exponential equations:
(a) $y = 10^x$,   (b) $y = 10^{-x}$.

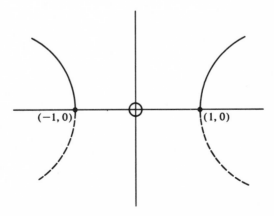

Figure 13.9

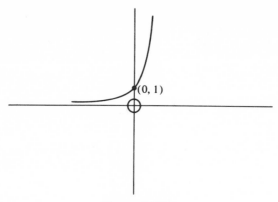

Figure 13.10

(a) For large positive x, 10^x clearly is large and positive. Therefore, the y coordinates are very large and positive. When x is large and negative, $y = 1/10^{|x|}$, which is close to zero. Finally, when $x = 0, y = 10^0 = 1$. Therefore, the graph (see Figure 13.10) has the negative x axis as an asymptote on the left, passes through the point $(0, 1)$, and continues in an upward direction to the right of $(0, 1)$. (At present we must assume that numbers like $10^{\sqrt{2}}$ and 10^π can be defined in a meaningful way later on.)

(b) For large positive x, $10^{-x} = 1/10^x$; hence, y is very close to zero. On the other hand, when x is large and negative, $10^{-x} = 10^{|x|}$; hence, y is a very large positive number. Also, when $x = 0, 10^{-x} = 1$. Therefore, this graph (see Figure 13.11) comes down from above and to the left of the point $(0, 1)$ and continues through $(0, 1)$ to approach the x axis on the right. Notice that the graph in Figure 13.11 is just a reflection or mirror image of the graph in Figure 13.10 in the y axis.

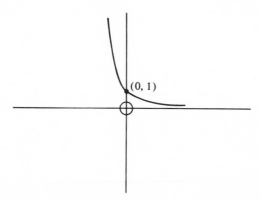

Figure 13.11

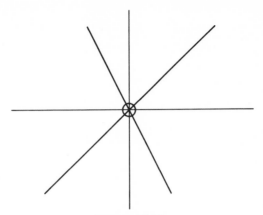

Figure 13.12

EXAMPLE 13.4 *The Exceptional Cases* (a) There are equations, such as $2x^2 + 3y^2 = -7$, that have graphs consisting of no points at all. For every real value of x and y, x^2 and y^2 are either positive or zero. Hence, $2x^2 + 3y^2 \geq 0$ for all real numbers x and y. Therefore, $2x^2 + 3y^2 = -7$ has no real solutions, and there are no points that have coordinates that satisfy this equation. Consequently, no graph exists.

(b) There are equations such as $(x - 1)^2 + (y + 2)^2 = 0$, that have as their graph a single point. For all real values of $x \neq 1$, $(x - 1)^2 > 0$; for all real values of $y \neq -2$, $(y + 2)^2 > 0$. Therefore, $(x - 1)^2 + (y + 2)^2 = 0$ if and only if $x = 1$ and $y = -2$. Hence, this equation has a graph consisting of the single point $(1, -2)$.

(c) There are equations of the second degree in x and y that look like equations of conics, yet turn out to have graphs that are one or more lines. The equation $4x^2 + 4xy + y^2 = 0$ can be factored into $(y + 2x)^2 = 0$. Therefore, the graph of this equation is the set of points (x, y) with the property that $y + 2x = 0$ or $y = -2x$. Hence, the graph is one line with equation $y = -2x$.

The equation $y^2 + xy - 2x^2 = 0$ can be factored into

$$(y - x)(y + 2x) = 0.$$

Therefore, the graph of this equation consists of the two lines $y = x$ and $y = -2x$ (see Figure 13.12).

PROBLEMS

Graph each of the following equations.

13.1 (a) $y = \sqrt{x}$
(b) $y = (x - 1)^{1/2}$

13.2 $x = \sqrt{y}$

13.3   (a) $y = x^3$

      (b) $y = (x + 1)^3$

13.4 $y = \sqrt[3]{x}$

13.5 $y = -\sqrt[3]{x}$

13.6   (a) $y = 1/x$

      (b) $y = 1/(x - 2)$

13.7 $y = -(1/x)$

13.8   (a) $y = 1/x^2$

      (b) $y = 1/(x + 1)^2$

13.9 $y = -(1/x^2)$

13.10 $y = |x|$

13.11 $y = -|x|$

13.12   (a) $y = |1 + x|$

       (b) $y = |x - 1|$

13.13   (a) $y = |1 - x^2|$

       (b) $y = |x^2 - x|$

       (c) $y = |x^2 - 2x|$

13.14   (a) $y = (4 - x^2)^{1/2}$

       (b) $y = -(4 - x^2)^{1/2}$

13.15   (a) $y = (x^2 - 4)^{1/2}$

       (b) $y = -(x^2 - 4)^{1/2}$

13.16   (a) $y = 2^x$

       (b) $y = 3^x$

13.17   (a) $y = 2^{-x}$

       (b) $y = 3^{-x}$

13.18 $y = a^x$ $a > 1$

13.19 $y = a^x$ $0 < a < 1$

13.20 $x^2 - y^2 = 0$

13.21 $x^2 + y^2 = 0$

13.22 $x^2 + y^2 = -1$

13.23 $x^2 + 2xy + y^2 = 0$

13.24* $|x + |y|| + ||x| + y| = 2$

Find the equation of each of the following graphs.

13.25   The portion of the circle $x^2 + y^2 = 4$ that is contained in the upper half plane.

13.26   The portion of the circle $x^2 + y^2 = 4$ that is contained in the lower half plane.

13.27   The portion of the parabola $y^2 = x$ that is contained in the upper half plane.

13.28   The portion of the ellipse $4x^2 + y^2 = 1$ that is contained in the lower half plane.

13.29   The portion of the hyperbola $y^2 - x^2 = 1$ that is contained in the upper half plane.

TANGENTS TO CONICS

We conclude our treatment of analytic geometry with a brief discussion of the notion of a tangent line to the graph of an *arbitrary conic*. We have already mentioned tangent lines to *circles* in Example 7.4 and Problem 7.3. In this case, we saw that the tangent line intersected the circle at only one point and was perpendicular to the radius line drawn to this point. We must now attempt to find a definition of a tangent line to an *arbitrary* conic that agrees with our *intuitive* notion of a tangent line. For example, in Figures 14.1 and 14.2, the lines through the points P intersect the graph at only one point, yet in each case only one of these lines (the one labeled T) fits our intuitive notion of a tangent line (you should recognize the graph in Figure 14.2 as part of the graph of $y = 1/x$). We now give a precise definition of the tangent line to the graph of a conic at a point P.

Definition 14.1 Let P be any point on the graph of an arbitrary conic (see Figure 14.3). Choose any sequence of points on the graph near P in the

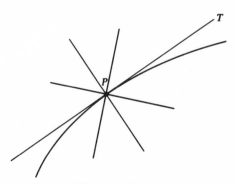

Figure 14.1

73

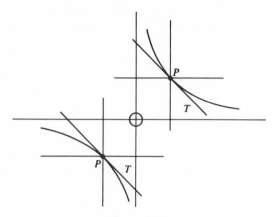

Figure 14.2

following way. P_1 is a point to the right and near P; P_2 is a point to the left and near P; P_3 is a point to the right closer to P than P_1; P_4 is a point to the left closer to P than P_2 ; and so on, where the odd-numbered points $P_1, P_3, P_5, \ldots$ approach P from the right, while the even-numbered points $P_2, P_4, P_6, \ldots$ approach P from the left.   From each point $P_i, i = 1, 2, 3, \ldots,$ draw the secant line to P.   Observe that, on either side of P, the shorter secant lines approach the line T where T is the line that best fits our intuitive idea of a tangent line at P.   Consequently, we *define* the tangent line at the point P to be the unique line T that is the "limiting line" of the secant lines from P_i to P for large values of i.   Note also that T does, in fact, intersect the graph only at the point P.

Now, if there exist no points of the graph on one side of the point P, we choose the sequence of points $P_1, P_2, P_3, \ldots$ on the other side only and still obtain a limiting line.   For example, $y = \sqrt{x}$ at $(0, 0)$ or $y = (x^2 - 1)^{1/2}$ at $(-1, 0)$ (see Figures 13.7 and 13.9).

Observe also that in all cases except that of a vertical tangent, the slopes of the secant lines from P_i to P for large values of i must closely approximate the slope m of the limiting line T.

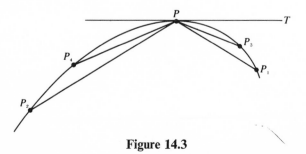

Figure 14.3

The preceding discussion leads to a technique for actually calculating the tangent line to a point on the graph of an arbitrary conic. We now describe the general technique and conclude this section with a few examples.

Once again, we use Figure 14.3. Let P_i, for some fixed positive integer i, have coordinates (x_i, y_i) and let P have coordinates (a, b). Then, the secant line S_i joining P_i to P has equation $y - b = m_i(x - a)$ where $m_i = (y_i - b)/(x_i - a)$. If the original graph is a conic, it will have a second-degree equation in x and y corresponding to it. If we solve the equation for S_i simultaneously with the equation of the conic (using the methods of Section 12), we obtain the distinct intersection points P_i and P. But the tangent line T is the limiting line of the secant lines S_i and intersects the graph only at P. Hence, T can be thought of as the secant line that intersects the graph in *two* points, *both of which are P.* If we assume that T has slope m, then T has equation $y - b = m(x - a)$. Solving simultaneously with the equation of the conic must yield a quadratic equation in x with terms in m, and since P is the only intersection point, this quadratic can have only one distinct root. This means its discriminant (which will depend on m) must be zero, which gives rise to a quadratic equation in m that has *one* distinct root. It is therefore this value of m that yields the slope of the required tangent line T.

EXAMPLE 14.1 Find the equation of the tangent line to the graph of the equation $y = x^2$ at the point $(2, 4)$.

SOLUTION The graphs appear in Figure 14.4. Suppose the slope of the required tangent line is m. Then the equation of the tangent line is $y - 4 = m(x - 2)$ or $y = mx + 4 - 2m$. The intersection of the line and the parabola $y = x^2$ is obtained by setting the two expressions in x equal to each other; that is, $x^2 = mx + 4 - 2m$. This leads to the quadratic

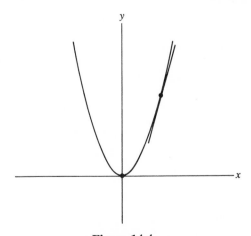

Figure 14.4

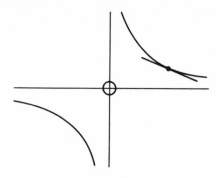

Figure 14.5

equation $x^2 - mx + (2m - 4) = 0$. But m is the slope of the tangent line; hence, we look for a value of m for which this quadratic will have only *one* distinct root. This happens when the discriminant

$$m^2 - 4(2m - 4) = 0.$$

This leads to the quadratic equation $m^2 - 8m + 16 = 0$, which factors into $(m - 4)^2 = 0$. Hence, $m = 4$ yields one distinct root in the preceding quadratic in x, namely, $x = 2$. Therefore, the equation of the required tangent line is $y = 4x - 4$.

As an exercise, prove that the slope of the tangent line to $y = x^2$ at $(0, 0)$ is zero. Also, find a point on this graph where the slope of the tangent line is 1.

EXAMPLE 14.2   Find the equation of the tangent line to the graph of the equation $xy = 1$ at the point $(2, \frac{1}{2})$.

SOLUTION   The graphs appear in Figure 14.5.   Let m be the slope of the required tangent line.   Then the equation of the tangent line is $y - \frac{1}{2} = m(x - 2)$ or $y = mx - 2m + \frac{1}{2}$.   To find the intersection of the tangent line with the hyperbola $y = 1/x$, we set $mx - 2m + \frac{1}{2} = 1/x$.   This leads to the quadratic equation $mx^2 + (\frac{1}{2} - 2m)x - 1 = 0$.   Once again, this equation will have only one distinct root if its discriminant $(\frac{1}{2} - 2m)^2 - 4m(-1) = 0$.   This reduces to the quadratic equation $4m^2 + 2m + \frac{1}{4} = 0$, which factors into $(2m + \frac{1}{2})^2 = 0$.   Hence, $m = -\frac{1}{4}$ is the slope of the tangent line and the equation is $y = -\frac{1}{4}x + 1$.

As an exercise, find the slope of the tangent line to $xy = 1$ at $(1, 1)$.   Also, find a point on the graph of $xy = 1$ where the slope of the tangent line is equal to -4.   Is there more than one such point?

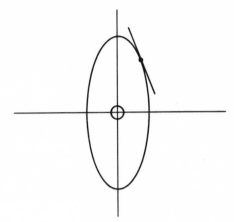

Figure 14.6

EXAMPLE 14.3 Find the equation of the tangent line to the graph of the equation $2x^2 + y^2 = 6$ at the point $(1, 2)$.

SOLUTION   The graphs appear in Figure 14.6.   Let the equation of the tangent line be given by $y - 2 = m(x - 1)$ or $y = mx + 2 - m$.   Substituting this value of y into the equation of the ellipse yields the equation $2x^2 + [m^2x^2 + 2m(2 - m)x + (2 - m)^2] = 6$.   This simplifies to $(2 + m^2)x^2 + 2m(2 - m)x + (m^2 - 4m - 2) = 0$.   The discriminant is $4m^2(2 - m)^2 - 4(2 + m^2)(m^2 - 4m - 2)$, and setting this equal to zero leads to (after some cancellation) $16m^2 + 32m + 16 = 0$, which factors into $16(m + 1)^2 = 0$.   Hence, $m = -1$ is the slope of the tangent line and the equation is $y = -x + 3$.

As an exercise, prove that the tangent to the graph of $2x^2 + y^2 = 6$ at $(\sqrt{3}, 0)$ is a *vertical* line.   Also, find a point on this graph where the slope of the tangent line is equal to -1.   Is there more than one such point?

PROBLEMS

14.1   Find the equation of the tangent line to the graph of each of the following equations at the indicated point.   Sketch the graphs and the tangent lines.
(a) $y = 5x^2$ $(1, 5)$
(b) $y = x^2 + x + 1$ $(-1, 1)$
(c) $xy = 2$ $(2, 1)$
(d) $xy = -1$ $(\tfrac{1}{2}, -2)$
(e) $2x^2 + y^2 = 3$ $(1, 1)$
(f) $x^2 - y^2 = 1$ $(1, 0)$
(g) $y^2 - x^2 = 1$ $(0, -1)$

14.2* Find the tangent of the *angle* between the tangent lines to the graphs of the parabola $y = x^2 - 1$ and the hyperbola $y^2 - x^2 = 1$ at each of their points of intersection. (See Example 12.7 and Theorem 4.1.)

14.3 In each of the following, find a point on the graph of the given equation where the tangent line has the given slope.

(a) $y = 5x^2$ slope 2

(b) $xy = 2$ slope -8

(c) $2x^2 + y^2 = 3$ slope -4

14.4* Using Definition 14.1, determine whether each of the following graphs has a tangent line at the given points.

(a) $y = |x|$ (1, 1) and (0, 0)

(b) $y = |x^2 - 1|$ $(-1, 0), (0, 1), (1, 0)$ (See Figure 13.6).

(Note that neither one of these equations yields the graph of a conic; hence, not all of the results discussed in this section apply.)

14.5 (a) Are either of the lines $x = 1$ or $y = 1$ tangent to the graph of $xy = 1$ at the point (1, 1)? If neither one is, find the tangent line at (1, 1). (b) May we define the tangent line to a point P on the graph of an arbitrary conic to be the unique line through P intersecting the graph at no other point? Explain your answer.

Part II

DIFFERENTIAL CALCULUS

FUNCTIONS

In this section, we explain the concept of a function. To do this precisely, we use the two basic methods of defining a function: first, the notion of a rule of correspondence or a mapping; and second, the notion of an ordered pair of numbers. We give examples and problems that require your becoming familiar with both definitions. Each one has its good and bad features and we attempt to draw on the good features of both definitions.

Definition 15.1 Let X and A be two sets of real numbers. A *function f* is a *rule* that assigns to each number x in X a *unique* number y in A. We usually indicate this assignment by writing $y = f(x)$ and we call y the *value* of the function f at the number x. We could also say that a function f is a *mapping* of the set X into the set A, denoting this by the expression $f: X \rightarrow A$. Then if $y = f(x)$, we say that f *maps* or *takes* the number x to the number y, denoting this by the expression $f: x \rightarrow y$, and we call y the *image* of x under f.

The set X is called the *domain* of the function f. If we let Y equal the subset of A consisting of all the numbers in A that are images of numbers in X, the set Y is called the *range* of the function f.

EXAMPLE 15.1 Let X be the set of all positive *odd* integers. Let A be the set of all integers. Let f be the rule that assigns to each number x in X the number y given by $x + 1$ in A. Then $f(x) = x + 1$ for each x in X, X is the domain of f, and the range Y is the subset of A consisting of all positive *even* integers. For example, in the particular case when $x = 1$, we say that f *assigns* the number 2 in Y to the number 1 in X, or that f maps the number 1 to the number 2, or that 2 is the image of 1 under f. We write $f(1) = 2$, $2 = f(1)$, or $f: 1 \rightarrow 2$. When $x = 3$, we have $f(3) = 4$ or $f: 3 \rightarrow 4$, and so on.

81

EXAMPLE 15.2 Let X and A both be the set of all real numbers. Let f be the function that assigns to each real number in X its square. Therefore, under this function, to each x in X we assign the number y given by x^2 and we may write $f(x) = x^2$. The range Y is the subset of A consisting of all the nonnegative real numbers. In the particular case when $x = 2$, we say that f assigns the number 4 to the number 2, or that f maps the number 2 to the number 4, or that 4 is the image of 2 under f. We write $f(2) = 4$, $4 = f(2)$, or $f: 2 \rightarrow 4$.

There is an alternate definition for a function that is equivalent to the preceding one. We give it now, along with some examples.

Definition 15.2 Let X and A be two sets of real numbers. A *function f* is a *subset* of the set of all ordered pairs (x, a) where x is in X and a is in A, with the additional property that each number x in X appears *exactly* once as the first member of an ordered pair in f. The set X is called the *domain* of f and the set Y, consisting of all the numbers that appear as the second member of an ordered pair in f, is called the *range* of f. The word " exactly " implies that every number x in X appears *once* and *only once* as the first member of an ordered pair in f. This rules out the possibility that two distinct ordered pairs in f have the same first member.

We now show the relationship between these two definitions of a function. Suppose we are given a function f with domain X and range Y, according to Definition 15.1. Then the set of ordered pairs that defines f according to Definition 15.2 would be the set of pairs (x, y) where x is in X and $y = f(x)$, or simply the set of pairs $(x, f(x))$ where x is in X. Now suppose we are given a function f defined by its set of ordered pairs (x, y) where x is in X and y is in Y. Each pair (x, y) in f can be used to associate the number y with the number x. Hence, each pair (x, y) assigns the number y in Y to the number x in X and this is what we mean by $y = f(x)$ in Definition 15.1. The additional condition in Definition 15.2, which states that each x in X appears *exactly* once as a first member of an ordered pair, guarantees that every number in X is actually assigned to a number in Y and that no number x is assigned to more than one number in Y. But this is what we meant by the uniqueness condition given in Definition 15.1 [viz., if $y = f(x)$ and $w = f(x)$, then $y = w$]. Conversely, given Definition 15.1, the fact that each x in X is assigned to a *unique* number y in Y implies that one and only one ordered pair (x, y) can be made with the number x as the first member.

For instance, in Example 15.1, the function f could have been defined by the set of all ordered pairs of the type $(x, x + 1)$ where x is in the set X of all positive odd integers. Observe that f does not contain *arbitrary* ordered pairs (m, n), where m is odd and n is even, such as $(1, 4)$, $(3, 2)$, or $(7, 10)$; but f contains only such pairs where the first member m is positive and odd and the second member is $m + 1$, such as $(11, 12)$, $(27, 28)$, and $(7, 8)$.

In the case of Example 15.2, the function f could have been defined as the set of all ordered pairs (x, x^2) where x is in the set X of all real numbers. Observe that the pair $(2, 4)$ is in f, which means that $4 = f(2)$ in the notation of Definition 15.1. Also, the pair $(-2, 4)$ is in f, which implies that $4 = f(-2)$. Note that there is no restriction on how many times a number in Y can appear as a second member of an ordered pair in f. This leads us to a particularly useful type of function that has the property that no two distinct numbers in its domain are mapped or assigned to the same number in its range. This type of function is called a *one-to-one* function. We now give a formal definition of such a function.

Definition 15.3 Let f be a function with domain X and range Y. Then if $f(x) = f(z)$ implies $x = z$, we say that the function f is a one-to-one function.

Note that if we use the ordered pair definition of a function, we could define a one-to-one function by stating that every number y in Y appears exactly once as the second member of an ordered pair in f.

Now observe that the function in Example 15.1 is one-to-one [since $f(x) = f(z)$ implies $x + 1 = z + 1$ and therefore $x = z$], whereas the function in Example 15.2 is not [since every positive real number r appears both in the pair $(\sqrt{r}, r)$ and in the pair $(-\sqrt{r}, r)$].

EXAMPLE 15.3   *Constant Functions*   Let X be the set of all real numbers. Suppose f is the function that assigns to every real number x in X a fixed real number k. Then $f(x) = k$ for all x in X and the range of f is simply the set that consists of the single real number k.

We could have defined this function by saying that it *maps* the entire set of real numbers to the constant k, or equivalently, that f is the set of all ordered pairs (x, k) where x is any real number.

The functions $f(x) = 2$, $g(x) = \pi$, $h(x) = -3$ are examples of specific constant functions.

EXAMPLE 15.4   *The Identity Function*   Let X be the set of all real numbers. Let f be the function that assigns to each real number x the number itself. Then $f(x) = x$ for all x in X and $Y = X$.

We could have defined this function by saying that it *maps* each real number x to itself, or equivalently, that f is the set of all ordered pairs (x, x) where x is a real number. Also, it is clear that f is a one-to-one function.

EXAMPLE 15.5   *The Absolute Value Function*   Let X be the set of all real numbers. Let f be the function that assigns to each real number x its absolute value $|x|$. Then $f(x) = x$ if $x \geq 0$ and $f(x) = -x$ if $x < 0$. For example, $f(2) = 2$, $f(-3) = 3$, $f(0) = 0$, and $f(-\pi) = \pi$. Consequently, Y is the set of all nonnegative real numbers. Is f a one-to-one function?

We could have defined f as the function that maps every nonnegative real number x to itself and every negative real number x to $-x$; equivalently, f is the set of all ordered pairs of the type (x, x) if $x \geq 0$ and $(x, -x)$ if $x < 0$.

EXAMPLE 15.6   *The Square Root Function*   Let X be the set of all nonnegative real numbers.   Let f be the function that assigns to each x in X its square root.   Then $f(x) = \sqrt{x}$ for each x in X and $Y = X$, since the convention is that $\sqrt{x}$ means the positive square root of x.   Is f a one-to-one function?

EXAMPLE 15.7   *The Greatest Integer Function*   Let X be the set of all real numbers.   Let f be the function that assigns to each real number x the greatest integer less than or equal to x. The symbol for the greatest integer less than or equal to x is $[x]$.   Hence, we write $f(x) = [x]$ for each x in X.   For example, $f(\pi) = 3, f(\sqrt{2}) = 1, f(5) = 5, f(-2) = -2,$ $f(-\pi) = -4$, $f(\frac{1}{2}) = 0$, $f(-\frac{1}{3}) = -1$, and in general, if $n \leq x < n + 1$, then $[x] = n$ where n is an integer.

Note that the *domain* X of this function is the set of all real numbers, while the *range* Y is the set of all integers.

EXAMPLE 15.8   Let X be the set of all nonzero real numbers. Let f be the function that assigns to each number x in X its reciprocal $1/x$. Then we may write $f(x) = 1/x$ for each x in X.   Note that the range of f is also the set of all nonzero real numbers.   Is f a one-to-one function?

EXAMPLE 15.9   *The Polynomial Functions*   A polynomial of degree n in the variable x with real number coefficients is an expression of the type $a_n x^n + a_{n-1} x^{n-1} + \cdots + a_2 x^2 + a_1 x + a_0$ where n is a nonnegative integer, a_i is a real number for each $i = 0, 1, 2, \ldots, n$, and $a_n \neq 0$.   Let X be the set of all real numbers.   Let f be the function that assigns to each real number x the value $a_n x^n + a_{n-1} x^{n-1} + \cdots + a_2 x^2 + a_1 x + a_0$. Then we write $f(x) = a_n x^n + a_{n-1} x^{n-1} + \cdots + a_2 x^2 + a_1 x + a_0$.

Some examples of polynomial functions are the following.

1. $f(x) = 3x^2 + 2x - 1$.   Then $f(0) = -1$, $f(1) = 4$, $f(\sqrt{2}) = 5 + 2\sqrt{2}$, and $f(\pi) = 3\pi^2 + 2\pi - 1$.   Is the number zero in the range of f?

2. $f(x) = x^3 - 2x^2 + x$.   Then $f(0) = 0$, $f(1) = 0$, $f(-1) = -4$, $f(\sqrt{2}) = 3\sqrt{2} - 4$, and $f(\pi) = \pi^3 - 2\pi^2 + \pi$.   Is the number 2 in the range of f?

3. Finally, observe that all constant functions of the form $f(x) = k$ are polynomial functions with $n = 0$, $a_0 = k$, and $a_i = 0$ for all $i \neq 0$.

(Note that for $k = 0$, we must drop the condition $a_n \neq 0$.) Also, the identity function $f(x) = x$ and the square function $f(x) = x^2$ are examples of polynomial functions. Is the function $f(x) = 1/x$ a polynomial function? Is the square root function a polynomial function?

EXAMPLE 15.10 Let X be the set of all rational numbers. Let f be the function that assigns to each number x the number 10^x. For example, $f(1) = 10$, $f(-1) = 1/10$, $f(\frac{1}{2}) = \sqrt{10}$, $f(0) = 1$, and $f(\frac{2}{3}) = (100)^{1/3}$. Then we write $f(x) = 10^x$ for each x in X. There are *three* questions that relate to this function. First, what is the range of f? Second, is f one-to-one? Finally, can the domain of this function be extended to the set of all real numbers? This would mean that a number would have to be assigned to expressions like $10^{\sqrt{2}}$, 10^π, and so on (see Example 13.3). We will answer these questions later on.

EXAMPLE 15.11 Let X be the set of real numbers greater than or equal to zero and less than or equal to π. Let f be the function that assigns to each number x in X the value $\sin x$ (see the Appendix). For example, $f(0) = 0$, $f(\pi/6) = 1/2$, $f(\pi/2) = 1$, and $f(\pi) = 0$. Then, we write $f(x) = \sin x$ for each x in X. What is the range of this function? Is this a one-to-one function? If we extend the domain to the set of real numbers between 0 and 2π, what is the range of the function? Finally, if we extend the domain to the set of all real numbers, what is the range of the function?

PROBLEMS

15.1 Find the function f such that $f(x)$ is the circumference of a circle of radius x. Determine the domain and range of f.

15.2 Find the function f such that $f(x)$ is the area of a circle of radius x. Determine the domain and range of f.

15.3 Find the function f such that $f(x)$ is the volume of a sphere of radius x. Determine the domain and range of f.

15.4 Let X be the set of integers from 1 through 20.
(a) Let f be the function that assigns to each x in X the number of distinct positive integral divisors of x. Find the range of f.
(b) Let g be the function that assigns to each x in X the number of positive primes less than x. Find the range of g.

15.5* Let X be the set of integers from 1 through 5.
(a) Let f be the function that assigns to each x in X the number of distinct positive integral divisors of 3^x. Find the range of f.
(b) Let g be the function that assigns to each x in X the number of distinct positive integral divisors of 10^x. Find the range of g.

In each of the following, find the largest set of real numbers that may serve as the domain of the given function. Then find the range and determine whether the function is one-to-one.

15.6 $f(x) = (x-1)^{1/2}$
15.7 $f(x) = x^3$
15.8 $f(x) = \sqrt[3]{x}$
15.9 $f(x) = 1/(x-2)$
15.10 $f(x) = 1/x^2$
15.11 $f(x) = |1+x|$
15.12 $f(x) = 1/(x+1)^2$
15.13 $f(x) = (4-x^2)^{1/2}$
15.14 $f(x) = -(4-x^2)^{1/2}$
15.15 $f(x) = |x|/x$
15.16 $f(x) = |x| + x$
15.17 $f(x) = 2x^2 - 12x + 19$
15.18 $f(x) = \cos x$ (See the Appendix.)
15.19 $f(x) = \tan x$ (See the Appendix.)

GRAPHS OF FUNCTIONS

In this section we define the graph of a function and determine the graphs of most of the functions we used as examples in Section 15.

Let f be a function with domain X and range Y. Then for each real number x in X, there is an assigned value $y = f(x)$ where y is in Y. This gives rise to an ordered pair (x, y) where $y = f(x)$. If we take the usual rectangular coordinate system, we may assign each of the real numbers in X to a point on the x axis in the usual way. In the same way, we may assign each of the real numbers in Y to a point on the y axis. Then the ordered pair (x, y), where $y = f(x)$, corresponds to a particular point in the *plane*, namely, the point with first coordinate x and second coordinate $y = f(x)$. This leads to the following definition.

Definition 16.1 Let the function f have domain X and range Y. Then the *graph* of f is the set of *all* points (x, y) in the plane where x is in X and $y = f(x)$. This statement can be made symbolically as follows. The graph of f is $\{(x, y) \mid x \in X \text{ and } y = f(x)\}$. (The expression $x \in X$ is a short way of writing x is in X.) This means that the graph of f is simply the graph of the equation $y = f(x)$. For example, the graph of $f(x) = x^2$ is the set of points (x, y) where $y = x^2$, but this is just the graph of the equation $y = x^2$.

Note that if we use Definition 15.2 as the definition of a function, we may define the graph of f to be *precisely* those points in the plane that correspond to the set of ordered pairs that belong to f.

EXAMPLE 16.1 The graph of the function given in Example 15.1 is simply the set of isolated points $(x, x + 1)$ where x is a positive odd integer. The graph for values of x from 1 through 9 is given in Figure 16.1.

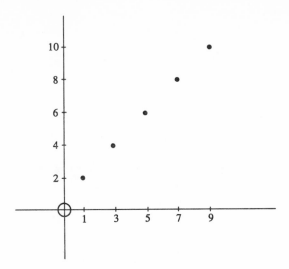

Figure 16.1

EXAMPLE 16.2   The square function of Example 15.2 has the parabola with equation $y = x^2$ as its graph.

EXAMPLE 16.3   The graph of the constant function $f(x) = k$ (of Example 15.3) is simply the set of all points having y coordinate k. This set of points is, of course, the horizontal line with equation $y = k$.

EXAMPLE 16.4   From the definition of the identity function mentioned in Example 15.4, the graph is the set of all points that have their x coordinate and y coordinate equal. This is just the graph of the equation $y = x$.

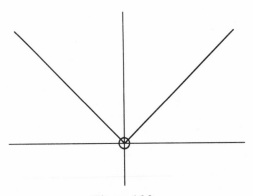

Figure 16.2

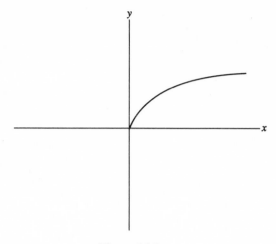

Figure 16.3

EXAMPLE 16.5   The absolute value function mentioned in Example 15.5 has the same graph as the equation $y = |x|$.   This graph is shown in Figure 16.2.   (Note that this graph consists of the graph of $y = x$ to the right of the origin and the graph of $y = -x$ to the left of the origin.)

EXAMPLE 16.6   The square root function in Example 15.6 has the same graph as the equation $y = \sqrt{x}$, namely, the upper half of the parabola with equation $y^2 = x$ (see Figure 16.3).

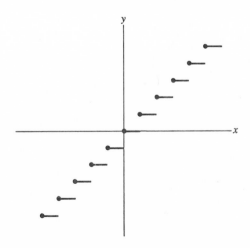

Figure 16.4

EXAMPLE 16.7   The greatest integer function $f(x) = [x]$ consists of infinitely many distinct horizontal line segments. For values of x such that $n \le x < n + 1$, where n is any integer, the graph coincides with the line having equation $y = n$.   See Figure 16.4 for the graph of this function for values of x from -5 up to and not including $+6$.

EXAMPLE 16.8   The graph of the reciprocal function mentioned in Example 15.8 is the hyperbola with equation $y = 1/x$.

EXAMPLE 16.9   The graph of a polynomial function of the type $f(x) = a_0$ falls into the category of Example 16.3 and its graph is the graph of the equation $y = a_0$.   The graph of a polynomial function of the type $f(x) = a_1 x + a_0$ is simply the line with equation $y = a_1 x + a_0$. The graph of a polynomial function of the type $f(x) = a_2 x^2 + a_1 x + a_0$ is a parabola with equation $y = a_2 x^2 + a_1 x + a_0$.   (Find the vertex and focus of this parabola.)   For polynomial functions of degree greater than 2, at present, we can only plot as many of the points $(x, f(x))$ as we need to trace the path of the graph.   For example, the graph of the function $f(x) = x^3 - 3x$ is given in Figure 16.5.   A rough sketch of this graph may be obtained by plotting such points as $(-2, -2)$, $(-1, 2)$, $(0, 0)$, $(1, -2)$, and $(2, 2)$.

We conclude this section with a few remarks about graphs of arbitrary functions.

REMARK 1   Suppose we have a function $y = f(x)$ with a graph like the one in Figure 16.6.   Note that we may use the graph to give us a pictorial representation of the function f as a *mapping* of points on the x axis to points on the y axis.   If we choose some arbitrary values of x, say x_1, x_2, and x_3,

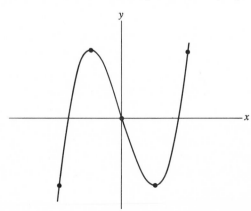

Figure 16.5

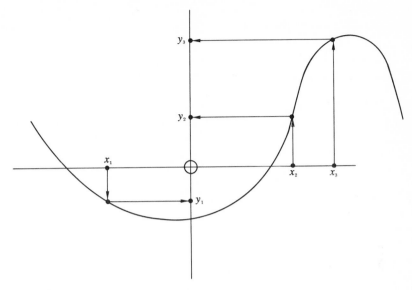

Figure 16.6

we may locate, on the y axis, the corresponding values $y_1 = f(x_1)$, $y_2 = f(x_2)$, and $y_3 = f(x_3)$ by simply drawing a vertical line from the point $(x_i, 0)$ for each $i = 1, 2, 3$ to the graph and then a horizontal line from the graph to the y axis. The horizontal line will meet the y axis at the point $(0, y_i)$ for each $i = 1, 2, 3$ where $f(x_i) = y_i$.

Note that if a function f is one-to-one, distinct points on the x axis must be mapped to distinct points on the y axis. This means that *any horizontal line intersects the graph in at most one point.* Conversely, if we have a graph of a function with the property that every horizontal line intersects the graph in at most one point, we can conclude that the function is one-to-one. Is the function having the graph shown in Figure 16.6 one-to-one?

REMARK 2   In Example 16.1, we saw a function that had a graph represented by infinitely many isolated points. The domain of this function was the set of positive odd integers. We would like to mention another function that has a graph that can only be represented partially by isolated points. In this case, the domain will be the whole set of real numbers, but the graph cannot be exhibited completely over any interval.

Let $f(x) = 0$ if x is rational and let $f(x) = 1$ if x is irrational. Then f is a well-defined function with domain X equal to the set of all real numbers and with range Y equal to the set of integers consisting of the numbers 0 and 1. For example, $f(1) = 0$, $f(0) = 0$, $f(\tfrac{1}{2}) = 0$, $f(-10) = 0$, $f(\sqrt{2}) = 1$, $f(2\sqrt{2}) = 1$, $f(\pi) = 1$, and $f(-\pi) = 1$.

For each rational point r on the x axis, the point $(r, 0)$ is on the graph of f and for each irrational point α on the x axis, the point $(\alpha, 1)$ is on the graph of f. Therefore, given any real number on the x axis, there is one and only one point corresponding to it on the graph of f.

We now observe that since the range consists of the numbers 0 and 1, the assumption that the graph can be represented completely over some interval must imply that the graph contains at least one line segment that lies on either the line $y = 1$ or the line $y = 0$ (the x axis). As an exercise, show that this is impossible. (*Hint:* Every interval on the x axis must contain both rational and irrational points.)

REMARK 3   We have seen that a function f has a graph that corresponds to the equation $y = f(x)$. It is also clear that if we have an equation in which y equals some expression in x, the graph of this equation is the graph of the function with rule of correspondence given by the same expression in x. For example, the equation $y = x^2 + 2x + 1$ has a graph that is the same as the graph of the function $f(x) = x^2 + 2x + 1$. Some equations, however, have graphs that are *not* graphs of functions; a few of these equations are the following.

(a) The equation $x^2 + y^2 = 1$ has a circle for a graph. We may solve for y, obtaining $y = \pm (1 - x^2)^{1/2}$. Now if there were a function f that had this graph, then $f(0)$ would equal 1 and $f(0)$ would also equal -1, which is impossible. It turns out that the upper half of the circle is the graph of $f(x) = (1 - x^2)^{1/2}$ and the lower half is the graph of $f(x) = -(1 - x^2)^{1/2}$ [see Example 13.2, part (b)].

(b) The equation $x^2 - y^2 = 1$ has a hyperbola for its graph. Solving for y, we obtain $y = \pm(x^2 - 1)^{1/2}$. If there were a function f that had this hyperbola for its graph, then $f(\sqrt{2})$ would equal 1 and $f(\sqrt{2})$ would also equal -1, which is impossible. It turns out that the portion of the hyperbola above the x axis is the graph of the function $f(x) = (x^2 - 1)^{1/2}$, while the portion of the hyperbola below the x axis is the graph of the function $f(x) = -(x^2 - 1)^{1/2}$ [see Example 13.2 (c)].

(c) For $x \le 0$, the equation $x^3 + y^3 = 3xy$ has a graph (see Figure 16.7) that corresponds to a function f. There is, however, no easy method for finding an explicit representation of f in the form $y = f(x)$. This equation is said to define a function f *implicitly*.

In general, the preceding examples (a), (b), and (c) are all examples of functions f defined implicitly by an equation in x and y. In parts (a) and (b), we are able to solve for y and arrive at two distinct functions of x that are defined implicitly; in part (c), however, this cannot easily be done. Consequently, after checking the graph of the equation, we assume that there is at least one, and perhaps more than one, function $y = f(x)$ that has a graph that is the same as the graph in Figure 16.7.

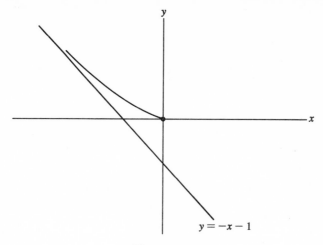

$$y = -x - 1$$

Figure 16.7

PROBLEMS

16.1 Find the graph of each of the functions mentioned in Examples 15.10 and 15.11.

16.2–16.19 Find the graph of each of the functions given in Problems 15.2 through 15.19.

16.20 Find the functions defined implicitly by each of the following equations.
(a) $4x^2 + y^2 = 4$
(b) $x^2 = y^3$
(c) $x^2 + 2xy + y^2 = 0$
(d) $xy + y^2 = 1$

16.21 Find the graph of each of the following functions.
(a) $f(x) = x - [x]$ $\quad x \geq 0$
(b) $f(x) = (x - [x])^{1/2}$ $\quad x \geq 0$

THE ALGEBRA
OF FUNCTIONS

In this section, we give a short treatment of the algebraic operations that are defined on the set of functions. We show that these operations, in most cases, satisfy the familiar properties of the basic operations defined on the set of real numbers.

Before discussing the operations defined on functions, we give a precise definition of what we mean when we say two functions f and g are equal.

Definition 17.1   Two functions f and g are *equal* if they have the same domain X and if $f(x) = g(x)$ for each $x \in X$.   If we use the ordered pair definition for f and g, we say that $f = g$ if they are the same set of ordered pairs.

In what follows, we assume that each function f and g has a common domain X, where X is some set of real numbers.   We also assume that the individual ranges of f and g, which may be distinct, are contained in the set of real numbers (we usually say simply that f and g are *real-valued* functions).

Definition 17.2   Let f and g be two arbitrary real-valued functions having a common domain X.   Then we define the *sum $f + g$* of the functions f and g to be the function F where for each $x \in X$, $F(x) = f(x) + g(x)$; thus, by definition, $(f + g)(x) = f(x) + g(x)$.   Consequently, to find the value of the *sum* of two functions at a particular number x, we simply add the individual functional values $f(x)$ and $g(x)$.

By replacing the plus sign with a minus sign in the preceding definition, we obtain the definition of the *difference* of two functions f and g; the difference is the function $F(x) = f(x) - g(x)$, written $(f - g)(x)$.

For example, if $f(x) = x^2$ and $g(x) = |x|$, then $(f + g)(x) = f(x) + g(x) = x^2 + |x|$ and $(f - g)(x) = f(x) - g(x) = x^2 - |x|$.   Observe that the common domain of f and g is the set of all real numbers. Find the range of $(f + g)$ and of $(f - g)$, and plot the graphs.

Definition 17.3 Let f and g be two arbitrary real-valued functions having a common domain X. Then we define the *product fg* of the functions f and g to be the function F where for each $x \in X$, $F(x) = f(x)g(x)$; thus, by definition, $(fg)(x) = f(x)g(x)$. Consequently, to find the *product* of two functions at a particular number x, we simply multiply the individual functional values $f(x)$ and $g(x)$.

In a similar way, we define the *quotient* of the functions f and g to be the function F such that $F(x) = f(x)/g(x)$ for each $x \in X$ for which $g(x) \neq 0$. (Note that in this case we must eliminate from X all the numbers x for which $g(x) = 0$.) Hence, by definition, $(f/g)(x) = f(x)/g(x)$.

For example, if $f(x) = x^2$ and $g(x) = |x|$, then $(fg)(x) = f(x)g(x) = x^2|x|$ and $(f/g)(x) = f(x)/g(x) = x^2/|x|$. Note that the domain of the product function $(fg)(x)$ is the set of all real numbers, whereas the domain of the quotient function $(f/g)(x)$ is the set of all *nonzero* real numbers. What is the range of the product function? What is the range of the quotient function? Plot the graphs. Can the quotient function actually be defined at $x = 0$?

Observe also that for each real number r there is a corresponding constant function $f(x) = r$. Consequently, if g is an arbitrary function with domain X, we can consider the product function formed by f and g, and we obtain the function $F(x) = rg(x)$. Therefore, we can consider the multiplication of an arbitrary function $g(x)$ by a real number r as the product of the *functions* $f(x) = r$ and $g(x)$. For example, if $r = 3$ and $g(x) = x^2$, we may consider the new function $h(x) = 3x^2$ as the product of $f(x) = 3$ and $g(x) = x^2$.

Summarizing, we have seen that two functions f and g with a common domain X can be added, subtracted, multiplied, and divided. Since these operations were defined in terms of the real numbers $f(x)$ and $g(x)$, we may conclude that the operations on functions satisfy the same familiar properties that they satisfy on the set of real numbers (e.g., the commutative laws for addition and multiplication $f + g = g + f$, $fg = gf$, etc.). Note that the associative law for real number addition $(a + b) + c = a + (b + c)$, for any three real numbers a, b, and c, can be stated in an analogous way for the addition of three functions f, g, and h as $(f + g) + h = f + (g + h)$. To prove that these functions are actually equal, we must show that for each x in the common domain X of the functions f, g, and h, $([f + g] + h)(x) = (f + [g + h])(x)$. But this follows from the fact that $[f(x) + g(x)] + h(x) = f(x) + [g(x) + h(x)]$ (from the associative law for real number addition).

In a similar way, verify that the set of functions under multiplication satisfies the associative law, namely, that $(fg)h = f(gh)$.

The familiar distributive law for real numbers, $a(b + c) = ab + ac$, is also satisfied in the set of functions since $f(g + h) = fg + fh$.   This result follows from the fact that for each x, $f(x)[g(x) + h(x)] = f(x)g(x) + f(x)h(x)$ (from the distributive law for real numbers).

Finally, the system of real numbers has the number 0 as its identity element under addition, since for any real number r, $0 + r = r + 0 = r$, and has the number 1 as its identity element under multiplication, since $1r = r1 = r$ for any real number r.   (Is there an identity element for the operation of subtraction?   For the operation of division?)   The constant functions $Z(x) = 0$ and $U(x) = 1$ do the analogous jobs in the system of functions, since for each function f, $f + Z = Z + f = f$ and $fU = Uf = f$.   The first equation can be verified as follows.   For each x, $(f + Z)(x) = f(x) + Z(x) = f(x) + 0 = f(x)$ and $(Z + f)(x) = Z(x) + f(x) = 0 + f(x) = f(x)$.   Verify the second equation.

It turns out that the set of functions has another operation that is quite different from the familiar operations on real numbers.   We now give a definition of this new operation along with some examples.

Definition 17.4   Let f and g be two real-valued functions with domains X and X', respectively.   Let Y and Y' be the respective ranges and let the range Y' of g be contained in the domain X of f.   Then, the *composition* of f with g, written $f \circ g$, is the function whose domain is X' and whose rule of correspondence is $(f \circ g)(x) = f(g(x))$.   The expression $f \circ g$ is usually read "f circle g" and the expression $f(g(x))$ is read "f of g of x."

In terms of mappings, $f \circ g$ takes a number x in X' (the domain of g) to a number y in Y (the range of f) by first taking it to a number z in Y' (by g) and then by taking z to the number y in Y (by f).   Pictorially, we describe the action of $f \circ g$ in Figure 17.1.

EXAMPLE 17.1   Let f be the square root function and let g be the function $g(x) = x^2 + 1$.   Then $(f \circ g)(x) = f(g(x)) = (x^2 + 1)^{1/2}$; the respective domains are X (the domain of f), the set of all nonnegative

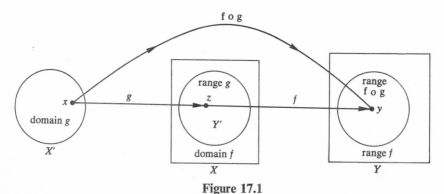

Figure 17.1

real numbers, and X' (the domain of g), the set of all real numbers. The corresponding ranges are Y (the range of f), the set of all non-negative real numbers, and Y' (the range of g), the set of all real numbers greater than or equal to 1. Clearly, Y' is contained in X; thus, $f(g(x)) = (x^2 + 1)^{1/2}$ is well defined. Since $Y' = $ set of all real numbers greater than or equal to 1, the range of $f \circ g$ is simply Y', which in this case is certainly contained in but not equal to Y. As an exercise, construct a diagram like the one in Figure 17.1.

In the particular case of the number $\sqrt{3}$ in X', $g(\sqrt{3}) = 3 + 1 = 4$ and consequently, $f(g(\sqrt{3})) = f(4) = \sqrt{4} = 2$. Therefore, $f \circ g$ takes $\sqrt{3}$ to 2 by first taking $\sqrt{3}$ to 4 by g and then by taking 4 to 2 by f. We represent this pictorially as

$$\sqrt{3} \xrightarrow{g} 4 \xrightarrow{f} 2.$$

What is $f(g(\sqrt{8}))$? Describe the intermediate steps. What is $f(g(-\sqrt{3}))$? Describe the intermediate steps.

Example 17.2 Let f be the square root function and let g be the function $g(x) = x^2 + 1$. Then $(g \circ f)(x) = g(f(x)) = x + 1$. Now the domain of $g \circ f$ is the domain of f, the domain of g contains the range of f, and finally, the range of $g \circ f$ is the set of all real numbers greater than or equal to 1. As an exercise, construct a diagram like the one in Figure 17.1.

By now, you have observed that in the preceding examples $(f \circ g)(x) = (x^2 + 1)^{1/2}$ and $(g \circ f)(x) = x + 1$. These functions are *not* equal since $(x^2 + 1)^{1/2} \neq x + 1$ for all choices of x. *Consequently, this new operation of composition is not a commutative one.*

However, given a set of compatible domains and ranges, the composition operation does satisfy some of the other properties, such as the associative law, namely, $(f \circ g) \circ h = f \circ (g \circ h)$ and the following right distributive laws.

1. $(f + g) \circ h = (f \circ h) + (g \circ h)$

2. $(fg) \circ h = (f \circ h)(g \circ h)$

The corresponding left distributive laws are not valid (see Problem 17.18).

There is also a function that is an identity element for this operation, namely, the function $f(x) = x$, which was defined in Example 15.4 and called, appropriately enough, the identity function. If we let $I(x) = f(x) = x$, then for any function g we have $I \circ g = g \circ I = g$ since $(I \circ g)(x) = I(g(x)) = g(x)$ and $(g \circ I)(x) = g(I(x)) = g(x)$.

Definition 17.5   *The Inverse of a Function*   (a) In the case of addition of functions, for each function $f(x)$ we may define the function $g(x) = -f(x)$. Then $f(x) + g(x) = 0$ for all x, and $f + g$ is therefore the *zero function* $Z(x) = 0$ (discussed earlier), which is the identity element for functions under *addition*. We call $-f$ the *additive* inverse of f.

(b) In the case of multiplication of functions, for each function $f(x)$ with the property that the number zero is not in the range of f, we may define the reciprocal function $g(x) = 1/f(x)$. Then $f(x)g(x) = 1$ for all x and fg is equal to the constant function $U(x) = 1$, which is the identity element for the set of functions under the *product* operation. We call $1/f$ the *multiplicative* inverse of f.

(c) In the case of the composition of functions, we may ask the same type of question. Given a function f, is there another function g with the property that $(f \circ g)(x) = (g \circ f)(x) = I(x) = x$ for all x, where $I(x)$ is the identity element for the operation of composition? If so, we call g the *inverse* of the function f under composition and we denote g by $f*$. In most cases, when we use the expression "the inverse of a function," we mean the inverse under composition. It turns out there is a rather large class of functions that possess inverses under composition, namely, the class of all one-to-one functions discussed in Definition 15.3.

Theorem 17.1   Let f be a one-to-one function with domain X and range Y. Define the function $f*$ on the set Y as follows. For each $y \in Y$, $f*(y) = x$ if $f(x) = y$. Then $f*$ is the inverse of f and the range of $f*$ is the set X.

PROOF   The domain of $f \circ f*$ is Y, so choose a number $y \in Y$ and compute $(f \circ f*)(y)$. Now $(f \circ f*)(y) = f(f*(y)) = f(x) = y$ [since $f*(y) = x$ implies $f(x) = y$]. On the other hand, $f* \circ f$ has domain X, so choose a number $x \in X$ and compute $(f* \circ f)(x)$. Now $(f* \circ f)(x) = f*(f(x)) = f*(y) = x$ [since $y = f(x)$ implies $f*(y) = x$]. Therefore, $f \circ f* = f* \circ f =$ the identity function. ∎

Note that if we use the *ordered pair* definition for f, then f one-to-one means that each $y \in Y$ appears *exactly* once as a second member of an ordered pair in f. Consequently, we may define the inverse function $f*$ as the set of all ordered pairs (y, x) such that (x, y) is in f. The fact that each y appears *exactly* once among the *second* members of the pairs in f implies that each y will appear *exactly* once as a *first* member of the pairs that belong to $f*$, and thus $f*$ is a well-defined function.

If f is *not* one-to-one, there are at least two distinct numbers x and z in X such that $f(x) = f(z) = y$, where $y \in Y$. Then, how would you define $f*(y)$? $f*(y)$ would not have a *unique* value; hence, $f*$ would not be a function. On the other hand, in the case of the ordered pairs that define f, the pairs (x, y) and (z, y) would both be in f. Therefore, $f*$ would have to

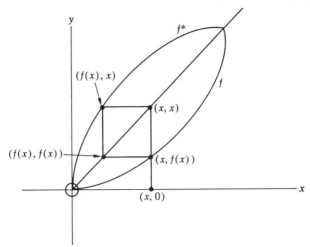

Figure 17.2

contain (y, x) and (y, z), which is impossible since y cannot appear as a first member more than once.

Now, if f^* is the inverse of f, we can construct the graph of f^* from the graph of f in the following way (Figure 17.2). Choose an arbitrary number x in the domain of f. Draw a vertical line through the point $(x, 0)$. This line intersects the graph of f at the point $(x, f(x))$ and intersects the line with equation $y = x$ at the point (x, x). Draw a horizontal line through the point $(x, f(x))$. This line intersects the line $y = x$ at the point $(f(x), f(x))$. Finally, the intersection point of the horizontal line through (x, x) and the vertical line through $(f(x), f(x))$ is precisely the point $(f(x), x)$ that must be a point on the graph of f^* [since $(x, f(x))$ is a point on the graph of f]. Therefore, the point $(f(x), x)$ is just the *reflection* of the point $(x, f(x))$ in the line $y = x$, and the set of all such reflection points will give us the graph of f^*.

Try this technique to obtain the graph of the inverse function of each of the following functions: $f(x) = 2x + 1, f(x) = \sqrt{x}, f(x) = 1/x$, and $f(x) = -1/x$.

Finally, we observe that if f^* is the inverse of f, the inverse of f^* must be f. The proof is clear from the ordered pair definition for f and f^*. (The inverse of f^* would have the same ordered pairs as the original function f had.)

EXAMPLE 17.3 Find the inverse function of $f(x) = 3x - 1$.

SOLUTION $y = 3x - 1$ if and only if $x = \frac{1}{3}(y + 1)$. Therefore, $f^*(x) = \frac{1}{3}(x + 1)$ is the inverse function of $f(x) = 3x - 1$. To verify this, choose any real number z; then

$$(f^* \circ f)(z) = f^*(f(z)) = f^*(3z - 1) = \frac{1}{3}[(3z - 1) + 1] = z$$

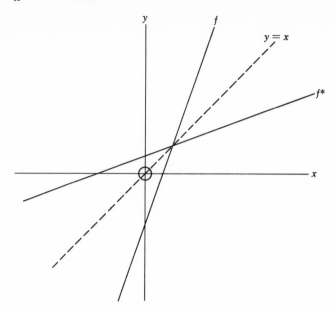

Figure 17.3

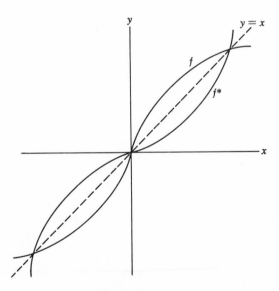

Figure 17.4

and

$$(f \circ f^*)(z) = f(f^*(z)) = f(\tfrac{1}{3}(z + 1)) = 3[\tfrac{1}{3}(z + 1)] - 1 = z$$

and this completes the proof. The graphs are in Figure 17.3.

EXAMPLE 17.4 Find the inverse function of $f(x) = \sqrt[3]{x}$.

SOLUTION $y = \sqrt[3]{x}$ if and only if $x = y^3$. Therefore, $f^*(x) = x^3$ is the inverse of $f(x) = \sqrt[3]{x}$. Verify this by choosing any real number z, and then prove that $(f \circ f^*)(z) = (f^* \circ f)(z) = z$. The graphs are shown in Figure 17.4.

We now present a shortcut for obtaining f^* from f. Recall that interchanging x and y in an equation results in the *reflection* of the original graph in the line $y = x$. Hence, we need only solve the equation $y = f(x)$ for x and then interchange x and y; this yields the inverse function f^*. For example, if $f(x) = 2x + 1$, we solve $y = 2x + 1$ for x, obtaining $x = \tfrac{1}{2}(y - 1)$. Then, interchanging x and y, we arrive at $y = \tfrac{1}{2}(x - 1) = f^*(x)$.

PROBLEMS

In Problems 17.1 through 17.14, let $f(x) = x^2 + 1$, $g(x) = \sqrt{x}$, and $h(x) = 1/x$. Find the required function and state its domain and range.

17.1 $f + g$
17.2 $f - g$
17.3 fg
17.4 fh
17.5 h/g
17.6 g/f
17.7 $f \circ g$
17.8 $g \circ f$
17.9 $f \circ h$
17.10 $h \circ f$
17.11 $g \circ (f + h)$
17.12 $(fh) \circ g$
17.13 $f(g + h)$
17.14 $f \circ g \circ h$
17.15   Show that every polynomial $a_n x^n + a_{n-1} x^{n-1} + \cdots + a_2 x^2 + a_1 x + a_0$ can be constructed from the constant functions and the identity function $I(x) = x$ by using the operations of addition and multiplication.
17.16   Sketch the graph of each of the functions mentioned in Problems 17.1, 17.2, 17.3, 17.5, 17.7, 17.9, and 17.14.
17.17   Find the inverse (if it exists) of each of the following functions and plot the graphs of both the given function and its inverse.
(a) $f(x) = x^5$
(b) $f(x) = 5x - 3$
(c) $f(x) = |x|$

(d) $f(x) = 1/x^2$
(e) $f(x) = \sin x$
(f) $f(x) = x^2 + 1, \qquad x \geq 0$

17.18 Show that the following two distributive laws are *not* valid for arbitrary functions f, g, and h.

(a) $h \circ (f + g) = (h \circ f) + (h \circ g)$
(b) $h \circ (fg) = (h \circ f)(h \circ g)$
    [*Hint:* Let $f(x) = x$, $g(x) = x$ and let $h(x) = k$, where k is any constant not equal to 0 or 1.]

17.19 (a)* If f and g are one-to-one functions, show that $f \circ g$ is one-to-one and that the inverse function of $f \circ g$ is equal to the composition of g^* with f^*, that is, $(f \circ g)^* = g^* \circ f^*$.

(b) Verify the result of part (a) in the special case where $f(x) = x + 1$ and $g(x) = \sqrt{x}$.

17.20 Let $f(x) = (1 - x^n)^{1/n}$ where $0 \leq x \leq 1$ and n is any positive integer. Show that f is its own inverse.

LIMITS OF SEQUENCES

In this section we discuss limits of sequences. We attempt a *conceptual* rather than *computational* approach, beginning with the definition of a sequence and some examples of sequences. We also require the definition of a neighborhood of a point on a line.

Definition 18.1 A *sequence* is a function f whose domain is *the set of natural numbers* and whose range is some set of real numbers. We may list the values that f takes on by simply writing $f(1), f(2), f(3), f(4), \ldots, f(n), \ldots$, where each value $f(n)$ is called a term or an *entry* of the sequence. We usually let $x_n = f(n)$ for each natural number n and write the entries in the form $x_1, x_2, x_3, x_4, \ldots, x_n, \ldots$. Since every sequence has its domain equal to the natural numbers, we distinguish between sequences simply by using the entries $x_1, x_2, x_3, \ldots, x_n, \ldots$ for one sequence, the entries $y_1, y_2, y_3, \ldots, y_n, \ldots$ for another, and the entries $z_1, z_2, z_3, \ldots, z_n, \ldots$ for a third sequence. We then denote these sequences by the expressions $\{x_n\}$, $\{y_n\}$, and $\{z_n\}$, respectively.

Observe that we must make a distinction between the terms or entries of a sequence and the numbers in the range of the sequence. All sequences have *infinitely* many terms, one for each natural number n; however, the range may not be an infinite set. For example, the sequence $1, 0, 1, 0, 1, 0, 1, 0, \ldots$, where each odd term is 1 and each even term 0 has infinitely many terms, but its range consists of only the integers 0 and 1.

EXAMPLE 18.1 *Examples of sequences*
(a) $\{x_n\} = \{1/n\} = 1, \frac{1}{2}, \frac{1}{3}, \frac{1}{4}, \ldots, 1/n, \ldots$
(b) $\{x_n\} = \{n/(n+1)\} = \frac{1}{2}, \frac{2}{3}, \frac{3}{4}, \frac{4}{5}, \ldots, n/(n+1), \ldots$
(c) $\{x_n\} = \{n\} = 1, 2, 3, 4, \ldots, n, \ldots$

Figure 18.1

(d) $\{x_n\} = 1, 0, 1, 0, 1, 0, 1, 0, \ldots$ where for odd n, $x_n = 1$ and for even n,
$x_n = 0$
(e) $\{x_n\} = 0.3, 0.33, 0.333, 0.3333, \ldots$
(f) $\{x_n\} = 1, -\frac{1}{2}, \frac{1}{3}, -\frac{1}{4}, \frac{1}{5}, -\frac{1}{6}, \ldots$

Since each term of a sequence is a real number, we may associate it with a point on the x axis or the y axis.   If, for the present, we associate the terms of a given sequence with points on the x axis, each sequence corresponds to a set of points on the x axis.   See Figure 18.1 for the set of points associated with the sequence $\{1/n\}$.   Also note that the sequence 1, 0, 1, 0 1, 0, ... can be associated with the points 0 and 1 alone, whereas the sequence 1, 2, 3, 4, 5, ... can be associated with the set of all points on the x axis with natural number coordinates.

Definition 18.2   *A Neighborhood of a Point*   Very often, when dealing with a fixed point c on the x axis, we become interested in other points on the x axis in the vicinity of c.   In fact, most of the time we are interested in *intervals* of the x axis that contain the point c (usually taking c to be the midpoint of the interval).   If the left endpoint of an interval containing c has coordinate a, while the right endpoint has coordinate b, the set of all points x such that $a < x < b$ (the *open interval* with endpoints a and b) is called a *neighborhood* of the point c.   If we exclude the point c from this interval, we obtain what is usually called a *deleted neighborhood* of the point c.   In Figure 18.2, we have depicted a neighborhood of c on the left and a deleted neighborhood of c on the right.

When dealing with neighborhoods of points, we will occasionally restrict ourselves to what we will call an *ε-neighborhood* of c.   This means a neighborhood of c with c as the midpoint and endpoints $c - \varepsilon$ and $c + \varepsilon$, that is, an open interval about c as center with radius ε (Figure 18.3).

In the following, we consider a given sequence as a set of points on the x axis.

Definition 18.3   We say that a point x_0 is the *limit* of a given sequence $\{x_n\}$ if, for any arbitrarily small ε neighborhood N of x_0, at most a finite number of terms of the sequence lie outside N.   We denote this by writing $\lim_{n \to \infty} x_n = x_0$

Figure 18.2

Figure 18.3

or simply $\{x_n\} \to x_0$. We also say that the sequence $\{x_n\}$ converges to the point x_0.

Alternate Definition   Given $\varepsilon > 0$, there exists a positive integer m (depending on ε) with the property that for all $n > m$, $|x_n - x_0| < \varepsilon$. This merely states that for any ε neighborhood of x_0 there exists a positive integer m with the property that all the terms x_n with $n > m$ lie *inside* the ε neighborhood. Therefore, *at most a finite number* of terms can lie outside, namely, the m terms $x_1, x_2, x_3, \ldots, x_m$.

EXAMPLE 18.2   *Examples of limits of sequences*
(a) $\{1/n\} \to 0$, since given an ε neighborhood of 0, we choose m such that $1/m < \varepsilon$. Then only the terms $1, \frac{1}{2}, \frac{1}{3}, \ldots, 1/(m-1)$ can lie outside this neighborhood.
(b) $\{n/(n+1)\} \to 1$, since given an ε neighborhood of 1, we choose m such that $1/(m+1) < \varepsilon$. Then for any $n > m$, the term $n/(n+1)$ is within the ε neighborhood of 1 [since $1 - n/(n+1) = 1/(n+1) < 1/(m+1) < \varepsilon$]; therefore, only the first m terms can lie outside this neighborhood.
(c) The sequence $\{n\}$ does not have a finite limit, but in this case we say that the limit is infinite and we write $\lim_{n \to \infty} x_n = \infty$ or $\{x_n\} \to \infty$. Note that this result is compatible with Definition 18.3 if we extend our notion of an ε neighborhood to include those of the type $a < x < \infty$ where a is any real number.
(d) The sequence $1, 0, 1, 0, 1, 0, \ldots$ does not have a limit. If 1 were the limit, then taking an ε neighborhood of 1 with $\varepsilon = \frac{1}{2}$, we would see that infinitely many terms of the sequence (all the even-numbered terms) lie outside this neighborhood. The same kind of argument applies to a similar neighborhood of 0, since we would then have infinitely many terms of the sequence (all the odd-numbered terms) outside the neighborhood. Finally, no other real number can be the limit, since any other number r must be a positive distance from both the points 0 and 1. Hence, for a small enough ε, *all* the terms of the sequence will lie outside the corresponding ε neighborhood of r.
(e) The sequence of Example 18.1(e) has the number $\frac{1}{3}$ as its limit. For if we express $\frac{1}{3}$ as the infinite decimal $0.3333333\cdots$ and if we take

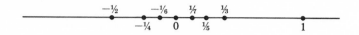

Figure 18.4

an ε neighborhood of $\frac{1}{3}$ with $\varepsilon = 1/10^n$, where n is any large natural number, we have

$$\frac{1}{3} - x_n = 0.\overbrace{00000 \cdots 03333}^{n} \cdots < \frac{1}{10^n}$$

Therefore, only the terms $x_1, x_2, \ldots, x_{n-1}$ lie outside this neighborhood and this sequence converges to $\frac{1}{3}$.

(f) If we plot the points (see Figure 18.4) corresponding to the sequence $1, -\frac{1}{2}, \frac{1}{3}, -\frac{1}{4}, \frac{1}{5}, -\frac{1}{6}, \ldots$, we see that the odd-numbered terms approach the point 0 from the right while the even-numbered terms approach the point 0 from the left. To prove 0 is the limit, take an ε neighborhood of 0. Choose the natural number m such that $1/m < \varepsilon$; then for all $n > m$, $|1/n| < 1/m < \varepsilon$ and only the first $m - 1$ terms can lie outside this neighborhood.

Theorem 18.1   If a sequence $\{x_n\}$ has a limit x_0, then x_0 is the *only* limit.

PROOF   Suppose α is another limit of the sequence $\{x_n\}$ where $\alpha \neq x_0$. Then let d be the distance between the points x_0 and α. Now, choose ε neighborhoods N and M of x_0 and α, respectively, with $\varepsilon = d/2$ (see Figure 18.5). Since x_0 is a limit of $\{x_n\}$, at most a finite number of terms of $\{x_n\}$ can lie outside N; consequently, infinitely many terms lie inside N. But this contradicts the fact that α is a limit, since it implies that *more* than a finite number of terms of $\{x_n\}$ lie outside M. Therefore, x_0 is the *unique* limit of the sequence $\{x_n\}$. This actually justifies the use of the expression "x_0 is *the limit* of the sequence $\{x_n\}$" since there can be no others. ∎

Finally, the preceding discussion would have been equally applicable to sequences considered as points on the y axis. In this event, the neighborhoods would have been open intervals on the y axis of the type $a < y < b$.

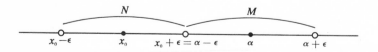

Figure 18.5

PROBLEMS

In each of the following, find the first six terms of the given sequence and compute the limit, if it exists.

18.1 $\{1/2^n\}$

18.2 $\{(n+1)/n\}$

18.3 $1, 0, -1, 1, 0, -1, 1, 0, -1, \ldots$

18.4 $1/3, -1/9, 1/27, -1/81, \ldots$

18.5 $0.6, 0.66, 0.666, 0.6666, \ldots$

18.6 $0.5, 0.55, 0.555, 0.5555, \ldots$

18.7 $\{3n/(n+2)\}$

18.8 $\{(n^2-1)/(n^2+1)\}$

18.9 $\{(n^2+1)/(2n-1)\}$

18.10 $\{(3n+1)/(n^2+1)\}$

18.11 $\{(n^4+n^2+1)/(n^4+n^3)\}$

18.12 $\{(n^4+n^2+1)/(n^3+1)\}$

18.13 $\{(n^4+n^2+1)/(n^5+1)\}$

THE ALGEBRA OF SEQUENCES

It turns out that within the set of sequences of real numbers, the operations of addition, subtraction, multiplication, and division can be defined. After all, by definition, a sequence is a real-valued function with domain equal to the set of all natural numbers. But in Section 17, these basic operations were defined on the set of *all* real-valued functions. Now, if we further restrict ourselves to the set of *convergent* sequences, we can also state some results concerning the limits of sequences that are sums, differences, products, and quotients of two given convergent sequences.

Definition 19.1 Let $\{x_n\} = x_1, x_2, x_3, \ldots, x_n, \ldots$ and $\{y_n\} = y_1, y_2, y_3, \ldots,$ $y_n, \ldots$ be two arbitrary sequences. We define the *sum* of $\{x_n\}$ and $\{y_n\}$ to be the sequence $\{z_n\} = x_1 + y_1, x_2 + y_2, x_3 + y_3, \ldots, x_n + y_n, \ldots$; that is, $\{x_n\} + \{y_n\} = \{x_n + y_n\}$.

We define the *difference* $\{x_n\} - \{y_n\}$ in a similar way as $\{z_n\} = x_1 - y_1,$ $x_2 - y_2, x_3 - y_3, \ldots, x_n - y_n, \ldots$; that is, $\{x_n\} - \{y_n\} = \{x_n - y_n\}$.

Definition 19.2 Let $\{x_n\} = x_1, x_2, x_3, \ldots, x_n, \ldots$ and $\{y_n\} = y_1, y_2, y_3, \ldots,$ $y_n, \ldots$ be two arbitrary sequences. We define the *product* of $\{x_n\}$ and $\{y_n\}$ to be the sequence $\{z_n\} = x_1y_1, x_2y_2, x_3y_3, \ldots, x_ny_n, \ldots$; that is, $\{x_n\}\{y_n\} = \{x_ny_n\}$.

We define the *quotient* $\{x_n\}/\{y_n\}$ in a similar way as $\{z_n\} = x_1/y_1, x_2/y_2,$ $x_3/y_3, \ldots, x_n/y_n, \ldots$ provided that $y_i \neq 0$ for each $i = 1, 2, 3, \ldots, n, \ldots$; that is $\{x_n\}/\{y_n\} = \{x_n/y_n\}$.

EXAMPLE 19.1 Let $\{x_n\} = 0, \frac{1}{2}, \frac{2}{3}, \frac{3}{4}, \ldots, (n-1)/n, \ldots$ and $\{y_n\} = 2,$ $\frac{3}{2}, \frac{4}{3}, \frac{5}{4}, \ldots, (n+1)/n, \ldots$. Then, $\{x_n\} + \{y_n\} = 2, 2, 2, 2, \ldots$; $\{x_n\}$

$- \{y_n\} = -2, \ -1, \ -\frac{2}{3}, \ -\frac{1}{2}, \ \ldots, \ -2/n, \ \ldots; \ \{x_n\}\{y_n\} = 0, \ \frac{3}{4}, \ \frac{8}{9}, \ \frac{15}{16},$
$\ldots, (n^2 - 1)/n^2, \ldots; \ \{x_n\}/\{y_n\} = 0, \ \frac{1}{3}, \ \frac{1}{2}, \ \frac{3}{5}, \ldots, (n-1)/(n+1), \ldots.$

EXAMPLE 19.2 Let $\{x_n\} = 1, 0, 1, 0, 1, 0, \ldots$ and $\{y_n\} = 0, -1, 0,$
$-1, 0, -1, \ldots.$ Then $\{x_n\} + \{y_n\} = 1, -1, 1, -1, 1, -1, \ldots; \ \{x_n\}$
$- \{y_n\} = 1, 1, 1, 1, 1, 1, \ldots; \ \{x_n\}\{y_n\} = 0, 0, 0, 0, 0, 0, \ldots; \ \{x_n\}/\{y_n\}$ is
undefined.

EXAMPLE 19.3 Let $\{x_n\} = 0, \ \frac{1}{2}, \ \frac{2}{3}, \ \frac{3}{4}, \ \ldots, \ (n-1)/n, \ \ldots$ and
$\{y_n\} = 1, 2, 3, 4, \ldots, n, \ldots.$ Then $\{x_n\} + \{y_n\} = 1, \ \frac{5}{2}, \ \frac{11}{3}, \ \frac{19}{4}, \ \ldots,$
$(n^2 + n - 1)/n, \ \ldots; \quad \{x_n\} - \{y_n\} = -1, \quad -\frac{3}{2}, \quad -\frac{7}{3}, \quad -\frac{13}{4}, \quad \ldots,$
$(-n^2 + n - 1)/n, \ldots; \ \{x_n\}\{y_n\} = 0, 1, 2, 3, 4, \ldots, n - 1, \ldots; \ \{x_n\}/\{y_n\}$
$= 0, \ \frac{1}{4}, \ \frac{2}{9}, \ \frac{3}{16}, \ \ldots, \ (n-1)/n^2, \ldots.$

In Example 19.1, you perhaps noticed that the sequences $\{x_n\}$ and $\{y_n\}$
each had limit 1, while $\{x_n\} + \{y_n\}$ had limit $1 + 1 = 2$, $\{x_n\} - \{y_n\}$ had limit
$1 - 1 = 0$, $\{x_n\}\{y_n\}$ had limit $1 \cdot 1 = 1$, and $\{x_n\}/\{y_n\}$ had limit $1/1 = 1$.
It turns out that for convergent sequences we can state the following results.

Theorem 19.1 If $\{x_n\} \to a$ and $\{y_n\} \to b$, then $\{x_n\} + \{y_n\} \to a + b$, $\{x_n\}$
$- \{y_n\} \to a - b$, $\{x_n\}\{y_n\} \to ab$, and $\{x_n\}/\{y_n\} \to a/b$, provided that $\{x_n\}/\{y_n\}$
is defined and $b \neq 0$.

PROOF We prove $\{x_n\} + \{y_n\} \to a + b$. The proofs for products and
quotients are more difficult and will be omitted.

Taking an arbitrary ε neighborhood of $a + b$, we must show that at most a
finite number of terms of the sequence $\{x_n + y_n\}$ lie outside this ε neighborhood; that is, there exists a positive integer m such that for all $n > m$, $x_n + y_n$
is in the ε neighborhood of $a + b$.

Choose $(\varepsilon/2)$ neighborhoods of a and b (see Figure 19.1). Then $\{x_n\} \to a$
implies that at most a finite number of terms of the sequence $\{x_n\}$ lie outside
the $(\varepsilon/2)$ neighborhood of a. Hence, there exists a positive integer j such
that for all $n > j$, $|x_n - a| < \varepsilon/2$. In a similar way, at most a finite number
of terms of $\{y_n\}$ lie outside the $(\varepsilon/2)$ neighborhood of b. Hence, there exists
a positive integer k such that for all $n > k$, $|y_n - b| < \varepsilon/2$. Now choose m
equal to the larger of the numbers j and k. Then for all $n > m$, $|(x_n + y_n)$
$- (a + b)| = |(x_n - a) + (y_n - b)| \leq |x_n - a| + |y_n - b| < \varepsilon/2 + \varepsilon/2 = \varepsilon$.
Consequently, at most m terms of the sequence $\{x_n + y_n\}$ lie outside the
ε neighborhood of $a + b$. Therefore, $\{x_n + y_n\} \to a + b$. ∎

Figure 19.1

EXAMPLE 19.4 Let $\{x_n\} = \{(2n + 1)/n\}$ and $\{y_n\} = \{(3n + 1)/n\}$. Then $\{x_n\} \to 2$ [since $(2n + 1)/n = 2 + 1/n$] and $\{y_n\} \to 3$ [since $(3n + 1)/n = 3 + 1/n$]. Using the results of Theorem 19.1, we may conclude that $\{x_n + y_n\}$ has limit 5, $\{x_n - y_n\}$ has limit -1, $\{x_n y_n\}$ has limit 6, and $\{x_n/y_n\}$ has limit $\frac{2}{3}$. Verify that these are the correct limits.

EXAMPLE 19.5 Observe that the sequences $\{x_n\}$ and $\{y_n\}$ of Example 19.2 are *not* convergent; however, the sequences $\{x_n - y_n\}$ and $\{x_n y_n\}$ are convergent. Consequently, the fact that two given sequences do *not* converge does not imply that none of the sequences $\{x_n + y_n\}$, $\{x_n - y_n\}$, $\{x_n y_n\}$, or $\{x_n/y_n\}$ converges. In fact, in general, it is possible to have either *one* or *both* of the sequences $\{x_n\}$ and $\{y_n\}$ not converge and yet have one or more of the sequences $\{x_n + y_n\}$, $\{x_n - y_n\}$, $\{x_n y_n\}$, or $\{x_n/y_n\}$ converge (see Problems 19.2 and 19.5).

PROBLEMS

In Problems 19.1–19.6, determine the limits (if they exist) of the sequences $\{x_n\}$, $\{y_n\}$, $\{x_n + y_n\}$, $\{x_n - y_n\}$, $\{x_n y_n\}$, and $\{x_n/y_n\}$. Also, write out the first five terms of each of these sequences.

19.1 $\{x_n\} = \{(-2n + 3)/(3n + 1)\}$
       $\{y_n\} = \{(2n^2 + 3n)/n^2\}$

19.2 $\{x_n\} = 1, -2, 3, -4, 5, -6, \ldots$
       $\{y_n\} = 1, -\frac{1}{2}, \frac{1}{3}, -\frac{1}{4}, \frac{1}{5}, -\frac{1}{6}, \ldots$

19.3 $\{x_n\} = \{(3n + 1)/(n^2 + 1)\}$
       $\{y_n\} = \{(n^2 + 1)/(2n - 1)\}$

19.4 $\{x_n\} = \{3n/(n + 2)\}$
       $\{y_n\} = \{(2n^2 - 1)/(n^2 + 1)\}$

19.5 $\{x_n\} = 1, -1, 1, -1, 1, -1, \ldots$
       $\{y_n\} = 1, -\frac{1}{2}, \frac{1}{3}, -\frac{1}{4}, \frac{1}{5}, -\frac{1}{6}, \ldots$

19.6 $\{x_n\} = 0, \frac{1}{2}, \frac{2}{3}, \frac{3}{4}, \frac{4}{5}, \ldots$
       $\{y_n\} = 1, 2, 3, 4, 5, \ldots$ (See Example 19.3.)

19.7* (a) Prove the following. If $\{x_n\} \to a$, $a \neq 0$, and $\{y_n\} \to \infty$, then $\{x_n + y_n\} \to \infty$, $\{x_n - y_n\} \to -\infty$, $\{x_n y_n\} \to \infty$ if $a > 0$, $\{x_n y_n\} \to -\infty$ if $a < 0$, and $\{x_n/y_n\} \to 0$.
       (b) Show that some of the results in (a) need not be true if $\{x_n\} \to 0$ or if $\{x_n\}$ is not a convergent sequence.

19.8* (a) Suppose a sequence $\{y_n\}$ satisfies the condition that $|y_n| \leq K$ for all n where K is a positive real number. Let $\{x_n\}$ be any sequence having limit zero. Prove that the product sequence $\{x_n y_n\}$ has limit zero.
       (b) Using part (a), find the limit of the product of the sequences $\{x_n\} = \{1/2^n\}$ and $\{y_n\} = 1, 0, -1, 0, 1, 0, -1, 0, \ldots$.

19.9* Using part (a) of the preceding problem, prove that if $\{x_n\} \to a$ and $\{y_n\} \to b$, then $\{x_n y_n\} \to ab$. (*Hint:* Write $x_n y_n - ab$ as $x_n y_n - a y_n + a y_n - ab$.)

LIMITS OF FUNCTIONS AND CONTINUITY

We now apply the results of the preceding two sections to the problem of computing limits of functions. We first define what we mean by the limit of a function *using sequences*, and then give numerous examples.

Recall the notion of a function as a mapping of points on the x axis to points on the y axis. This approach yields the most meaningful definition of a limit of a function. In what follows, we assume that the sequence $\{x_n\}$ corresponds to a set of points on the x axis in the domain of f. Then, for each term x_n, we let $y_n = f(x_n)$ and we form the sequence $\{y_n\}$, which corresponds to a set of points on the y axis in the range of f. Therefore, the function f maps each term x_i of the sequence $\{x_n\}$ to the term $y_i = f(x_i)$ of the sequence $\{y_n\}$.

Definition 20.1 Let f be a function with domain X and range Y. If for each sequence $\{x_n\}$ in X having limit a with $x_n \neq a$ for any n, the corresponding sequence $\{y_n\} = \{f(x_n)\}$ in Y has limit b, we say that *the limit of the function f as x approaches a is equal to b* and we denote this by $\lim_{x \to a} f(x) = b$.

Also, recall that the limit of a sequence is unique (Theorem 18.1); therefore, if the limit of a function f as x approaches a exists, it is unique.

Alternate Definition Given $\varepsilon > 0$, there exists a $\delta > 0$, δ depending on ε, with the property that for all x such that $|x - a| < \delta$, we have $|f(x) - b| < \varepsilon$. This means that given an ε neighborhood M of b on the y axis, there is a corresponding δ neighborhood N of a on the x axis such that every point of N is mapped to a point of M.

111

Note that geometrically (see Figure 20.1), both of these definitions merely state that points on the x axis that are close to a are mapped by f to points on the y axis close to b.

For sequences that have infinite limits or no limits, we use the following terminology and notation.

If for each sequence $\{x_n\} \to a$, the sequence $\{y_n\} \to \infty$, we write $\lim_{x \to a} f(x) = \infty$. If for at least one sequence $\{x_n\} \to a$, the sequence $\{y_n\}$ does not have either a finite or an infinite limit, we say that the limit of f as x approaches a does not exist. Also, if for each sequence $\{x_n\} \to \infty$, $\{y_n\}$ has limit b, we write $\lim_{x \to \infty} f(x) = b$. Finally, if for each sequence $\{x_n\} \to \infty$, $\{y_n\} \to \infty$, we write $\lim_{x \to \infty} f(x) = \infty$.

REMARK   Observe that to prove that the limit of a function f exists as x approaches a, we must show that for *any* choice of a sequence $\{x_n\} \to a$, the corresponding sequence $\{y_n\} = \{f(x_n)\}$ has the same limit. On the other hand, to prove that the limit of a function does not exist as x approaches a, we need only exhibit *one* sequence $\{x_n\} \to a$ yielding a corresponding $\{y_n\}$ sequence having *no* limit, *or* exhibit *two* sequences $\{x_n\} \to a$ and $\{x_n'\} \to a$ yielding corresponding sequences $\{y_n\}$ and $\{y_n'\}$ having *unequal* limits (see Example 20.5). In general, to obtain an arbitrary sequence $\{x_n\} \to a$, we may

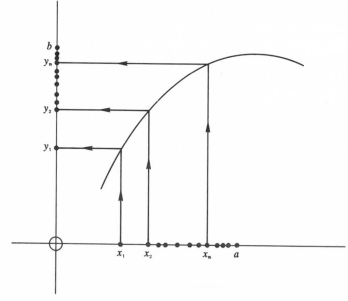

Figure 20.1

use a sequence $\{a + \varepsilon_n\}$ where $\{\varepsilon_n\}$ represents any sequence that approaches zero. We then compute the limit b of the sequence $\{y_n\} = \{f(a + \varepsilon_n)\}$. Then b is the limit of f as x approaches a. In the following three examples, we use this method to find the required limit.

EXAMPLE 20.1 Let $f(x) = x^2 - 1$. Find $\lim\limits_{x \to 2} f(x)$.

SOLUTION Let $\{\varepsilon_n\} = \varepsilon_1, \varepsilon_2, \varepsilon_3, \ldots$ be an arbitrary sequence converging to 0. Then $\{x_n\} = \{2 + \varepsilon_n\}$ represents an arbitrary sequence approaching 2. The corresponding sequence $\{y_n\} = \{(2 + \varepsilon_n)^2 - 1\} = \{4 + 4\varepsilon_n + \varepsilon_n^2 - 1\} = \{3 + 4\varepsilon_n + \varepsilon_n^2\}$. Hence, $\{y_n\} \to 3$ (since $\{\varepsilon_n\} \to 0$) and we have $\lim\limits_{x \to 2} (x^2 - 1) = 3$.

EXAMPLE 20.2 Let $f(x) = 1/x$. Find $\lim\limits_{x \to 2} f(x)$.

SOLUTION Let $\{x_n\} = \{2 + \varepsilon_n\}$, just as in the preceding example. Then the corresponding sequence is $\{y_n\} = \{1/(2 + \varepsilon_n)\}$ and has limit $\frac{1}{2}$ (since $\{\varepsilon_n\} \to 0$). Therefore, we have $\lim\limits_{x \to 2} 1/x = \frac{1}{2}$.

EXAMPLE 20.3 Let $f(x) = 1/x$, $x > 0$. Find $\lim\limits_{x \to 0} f(x)$.

SOLUTION Let $\{\varepsilon_n\}$ be an arbitrary sequence approaching zero through positive values. The corresponding sequence is $\{y_n\} = \{1/\varepsilon_n\}$ and has limit ∞; therefore, $\lim\limits_{x \to 0} 1/x = \infty$. Observe that in this example, we ask for the limit of $f(x)$ as x approaches 0, even though 0 is not in the domain of f. Reading Definition 20.1 again, we notice that only the terms of the sequences $\{x_n\}$ and $\{y_n\}$ must lie in the domain X and the range Y, respectively. There is no stipulation whether the limits a and b are in X and Y, respectively. Finally, looking at the graph of $y = 1/x$ (see Figure 11.8 with $a = 1$), observe that for positive values of x close to zero, the corresponding y values are extremely large. This substantiates geometrically what we have just proved analytically.

Finally, as an exercise, let $f(x) = 1/x$, $x < 0$. What is the limit of f as x approaches zero?

EXAMPLE 20.4 Let $f(x) = 1/x$. Find $\lim\limits_{x \to \infty} f(x)$.

SOLUTION Let $\{x_n\}$ be an arbitrary sequence approaching ∞. The corresponding sequence $\{y_n\} = \{1/x_n\}$ must approach 0; hence, $\lim\limits_{x \to \infty} 1/x = 0$. Once again, looking at the graph of $y = 1/x$, for large positive x, we observe that $1/x$ is positive and very close to zero, which confirms our result.

EXAMPLE 20.5   Let $f(x) = [x]$, the greatest integer function (see Examples 15.7 and 16.7).   Let p be any positive integer.   Find $\lim\limits_{x \to p} f(x)$.

SOLUTION   In this situation, if we choose the sequence $\{x_n\}$ to be $\{p - 1/n\}$, the sequence $\{y_n\} = p - 1, p - 1, p - 1, p - 1, \ldots$ and therefore has limit $p - 1$.   If, however, we choose $\{x_n\} = \{p + 1/n\}$, the corresponding sequence $\{y_n\} = p, p, p, p, p, \ldots$ and has limit p.   Since $p \neq p - 1$, it is *not* true that for each sequence $\{x_n\}$ that approaches p the corresponding sequence $\{y_n\}$ approaches a fixed number b.   Therefore, by definition, $\lim\limits_{x \to p} [x]$ *does not* exist (see the Remark preceding Example 20.1).   However, the limit $p - 1$ is called the *left-hand limit* of $f(x)$ at $x = p$, whereas the limit p is called the *right-hand limit* of $f(x)$ at $x = p$.

Definition 20.2   In general, if $\{\varepsilon_n\}$ is any sequence approaching zero through positive values, we can define the *left-hand limit* of $f(x)$ at $x = a$ to be the limit of the sequence $\{y_n\}$ corresponding to the sequence $\{x_n\} = \{a - \varepsilon_n\}$, while the right-hand limit of $f(x)$ at $x = a$ is the limit of the sequence $\{y_n\}$ corresponding to the sequence $\{x_n\} = \{a + \varepsilon_n\}$.

Finally, we say that $\lim\limits_{x \to a} f(x)$ exists and equals the number b if and only if the left-hand limit *equals* the right-hand limit *equals* b.   This is an *equivalent definition* for the *limit of a function* $f(x)$ as x approaches a.

Observe that in Example 20.5 we could have chosen $\{x_n\} = \{p + (-1)^{n-1}/n\}$. Then the sequence $\{y_n\} = p + 1, p - 1, p, p - 1, p, p - 1, \ldots$ and does not have a limit.   Hence, the limit of the function $f(x) = [x]$ as x approaches p does not exist (see the Remark preceding Example 20.1).

Definition 20.3   We say that a function f is *continuous* at $x = a$ if $\lim\limits_{x \to a} f(x) = f(a)$.   This means that for every sequence $\{x_n\}$ that has limit a, the corresponding sequence $\{y_n\}$ must not only have a limit, but the limit must be the functional value of f at $x = a$, that is, $f(a)$.   If a function f is *continuous* at every point x in an interval $c \leq x \leq d$, we say that f is *continuous on the interval* $c \leq x \leq d$.

An *alternate definition of continuity* at $x = a$ can be stated as follows.   If at $x = a$ the left-hand limit of f *equals* the right-hand limit *equals* $f(a)$, then f is continuous at $x = a$.   Another definition may be obtained by replacing b with $f(a)$ in the alternate definition following Definition 20.1.

Finally, observe that if f is continuous at $x = a$, then f maps points on the x axis close to a to points on the y axis close to $f(a)$ [see Figure 20.1 with $b = f(a)$].

EXAMPLE 20.6   In Example 20.1, $\lim\limits_{x \to 2} (x^2 - 1) = 3 = f(2)$.   Therefore, $f(x) = x^2 - 1$ is continuous at $x = 2$.

EXAMPLE 20.7 In Example 20.2, $\lim_{x \to 2} 1/x = \frac{1}{2} = f(2)$. Therefore, $f(x) = 1/x$ is continuous at $x = 2$. In fact, if we choose $x = a$ where a is any positive real number, we may replace 2 by a and repeat the argument used in Example 20.2. This leads us to the conclusion that $f(x) = 1/x$ is continuous at $x = a$ for every positive real number a. If we consider $f(x) = 1/x$ for $x < 0$, we can also show that $f(x) = 1/x$ is continuous at every negative real number a. Therefore, $f(x) = 1/x$ is continuous for all $x \neq 0$.

EXAMPLE 20.8 In Example 20.5, we proved that for any positive integer p, $\lim_{x \to p} [x]$ did not exist. Consequently, this limit cannot equal $f(p) = p$. Therefore, $f(x) = [x]$ is *not continuous at any positive integral value of x*. If, however, we take any positive real number a, a not an integer, and if we choose $\{x_n\} = \{a + \varepsilon_n\}$ where $\{\varepsilon_n\}$ is any sequence approaching zero, then for large enough n, all the terms of the sequence $\{y_n\}$ will be equal to the greatest integer less than a, that is, $[a]$. Therefore, $\{y_n\} \to [a]$ and $\lim_{x \to a} f(x) = f(a)$. Hence, the greatest integer function is *continuous at all positive nonintegral values of x*.

As an exercise, prove that the greatest integer function is also *continuous* for all negative nonintegral values of x and *not continuous* at $x = 0$ or at any negative integral value of x.

PROBLEMS

20.1 In each of the following, compute the limits of the given function as x approaches each of the numbers 0, 1, -1, and 2.
(a) $f(x) = x^3 - 2x^2 + 3x + 2$
(b) $f(x) = x^5 + 1$
(c) $f(x) = |1 - x|$
(d) $f(x) = (1 + x)^{1/2}$
(e) $f(x) = 2^x$
(f) $f(x) = 2^{1/x}$
(g) $f(x) = (x^2 + 1)/(x^3 + 2)$

20.2 In each of the following, compute the limit of the given function as x approaches ∞.
(a) $f(x) = (x - 1)/(x^2 + 1)$
(b) $f(x) = (x^3 - 1)/(x^3 + 1)$
(c) $f(x) = (2x^4 - 4x^2 + 1)/(3x^4 + 2x^3 - 3x^2 + x + 1)$

20.3 Show that $f(x) = x^2 - 1$ is continuous at every real number a by imitating the argument of Example 20.1 after replacing 2 with a.

20.4 Let $f(x) = |x|/x$, $x \neq 0$, and $f(0) = 0$.
(a) Find the *left-hand limit* of $f(x)$ at $x = 0$ and find the *right-hand limit* of $f(x)$ at $x = 0$ (see Definition 20.2).
(b) Discuss the continuity of this function.

20.5 Prove that $f(x) = |x|$ is continuous at every real number a.

20.6   Prove that if $f(x)$ is continuous at $x = a$ and $f(a) \neq 0$, the function $1/f(x)$ is continuous at $x = a$.

20.7*   Discuss the continuity of the following functions.

(a) $f(x) = \begin{cases} 0 & \text{if } x \text{ is rational} \\ 1 & \text{if } x \text{ is irrational} \end{cases}$

   (See Remark 2 at the conclusion of Section 16.)

(b) $f(x) = \begin{cases} 0 & \text{if } x \text{ is irrational or if } x = 0 \\ 1/q, & \text{if } x = p/q, q > 0, \text{ and } p/q \text{ is reduced to lowest terms} \end{cases}$

   (*Hint:* Use the fact that every irrational number has a sequence of rational numbers approaching it. For example, $\sqrt{2}$ is the limit of the sequence 1, 1.4, 1.41, 1.414, 1.4142, 1.41421, ...)

20.8*   Prove that the following is an *alternate definition of continuity*. We say that a function f is continuous at a point a if $\lim_{h \to 0} f(a + h) = f(a)$.

20.9*   If f is a continuous function defined for all real numbers x with the property that $f(x + y) = f(x) + f(y)$ for any real numbers x and y, then $f(x) = kx$ where $k = f(1)$.

BASIC THEOREMS ON LIMITS OF FUNCTIONS

We now state the basic theorems on limits of sums, differences, products, quotients, and compositions of functions. The proofs depend on the results obtained for sequences in Theorem 19.1.

Theorem 21.1 If $\lim_{x \to a} f(x) = b$ and $\lim_{x \to a} g(x) = c$, then we have the following results.

(a) $\lim_{x \to a} [f(x) + g(x)] = b + c$; that is, the limit of the sum of two functions equals the sum of the individual limits.

(b) $\lim_{x \to a} [f(x) - g(x)] = b - c$; that is, the limit of the difference of two functions equals the difference of the individual limits.

(c) $\lim_{x \to a} [f(x)g(x)] = bc$; that is, the limit of the product of two functions equals the product of the individual limits.

(d) $\lim_{x \to a} [f(x)/g(x)] = b/c$ provided that $c \neq 0$; that is, the limit of the quotient of two functions equals the quotient of the individual limits provided that $c \neq 0$.

PROOF (a) Let $\{x_n\}$ be any sequence with limit a. We must show that the sequence $\{y_n\} = \{f(x_n) + g(x_n)\}$ has limit $b + c$. But $\{f(x_n) + g(x_n)\} = \{f(x_n)\} + \{g(x_n)\}$, and from Theorem 19.1 we know that the sequence $\{f(x_n)\} + \{g(x_n)\}$ has a limit equal to the sum of the limits of the sequences $\{f(x_n)\}$ and $\{g(x_n)\}$. But $\lim_{x \to a} f(x) = b$ implies that $\{f(x_n)\} \to b$ and $\lim_{x \to a} g(x) = c$ implies that $\{g(x_n)\} \to c$. Therefore, $\{f(x_n) + g(x_n)\} \to b + c$ and we have completed the proof of part (a). ∎

117

The proofs of parts (b), (c), and (d) all depend on the analogous results given for sequences in Theorem 19.1. Write out these proofs as an exercise. The following are a few final remarks regarding this theorem.

REMARK 1   Statement (c) includes the possibility that $f(x) = k$ where k is any real constant. The statement would then read $\lim_{x \to a} [kg(x)] = kc$ since, in this event, $b = k$. Also, in the proof, we would use the fact that the limit of the sequence $k, k, k, \ldots$ is certainly k.

REMARK 2   These results may be extended to include the limits of more than two functions. For example, if the limit as x approaches a of each of the functions f, g, and h exists and equals b, c, and d, respectively, then $\lim_{x \to a} [f(x) + g(x) + h(x)] = b + c + d$ and $\lim_{x \to a} [f(x)g(x)h(x)] = bcd$, and so on for any finite number of functions.

Finally, if we assume that $f(x)$ and $g(x)$ are *continuous* at a, then $b = f(a)$, $c = g(a)$, and we get the following corollary to Theorem 21.1.

Corollary f and g *continuous* at a imply that $f + g$, $f - g$, fg, and f/g [provided that $g(a) \neq 0$] are all continuous at a. If we also assume that f and g are *continuous* on some interval J, we may say that the sum, difference, and product of f and g are also continuous on J, while the quotient will be continuous at all points x of J for which $g(x) \neq 0$.

EXAMPLE 21.1   Let $f(x) = x^2 - 1$ and let $g(x) = 1/x$. We know from Examples 20.1 and 20.2 that $\lim_{x \to 2} f(x) = 3$ and $\lim_{x \to 2} g(x) = \frac{1}{2}$. Therefore, from the preceding theorem, we obtain the following results.

(a) $\lim_{x \to 2} (x^2 - 1 + 1/x) = 3 + \frac{1}{2} = \frac{7}{2}$

(b) $\lim_{x \to 2} (x^2 - 1 - 1/x) = 3 - \frac{1}{2} = \frac{5}{2}$

(c) $\lim_{x \to 2} [(x^2 - 1)(1/x)] = 3(\frac{1}{2}) = \frac{3}{2}$

(d) $\lim_{x \to 2} [(x^2 - 1)/(1/x)] = 3/(\frac{1}{2}) = 6$

Theorem 21.2   If $\lim_{x \to a} g(x) = b$ and f is continuous at b, then $\lim_{x \to a} f(g(x)) = f(b)$.

Before we prove this theorem, you should observe that the statement above could have been put in the following way. If $\lim_{x \to a} g(x) = b$ and $\lim_{x \to b} f(x) = f(b)$, then $\lim_{x \to a} f(g(x)) = f(\lim_{x \to a} g(x))$. For example, if $f(x) = 1/x$ and $g(x) = x^2 - 1$, then $f(g(x)) = 1/(x^2 - 1)$ and $\lim_{x \to 2} f(g(x)) = f(\lim_{x \to 2} g(x)) = f(3) = \frac{1}{3}$. Verify this result by directly computing $\lim\limits_{x \to 2} [1/(x^2 - 1)]$ by using Definition 20.1 with $\{x_n\} = \{2 + \varepsilon_n\}$ or by using part (d) of Theorem 21.1.

PROOF OF THEOREM 21.2   Let $\{x_n\}$ be any sequence in the domain of g with limit a.   We must show that the sequence $\{y_n\} = \{f(g(x_n))\}$ has limit $f(b)$. Now $\{x_n\} \to a$ implies that the sequence $\{z_n\} = \{g(x_n)\} \to b$ [since $\lim\limits_{x \to a} g(x) = b$].   Therefore, $\{y_n\} = \{f(z_n)\}$ where $\{z_n\} \to b$.   But f continuous at b implies that for any sequence $\{z_n\} \to b$, the sequence $\{f(z_n)\} \to f(b)$. Therefore, $\{y_n\} = \{f(g(x_n))\}$ has limit $f(b)$ and the proof is completed. ∎

If in Theorem 21.2 we assume that $g(x)$ is *continuous* at $x = a$, we may conclude that $\lim\limits_{x \to a} f(g(x)) = f(g(a))$ and therefore, the function $f(g(x))$ is also *continuous* at $x = a$.   This leads to the following corollary to Theorem 21.2.

Corollary   If $g(x)$ is continuous at $x = a$ and $f(x)$ is continuous at $x = g(a)$, the composition $f(g(x))$ is continuous at $x = a$.
   For example, $g(x) = x^2 - 1$ is continuous at $x = 2$; $f(x) = 1/x$ is continuous at $g(2) = 3$; therefore, the function $f(g(x)) = 1/(x^2 - 1)$ must be continuous at $x = 2$.   This implies $\lim\limits_{x \to 2} f(g(x)) = f(g(2)) = \frac{1}{3}$, a fact we already verified in the example immediately following the statement of Theorem 21.2.

EXAMPLE 21.2   Let $h(x) = (x^2 - 1)^5$.   Find $\lim\limits_{x \to 2} h(x)$.

SOLUTION   By the preceding theorem with $f(x) = x^5$ and $g(x) = x^2 - 1$, we must simply compute $f(\lim\limits_{x \to 2} g(x)) = f(3) = 3^5$.

EXAMPLE 21.3   Let $h(x) = |1 - x^2|$.   Find $\lim\limits_{x \to 3} h(x)$.

SOLUTION   We let $f(x) = |x|$, $g(x) = 1 - x^2$, and we compute $f(\lim\limits_{x \to 3} g(x)) = |\lim\limits_{x \to 3} (1 - x^2)| = |-8| = 8$.

We now state a theorem, the proof of which depends on the results obtained in Theorem 21.1.

Theorem 21.3   Every polynomial

$$f(x) = a_n x^n + a_{n-1} x^{n-1} + \cdots + a_1 x + a_0$$

is continuous at every real number r.   This means, in effect, that for any real number r, $\lim\limits_{x \to r} f(x) = f(r)$.

This result, together with Theorem 21.1 [part (d)] and its Corollary, also implies that every *rational function* (i.e., a function that is a quotient of two polynomials) is *continuous* at every real number that is not a root of the denominator. This means that if $h(x)$ is the quotient of two polynomials $f(x)$ and $g(x)$, then $\lim_{x \to r} [f(x)/g(x)] = f(r)/g(r)$ provided that $g(r) \neq 0$. Observe also that a rational function is not defined at a point that is a root of the denominator. Consequently, we may say that a rational function is continuous at every point in its domain.

PROOF OF THEOREM 21.3   Every polynomial can be built up from the identity function $I(x) = x$ and the constant functions simply by taking sums and products. For example, $f(x) = 3x^2 - 2x + 1$ can be built by multiplying $I(x)$ by itself and then by the constant function $k(x) = 3$, which yields the term $3x^2$. We then subtract the product of the functions $k(x) = -2$ and $I(x) = x$; finally, we add the function $k(x) = 1$. In general, if we want to build the term $a_i x^i$ for $i > 0$, we multiply $I(x) = x$ by itself $i - 1$ times and multiply the result by $k(x) = a_i$. Having done this for each $i = 1, 2, 3, \ldots, n$, we add the results and conclude the process by adding the function $k(x) = a_0$. We then have the polynomial $a_n x^n + a_{n-1} x^{n-1} + \cdots + a_1 x + a_0$.

Therefore, we need only show that any constant function is continuous and that the identity function $I(x) = x$ is continuous. Then we apply the Corollary to Theorem 21.1 and conclude that every polynomial is continuous since it can be built up by sums and products of continuous functions.

Certainly every constant function $k(x) = a$ is continuous at every real number r since $\lim_{x \to r} k(x) = a = k(r)$. On the other hand, $\lim_{x \to r} I(x) = r = I(r)$; consequently, $I(x)$ is continuous at every real number r and we have completed the proof. ■

Finally, as a consequence of this theorem, we see that all functions of the form $f(x) = mx + b$ must be continuous for all real numbers r; hence, every function that has a nonvertical line for its graph is continuous. Every function $f(x) = ax^2 + bx + c$ must also be continuous for all real numbers r; hence, every function that has a parabola with a vertical axis of symmetry for its graph is continuous. All functions of the type $f(x) = x^n$ for any positive integer n must be continuous. Rational functions like $f(x) = (x^3 + 3x^2 - 2x + 1)/(x^5 + 9x^4 - 7x^3 + x^2 - 3x + 1)$ are continuous at every real number r that is not a root of the denominator. There are many other examples of continuous functions mentioned in the problems that follow this section, and we now conclude the section with a few more.

EXAMPLE 21.4   Compute each of the following limits.

(a) $\lim_{x \to 2} (x^2 + x - 2)$

(b) $\lim_{x \to 1} [(x^{17} + x^7 - x^4)/(x^{13} + x^3)]$

(c) $\lim_{x \to 1} [(x^2 + x - 2)/(x + 1)]$

SOLUTIONS (a) $f(x) = x^2 + x - 2$ is a polynomial; therefore, it is continuous and the limit is simply $f(2) = 4$.

(b) $f(x) = (x^{17} + x^7 - x^4)/(x^{13} + x^3)$ is a rational function; 1 is not a root of the denominator; therefore, the limit is $f(1) = \frac{1}{2}$.

(c) $f(x) = (x^2 + x - 2)/(x + 1)$ is also a rational function; 1 is not a root of the denominator; therefore, the limit is $f(1) = 0$.

EXAMPLE 21.5 Let $f(x) = (x^2 + x - 2)/(x - 1)$. Find $\lim_{x \to 1} f(x)$.

SOLUTION $f(x) = (x^2 + x - 2)/(x - 1)$ is a rational function; however, 1 *is* a root of the denominator and $f(1)$ is not defined; that is, 1 is not in the domain of f. In this case, the numerator is factored into the product $(x - 1)(x + 2)$ and the term $(x - 1)$ is canceled by the denominator, leaving simply $(x + 2)$ in the numerator. Now, the limit of the function $(x + 2)$ as x approaches 1 is 3; therefore, $\lim_{x \to 1} f(x) = 3$.

We now justify this method in the following manner. Let $\{x_n\}$ be any sequence such that $\{x_n\} \to 1$. We must determine whether the sequence $\{y_n\} = \{f(x_n)\}$ has a limit. Now, $f(x_n) = (x_n^2 + x_n - 2)/(x_n - 1) = (x_n - 1)(x_n + 2)/(x_n - 1)$. Moreover, $x_n \neq 1$ for any n (see Definition 20.1); therefore, $x_n - 1 \neq 0$ and it can be canceled, leaving $f(x_n) = x_n + 2$. Therefore, $\{f(x_n)\} = \{x_n + 2\}$ and has limit 3 (since $\{x_n\} \to 1$).

In the *general* situation of a rational function $f(x) = (x - a)p(x)/(x - a) \times q(x)$, where $q(a) \neq 0$, if we want to find the limit of $f(x)$ as x approaches a, we may imitate the preceding argument by choosing an arbitrary sequence $\{x_n\} \to a$. Then the sequence $\{y_n\} = \{f(x_n)\}$ reduces to $\{p(x_n)/q(x_n)\}$, which has limit $p(a)/q(a)$ [since $p(x)$ and $q(x)$ are continuous functions]. This completes one justification of the cancellation method.

Another way of looking at this general problem involves the comparison of the two functions $f(x) = (x - a)p(x)/(x - a)q(x)$ and $g(x) = p(x)/q(x)$. As x approaches a, x is never equal to a; hence, the limit of both f and g as x approaches a depends only on values of $x \neq a$. But for all $x \neq a$, $f(x) = g(x)$; hence, $\lim_{x \to a} f(x) = \lim_{x \to a} g(x) = p(a)/q(a)$.

PROBLEMS

21.1 In each of the following, compute the limits of the given function as x approaches each of the numbers 0, 1, -1, and 2.

(a) $f(x) = 7x^3 - 2x^2 + 3x + 2$

(b) $f(x) = x^5 + 2x^4 - 3x^3 + 2x^2 - 5x + 1$

(c) $f(x) = (x^{10} - 1)/(x^5 + 1)$

(d) $f(x) = (x^2 - 3x + 1)(x^5 - 3x^3 + 2x^2 + 1)$

(e) $f(x) = |3x^3 + 4x^2 + 2x - 7|$

(f) $f(x) = 2^{(x^2+1)}$ (You may assume that $g(x) = 2^x$ is continuous for all real x.)

(g) $f(x) = 2^{1/(x-1)}$ (You may assume that $g(x) = 2^x$ is continuous for all real x.)

(h) $f(x) = (x^7 + x^5 - x^3 + 1)^5$

(i) $f(x) = (x^3 - 2x^2 + x - 2)/(x^3 + 2x^2 - 9x - 18)$

(j) $f(x) = (x^2 + 3x - 1)/(x^5 + 3x^3 - 2x^2 + 1)$

(k) $f(x) = (x^3 - 2x^2 + x - 2)/(x^5 + 3x^3 - 2x^2 + 1)$

(l) $f(x) = (x^2 + 3x + 2)(3x^3 - 2x^2 + 1)$

(m) $f(x) = (x^3 + 3x + 1)(x^4 - 2x^2 + 1)$

21.2 (a) Prove that $f(x) = \sqrt{x}$ is continuous at every nonnegative real number a.

   (b) Compute the limit of the function $f(x) = (5x^2 + 3x + 1)^{1/2}$ as x approaches each of the numbers 0, 1, and -1.

21.3 (a) If $\lim_{x \to a} f(x) = b$, $b \neq 0$, prove that $\lim_{x \to a} (f(x))^n = b^n$ and that $\lim_{x \to a} (f(x))^{1/n} = b^{1/n}$ for *any* integer $n \neq 0$.

   (b) Prove that if $f(x)$ is continuous at the point a, $f(a) \neq 0$, the functions $(f(x))^n$ and $(f(x))^{1/n}$ are also continuous at the point a, where n is *any* nonzero integer.

21.4 (a) If $\lim_{x \to a} f(x) = b$, does $\lim_{x \to a} |f(x)|$ always exist? If so, prove it; if not, give a counterexample.

   (b) If $\lim_{x \to a} |f(x)| = b$, does $\lim_{x \to a} f(x)$ always exist? If so, prove it; if not, give a counterexample.

21.5 (a)* Prove that the functions $f(x) = \sin x$ and $f(x) = \cos x$ (see the Appendix) are continuous at every real number a.   [*Hint:* Use the result of Problem 20.8 and the formula for the $\sin(a + h)$ and $\cos(a + h)$.]

   (b) Compute the limit of the function $f(x) = x \sin x$ as x approaches each of the numbers $\pi/4$, $\pi/2$, and π.

   (c) Compute the limit of the function $f(x) = (1/x)\cos x$ as x approaches each of the numbers $\pi/4$, $\pi/2$, and π.

   (d) Compute the limit of the function $f(x) = \sin(2x + \pi/2)$ as x approaches each of the numbers 0, $\pi/4$, $\pi/2$, and π.   [*Hint:* $\sin(2x + \pi/2)$ is a composition of the functions $f(x) = \sin x$ and $g(x) = 2x + \pi/2$. Therefore, simply apply Theorem 21.2.]

   (e) Compute the limit of the function $f(x) = \cos(2x)$ as x approaches each of the numbers 0, $\pi/4$, $\pi/2$, and π.

   (f) Compute $\lim_{x \to 0} \sin(\cos x)$.

   (g)* Compute $\lim_{x \to 0} \sin(\cos(2x + \pi))$.

21.6* (a) Prove the following theorem. Suppose the functions f, g, and h are defined in some deleted neighborhood N of a point a with the property that for all $x \in N, f(x) \leq g(x) \leq h(x)$. Then if $\lim_{x \to a} f(x) = \lim_{x \to a} h(x) = b$, prove that $\lim_{x \to a} g(x) = b$.

   (b) Use the result of part (a) to prove that $\lim_{x \to 0} (\sin x)/x = 1$.

   (c) Prove that $\lim_{x \to 0} (1 - \cos x)/x = 0$.

21.7* In each of the following, discuss the continuity of the given functions and sketch the graph of each function.

(a) $f(x) = (\sin x)/x$, $x \neq 0$, $f(0) = 1$

(b) $f(x) = (1 - \cos x)/x$, $x \neq 0$, $f(0) = 0$

(c) $f(x) = (\tan x)/x$, $x \neq 0$, $f(0) = 1$

(d) $f(x) = \sin(1/x)$, $x \neq 0$, $f(0) = 0$

(e) $f(x) = x \sin(1/x)$, $x \neq 0$, $f(0) = 0$

THE DEFINITION
OF THE DERIVATIVE

We now deal with the limit of a special kind of function. We first give a few examples and then proceed to a discussion of two practical problems, both of which are resolved by the computation of this limit.

Suppose we choose a polynomial function such as $f(x) = x^2 + x + 1$. Let a be any real number. Form the quotient $[f(x) - f(a)]/(x - a)$. This quotient, for a fixed real number a, is still a function of x defined for all $x \neq a$; call it $Q(x)$. Then,

$$Q(x) = \frac{x^2 + x + 1 - a^2 - a - 1}{x - a} = \frac{(x^2 - a^2) + (x - a)}{x - a}$$

Now, for all values of $x \neq a$, $Q(x)$ equals $(x + a) + 1$. Since $(x + a) + 1$ is a polynomial function of x, we may compute $\lim_{x \to a} Q(x) = Q(a) = 2a + 1$ (see Theorem 21.3). This limit is defined later as the *derivative* of the function $f(x) = x^2 + x + 1$ at $x = a$.

Observe that in the preceding discussion we could have replaced the *non-zero* quantity $x - a$ by the *nonzero* quantity h, thus making $Q(x)$ "a function of the variable h." Then, $x^2 - a^2 = (x - a)(x + a) = h(h + 2a)$ and $Q(x)$ is changed into the function of h given by

$$R(h) = \frac{h(h + 2a) + h}{h} = h + 2a + 1$$

Now as x approaches a, $h = x - a$ approaches 0; therefore, $\lim_{x \to a} Q(x) = \lim_{h \to 0} R(h)$. Hence, we have an equivalent method of computing the derivative of $f(x)$.

124

Definition 22.1 In general, if we have an arbitrary function f with domain X, we may choose any real number a in X and form the function $Q(x) = [f(x) - f(a)]/(x - a)$, which is defined for all $x \neq a$. Then, we define the *derivative of f at x = a*, denoted by $f'(a)$, to be $\lim_{x \to a} Q(x)$, if this limit exists.

Also, in the general case, if we let $h = x - a$, we may substitute $a + h$ for x and we obtain the function of h

$$R(h) = \frac{f(a + h) - f(a)}{h}$$

which is defined for all $h \neq 0$. Then, we may also define the *derivative of f at x = a* to be $\lim_{h \to 0} R(h)$, if this limit exists.

Finally, if f has a derivative at every number a in X, we say that f is *differentiable* on X.

EXAMPLE 22.1 Find the derivative of the function $f(x) = x^2$ at $x = 2$.

SOLUTION 1 $Q(x) = [f(x) - f(2)]/(x - 2) = (x^2 - 4)/(x - 2) = x + 2$. Now $\lim_{x \to 2} (x + 2) = 4$; therefore, $f'(2) = 4$.

SOLUTION 2 $R(h) = [f(2 + h) - f(2)]/h = (4 + 4h + h^2 - 4)/h = 4 + h$. Now $\lim_{h \to 0} (4 + h) = 4$, therefore, $f'(2) = 4$.

EXAMPLE 22.2 Let $f(x) = c$ where c is any constant. Compute $f'(a)$ for any real number a.

SOLUTION $Q(x) = [f(x) - f(a)]/(x - a) = (c - c)/(x - a) = 0$;   hence, $\lim_{x \to a} Q(x) = 0$ and $f'(a) = 0$ for every real number a.

EXAMPLE 22.3 Let $f(x) = kx$ where k is any constant. Compute $f'(a)$ for any real number a.

SOLUTION 1

$$Q(x) = [f(x) - f(a)]/(x - a) = (kx - ka)/(x - a)$$
$$= k(x - a)/(x - a) = k$$

Now $\lim_{x \to a} k = k$; therefore, $f'(a) = k$ for every real number a.

SOLUTION 2

$$R(h) = [f(a + h) - f(a)]/h = [k(a + h) - ka]/h$$
$$= (kh)/h = k$$

Now $\lim_{h \to 0} k = k$; therefore, $f'(a) = k$ for every real number a.

EXAMPLE 22.4   Let $f(x) = kx^n$ where k is any constant and n is any positive integer.   Compute $f'(a)$ for any real number a.

SOLUTION

$$Q(x) = (kx^n - ka^n)/(x - a) = k(x^n - a^n)/(x - a)$$
$$= k(x^{n-1} + x^{n-2}a + x^{n-3}a^2 + \cdots + x^2a^{n-3} + xa^{n-2} + a^{n-1}),$$

since $(x^n - a^n)$ has the factor $(x - a)$.   Now since $Q(x)$ is a polynomial function, it is continuous at every real number a.   Therefore, $\lim\limits_{x \to a} Q(x) =$ $Q(a) = k(a^{n-1} + a^{n-2}a + a^{n-3}a^2 + \cdots + a^2a^{n-3} + aa^{n-2} + a^{n-1}) = kna^{n-1}$ and $f'(a) = kna^{n-1}$.

As an exercise, find the derivative of $f(x) = kx^n$ at $x = a$ by computing $\lim\limits_{h \to 0} R(h)$ where $R(h) = [k(a + h)^n - ka^n]/h$.

EXAMPLE 22.5   Let $f(x) = 1/x$.   Compute $f'(a)$ for any real number $a \neq 0$.

SOLUTION

$$R(h) = [f(a + h) - f(a)]/h = [1/(a + h) - 1/a]/h$$
$$= -h/a(a + h)h = -1/a(a + h) = -1/(a^2 + ah).$$

Now $R(h) = -1/(a^2 + ah)$ is a rational function of h and $h = 0$ is *not* now a root of the denominator.   Therefore, $R(h)$ is continuous at $h = 0$ (see Theorem 21.3) and $\lim\limits_{h \to 0} = R(h) = R(0) = -1/a^2$.   Hence, $f'(a) = -1/a^2$ for every nonzero real number a.

Observe that in Example 22.4 the derivative of f at $x = a$ assigned the value kna^{n-1} to each real number a in its domain, whereas in Example 22.5, the derivative of f at $x = a$ assigned the value $-1/a^2$ to each real number a in its domain.   Consequently, the derivative of a function associates a real number $f'(a)$ to each a in its domain for which $f'(a)$ exists.   Therefore, *the derivative of a function is itself a function with domain contained in or equal to the domain of f*.   The derivative of the function $f(x) = kx^n$ can therefore be written $f'(x) = knx^{n-1}$ for each x in the domain of f and the derivative of the function $f(x) = 1/x$ can be written $f'(x) = -1/x^2$ for each x in the domain of f.   As we continue with the discussion of the derivative, we will usually write $f'(x)$ for the derivative of an arbitrary function $f(x)$ at any point x in the domain of f and $f'(a)$ for the derivative of an arbitrary function $f(x)$ at the particular point $x = a$.

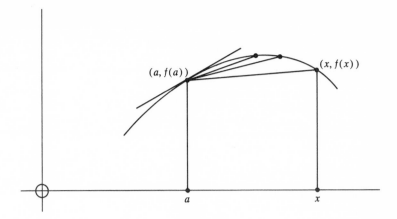

Figure 22.1

In the preceding, we have attempted to give the definition of the derivative in a purely *analytic* manner. We now give two examples of problems where the particular limit that is the derivative comes about in a natural way. Either of these problems could have been used as an introduction to the discussion of the derivative.

PROBLEM I Find the *slope of the tangent* to the graph of $y = f(x)$ at the point $(a, f(a))$.

SOLUTION Choose an arbitrary point x on the x axis near a (see Figure 22.1). Then draw the secant line joining the points $(a, f(a))$ and $(x, f(x))$. The slope of this line is $[f(x) - f(a)]/(x - a)$, which is the function of x called $Q(x)$ earlier. As x approaches a along the x axis, the point $(x, f(x))$, moving along the graph of f, will approach the point $(a, f(a))$. Meanwhile, the secant lines from $(a, f(a))$ to $(x, f(x))$ get shorter and shorter, and they become *almost parallel* to the tangent line at $(a, f(a))$ (see Definition 14.1 and Figure 14.3). Therefore, the analytic counterpart of this geometrical notion involves the computation of the numbers that represent the slopes of the secant lines for values of x that approach a. But this is *exactly* what we mean when we write $\lim_{x \to a} Q(x)$. Consequently, it is very natural to *define* the *slope of the tangent line* to the graph of $y = f(x)$ at $x = a$ to be $f'(a)$, which is the limit of the sequence of slopes obtained from $Q(x)$ as x approaches a along the x axis.

Observe that if we had chosen a small value $h \neq 0$, a point near to a on the x axis could have been denoted by $a + h$ (see Figure 22.2). The slope of the secant line in such a case would be $[f(a + h) - f(a)]/h$, which is the function of h called $R(h)$ earlier. Then as h approaches 0, the point $a + h$ approaches a along the x axis, while the slopes of the secant lines joining the points $(a, f(a))$ and $(a + h, f(a + h))$ approach the slope of the tangent line to the graph at the point $(a, f(a))$. In this case, the analytic counterpart would be the computation of $\lim_{h \to 0} R(h)$, which could also be used to define the slope of the tangent to the graph of $y = f(x)$ at $x = a$. (Again, refer to Definition 14.1 and Figure 14.3.)

EXAMPLE 22.6   Find the equation of the tangent line to the graph of $y = f(x) = x^2$ at the point $(2, 4)$.

SOLUTION   The required slope is the value of the derivative of $f(x) = x^2$ at $x = 2$. This was calculated in Example 22.1 and equals 4. Consequently, the required line is the line through $(2, 4)$ with slope 4. Hence, the line has equation $y - 4 = 4(x - 2)$ or $y = 4x - 4$. The graph with its tangent line is shown in Figure 22.3. Compare this method with the one given in Example 14.1.

EXAMPLE 22.7   Find the slope of the tangent line to the graph of $y = f(x) = 1/x$ at the point $(2, \frac{1}{2})$.

SOLUTION   In Example 22.5, we showed that the derivative of $f(x) = 1/x$ at $x = a$ is $f'(a) = -1/a^2$. Therefore, $f'(2) = -\frac{1}{4}$ is the required slope. Compare this solution with the one presented in Example 14.2.

PROBLEM II   Suppose $s(t)$ is a function that gives at time t the position of a particle P that is moving along a horizontal coordinate line L (see Figure 22.4). For example, if $s(t) = t^2 - t$, then at $t = 1$, P is

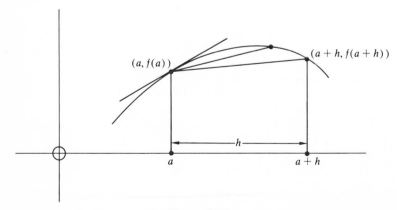

Figure 22.2

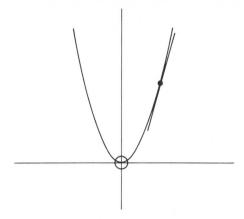

Figure 22.3

at the origin; at $t = 2$, P is at the point 2; at $t = 3$, P is at the point 6; and so on.

If we want to find the *average speed or velocity* of the particle from time $t = a$ to time $t = a + h$ where h is a small nonzero number, we need only compute the distance traveled during this time span and divide by the length of the time span, namely, h. But the distance traveled is simply $s(a + h) - s(a)$; therefore, the average velocity is precisely $[s(a + h) - s(a)]/h$, which is the function of h called $R(h)$ earlier (with s substituted for f).

Suppose we are interested in computing the velocity of P at the *exact* time $t = a$. We may certainly approximate this velocity by computing $R(h)$ for very small values of h, say, $h = 0.1$, $h = 0.01$, $h = 0.001$, and so on. This gives us the average velocity over the time spans from $t = a$ to $t = a + 0.1$, from $t = a$ to $t = a + 0.01$, and from $t = a$ to $t = a + 0.001$, and so on. Clearly, the smaller the time span, the closer our *average velocity* will be to the velocity at *exactly* $t = a$ (the *instantaneous* velocity). The analytic counterpart of this physical situation involves the simple calculation of $\lim_{h \to 0} R(h)$ where $R(h) = [s(a + h) - s(a)]/h$. Hence, we define the *instantaneous* velocity of the particle P at time $t = a$ to be the value of the derivative of the function $s(t)$ at $t = a$.

As an exercise, attempt the preceding development, arriving at the equivalent $\lim_{t \to a} Q(t)$ where $Q(t) = [s(t) - s(a)]/(t - a)$.

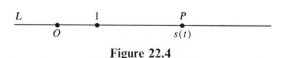

Figure 22.4

EXAMPLE 22.8   From physics, we know that under certain conditions a falling object will travel approximately $16t^2$ feet in t seconds.   Compute the instantaneous velocity at $t = 2$.

SOLUTION   The instantaneous velocity at $t = 2$ is simply the value of the derivative of the function $s(t) = 16t^2$ at $t = 2$.   But from Example 22.4, with $f = s$, $t = x$, and $k = 16$, we find that $s'(a) = 32a$ for any value a.   Therefore, setting $a = 2$, we obtain the required velocity $s'(2) = 64$ ft/sec.

Finally, the velocity of a moving particle or object is simply the *rate of change* of *distance* with respect to *time*.   The quotient $[s(a + h) - s(a)]/h$ measures the *average* rate of change of distance with respect to time, while the limit of the quotient as h approaches 0 yields the rate of change of distance with respect to time at the *instant* $t = a$.   There are other practical problems that involve rates of change of quantities other than distance and time.   We have included some of them among the following problems (Problems 22.3 and 22.4).

PROBLEMS

22.1   Find the slope of the tangent line to each of the following graphs at the given points.
(a) $y = x^2$ $(-1, 1)$, $(3, 9)$, $(4, 16)$
(b) $y = x^3$ $(-1, -1)$, $(1, 1)$, $(2, 8)$
(c) $y = x^3 + x^2 - 2x + 1$ $(0, 1)$, $(1,1)$, $(-1, 3)$
22.2   (a)   Find the velocity at $t = 1$ and $t = 2$ of a particle P that moves along a horizontal coordinate line with position function $s(t) = t^2 - t$.
(b)   Find the velocity of a falling object 3 sec after it is dropped.   At what time is the velocity equal to 160 ft/sec?
22.3   (a)   What is the *average* rate of change of the area of a circle with respect to its diameter as the diameter increases from 2 to 4 inches?   From 2 to 3 inches?
(b)   What is the *instantaneous* rate of change of the area with respect to the diameter when the diameter is exactly 2 inches?
22.4   (a)   What is the *average* rate of change of the volume of a sphere with respect to its radius as the radius increases from 2 to 4 inches?   From 2 to 3 inches?
(b)   What is the *instantaneous* rate of change of the volume with respect to the radius when the radius is exactly 2 inches?
22.5   (a)   Let $y = f(x) = mx + b$ be the equation of an arbitrary nonvertical line.   Using the definition of the derivative, find the slope of the tangent line to this graph at any point $(x_0, f(x_0))$.
(b)   Let $y = f(x) = ax^2 + bx + c$ be the equation of an arbitrary parabola having a vertical axis of symmetry.   Using the definition of the derivative, find the slope of the tangent line to this graph at any point $(x_0, f(x_0))$.   Compare this method with the one used in Example 14.1.

(c) Find the equations of both the *tangent* line and the *normal* line to the graph mentioned in (b) at any point $(x_0, f(x_0))$. (The *normal* line at a point is simply the line through the point that is perpendicular to the tangent line at the point.)

22.6 (a) Prove that if $f'(a)$ exists, f is continuous at a.

(b) Give an example showing that the converse of the result in (a) is not true.

22.7* (a) Prove that the derivative of the function $f(x) = \sqrt{x}$ at $x = a$ is given by $1/(2\sqrt{a})$.

(b) Let n be any positive integer. Prove that the derivative of the function $f(x) = x^{1/n}$ at $x = a$, $a \neq 0$, is given by $(1/n)a^{(1/n)-1}$.

22.8 Let $f(x) = |x|/x$, $x \neq 0$, and $f(0) = 0$. Find the derivative of f (if possible) at $x = -1, 0$, and 1.

22.9* Let f be defined on an open interval J containing the point a.

(a) What do the expressions "left-hand derivative" and "right-hand derivative" of f at $x = a$ mean? (See Definition 20.2.)

(b) Prove that $f'(a)$ exists if and only if f has both a left-hand and a right-hand derivative at $x = a$ and they are equal.

(c) Compute the left-hand and right-hand derivatives of $f(x) = [x]$ at $x = p$ where p is any positive integer (see Example 20.5).

22.10* We know that the derivative $f'(x)$ of a function $f(x)$ is also a function of x, say, $g(x)$. Let a, b be two constants and let x_0 be a point in the domain of f with the property that f' exists at the point $ax_0 + b$. Prove that the derivative of the function $f(ax + b)$ at $x = x_0$ is given by $ag(ax_0 + b)$.

THE DERIVATIVE FORMULAS

In this section, we state the basic theorems that give formulas for the calculation of the derivatives of sums, products, quotients, and compositions of functions. The proofs of most of these depend on the basic results for limits of functions contained in Theorems 21.1 and 21.2 Many examples of the applications of these results are also included.

Theorem 23.1   (a)   If $f(x) = c$ where c is any constant, $f'(x) = 0$ for all x in the domain of f (see Example 22.2).

(b)   If $f(x) = kx^n$, where k is any constant and n is a positive integer, $f'(x) = knx^{n-1}$ (see Example 22.4).

(c)   If $f'(x)$ exists on some interval J and k is any constant, the derivative of the function $kf(x)$ is equal to $kf'(x)$; that is, the derivative of a constant times a function is equal to the constant times the derivative of the function.

(d)   If f and g are differentiable on some interval J, the sum $f + g$ is also differentiable on J and $(f+g)' = f' + g'$; that is, the derivative of the sum of two functions is equal to the sum of the derivatives of the individual functions. Observe that the *converse* of this result is *not true* since it is possible for $f + g$ to be differentiable on some interval J while neither f nor g are differentiable on J (see Problem 23.7).

Theorem 23.2   Every polynomial function $f(x) = a_n x^n + a_{n-1}x^{n-1} + \cdots + a_2 x^2 + a_1 x + a_0$ has a derivative $f'(x) = na_n x^{n-1} + (n-1)a_{n-1}x^{n-2} + \cdots + 2a_2 x + a_1$.

PROOF   Simply apply parts (b) and (d) of the preceding theorem.   ∎

Theorem 23.3 If f and g are differentiable on some interval J, the product g is also differentiable on J and $(fg)' = fg' + f'g$. (The *converse* of this theorem is *not true*; see Problem 23.7.)

Theorem 23.4 If g is differentiable on some interval J with the property that $g(x) \neq 0$ for all x in J, the reciprocal function $1/g$ is also differentiable on J and $(1/g)' = -g'/g^2$.

PROOF Let a be any point in J. Then

$$\left(\frac{1}{g}\right)'(a) = \lim_{h \to 0} \frac{1/g(a+h) - 1/g(a)}{h} = \lim_{h \to 0} \frac{g(a) - g(a+h)}{hg(a)g(a+h)}$$

$$= \lim_{h \to 0} \left[\frac{-1}{g(a)g(a+h)} \cdot \frac{g(a+h) - g(a)}{h}\right]$$

$$= \lim_{h \to 0} \left(\frac{-1}{g(a)g(a+h)}\right) \lim_{h \to 0} \frac{g(a+h) - g(a)}{h}.$$

(since the limit of the product equals the product of the limits). The limit on the right is simply $g'(a)$ (by definition of the derivative). The limit on the left can be computed as follows.

$$\lim_{h \to 0} \left(\frac{-1}{g(a)g(a+h)}\right) = \frac{-1}{\lim_{h \to 0} [g(a)g(a+h)]} = \frac{-1}{\lim_{h \to 0} g(a) \lim_{h \to 0} g(a+h)}.$$

But $\lim_{h \to 0} g(a) = g(a)$ [since $g(a)$ is a constant]. Also, since g is differentiable at a, g is continuous at a, and this implies that $\lim_{h \to 0} g(a+h) = g(a)$ (see Problem 20.8). Therefore, the limit on the left is simply $-1/g(a)g(a) = 1/g^2(a)$. Consequently, $(1/g)'(a) = (-1/g^2(a))(g'(a)) = -g'(a)/g^2(a)$ and this completes the proof. ∎

Theorem 23.5 If f and g are differentiable on an interval J with the property that $g(x) \neq 0$ for all x in J, the quotient f/g is also differentiable on J and $(f/g)' = (gf' - fg')/g^2$.

PROOF Write f/g as $(f)(1/g)$ and then use Theorems 23.3 and 23.4. (Observe that the *converse* of this theorem is *not true*; see Problem 23.7.) ∎

EXAMPLE 23.1 Let $f(x) = x^2 - 1$ and let $g(x) = x^3 + 1$. Compute $f', g', (f+g)', (fg)', (1/g)',$ and $(f/g)'$.

Solution

$$f'(x) = 2x \text{ and } g'(x) = 3x^2$$
$$(f+g)'(x) = f'(x) + g'(x) = 2x + 3x^2$$
$$(fg)'(x) = f(x)g'(x) + f'(x)g(x) = (x^2 - 1)(3x^2) + (2x)(x^3 + 1)$$
$$(1/g)'(x) = -g'(x)/g^2(x) = -3x^2/(x^3 + 1)^2$$
$$(f/g)'(x) = [g(x)f'(x) - f(x)g'(x)]/g^2(x)$$
$$= [(x^3 + 1)(2x) - (x^2 - 1)(3x^2)]/(x^3 + 1)^2$$

THE CHAIN RULE

We now come to the formula for the derivative of the *composition* of two functions f and g. For example, functions like $(x^2 + x + 1)^{1/2}$ and $(x^2 + 1)^{17}$ would be extremely difficult to differentiate if we were allowed to use only the definitions and theorems we have discussed up to now. The following theorem gives a fomula that enables us to easily differentiate functions similar to the two previously mentioned ones.

Theorem 23.6 If g is differentiable on an interval J and f is differentiable on the range of g, then $f(g(x))$ is differentiable on J and $(f(g(x)))' = f'(g(x))g'(x)$. Using the circle notation for the composition of f with g, we may state the result as $(f \circ g)' = (f' \circ g)g'$.

Note that we use $(f(g(x)))'$ to denote the derivative of the function $f(g(x))$. In general, whenever we write a function $F(x)$ inside parentheses, such as $(F(x))'$, we simply mean the derivative of $F(x)$. Hence, $(x^2 + 2x)'$ and $(\sin x)'$ represent the respective derivatives.

EXAMPLE 23.2 Compute the derivatives of the functions (a) $(x^2 + x + 1)^{1/2}$ and (b) $(x^2 + 1)^{17}$.

Solution (a) Let $f(x) = \sqrt{x}$ and $g(x) = x^2 + x + 1$; then $f'(x) = 1/(2\sqrt{x})$ and $g'(x) = 2x + 1$. Now $f(g(x)) = (x^2 + x + 1)^{1/2}$ and from the preceding theorem, $(f(g(x)))' = f'(g(x))g'(x) = [\frac{1}{2}(x^2 + x + 1)^{-1/2}] \times (2x + 1)$.

Solution (b) Let $f(x) = x^{17}$ and $g(x) = x^2 + 1$; then $f'(x) = 17x^{16}$ and $g'(x) = 2x$. Now $f(g(x)) = (x^2 + 1)^{17}$ and from the preceding theorem, $(f(g(x)))' = f'(g(x))g'(x) = 17(x^2 + 1)^{16}(2x)$.

Finally, we also observe that the preceding theorem can be *extended* to the composition of more than two functions. For example, $(f(g(h(x))))' = [f'(g(h(x)))][g'(h(x))]h'(x)$; $(f(g(h(k(x)))))' = [f'(g(h(k(x))))][g'(h(k(x)))] \times [h'(k(x))]k'(x)$; and so on.

EXAMPLE 23.3 Let f be an arbitrary function that is differentiable on an interval J. Let $g(x) = ax + b$ where a and b are constants. Let $F(x) = f(g(x))$. Compute $F'(x)$ (a) by using the chain rule and (b) by using the definition of the derivative (see Problem 22.10).

SOLUTION (a) $F'(x) = f'(g(x))g'(x) = f'(ax + b)a$ [since $(ax + b)' = a$].

SOLUTION (b) This solution is a special case of the general technique that is used to prove Theorem 23.6 in most calculus texts. We feel that, instead of repeating the general proof, it would be advantageous to apply the general technique in a particular case. Now

$$F'(x) = \lim_{h \to 0} \left[\frac{f(g(x + h)) - f(g(x))}{h} \right]$$

$$= \lim_{h \to 0} \left[\frac{f(a(x + h) + b) - f(ax + b)}{h} \right].$$

If we multiply the numerator and denominator by the expression $(a(x + h) + b) - (ax + b)$, we obtain

$$\lim_{h \to 0} \left[\frac{f(a(x + h) + b) - f(ax + b)}{(a(x + h) + b) - (ax + b)} \frac{(a(x + h) + b) - (ax + b)}{h} \right]$$

$$= \lim_{h \to 0} \left[\frac{f(a(x + h) + b) - f(ax + b)}{ah} \frac{ah}{h} \right]$$

$$= \lim_{h \to 0} \left[\frac{f(a(x + h) + b) - f(ax + b)}{ah} \right] a$$

$$= a \lim_{h \to 0} \left[\frac{f(a(x + h) + b) - f(ax + b)}{ah} \right]$$

Let $z = ax + b$ and let $k = ah$; then the preceding limit can be expressed as

$$a \lim_{h \to 0} \frac{f(z + k) - f(z)}{k}$$

But as $h \to 0$, so does $ah = k$. Therefore, we may rewrite the limit in the form

$$a \lim_{k \to 0} \frac{f(z + k) - f(z)}{k} = af'(z)$$

(by the definition of the derivative of f at z). Replacing z by $ax + b$, we obtain $af'(ax + b)$, which is the required answer.

EXAMPLE 23.4   Using the chain rule, prove that if $f(x) = x^{-n}$ where n is any *positive* integer, then $f'(x) = -nx^{-n-1}$. (Observe that this result together with Example 22.4 yields the formula $(x^n)' = nx^{n-1}$ where n is *any* nonzero integer.)

SOLUTION x^{-n} is the composition of the function $g(x) = x^n$ and $h(x) = 1/x = x^{-1}$. Therefore, $x^{-n} = g(h(x)) = (1/x)^n$. From Example 22.5 we know that the derivative of $1/x$ is $-1/x^2$, and from Example 22.4 we know that the derivative of x^n is nx^{n-1}. Applying the chain rule, we obtain $(g(h(x)))' = g'(h(x))h'(x) = n(1/x)^{n-1}(-1/x^2) = -n(1/x^{n-1+2}) = -n(1/x^{n+1}) = -nx^{-n-1}$ and we are done. Try to prove this result by the direct application of Theorem 23.4.

EXAMPLE 23.5   Using the chain rule, prove that if $f(x) = x^{p/q}$ where p/q is a positive rational number reduced to lowest terms, then $f'(x) = (p/q)x^{(p/q)-1}$.

SOLUTION $x^{p/q}$ is the composition of the functions $g(x) = x^p$ and $h(x) = x^{1/q}$. We know that $g'(x) = px^{p-1}$ while $h'(x) = (1/q)x^{(1/q)-1}$ [from Problem 22.7(b)]. Therefore, $(g(h(x)))' = g'(h(x))h'(x) = p(x^{1/q})^{p-1}((1/q)x^{(1/q)-1}) = (p/q)(x^{(p-1)/q}x^{(1-q)/q}) = (p/q)x^{(p/q)-1}$.

EXAMPLE 23.6   Prove that if n is any nonzero integer and f' is differentiable on some interval J, then $(f^n(x))' = nf^{n-1}(x)f'(x)$.

SOLUTION $f^n(x) = (f(x))^n$; hence, $f^n(x)$ is a composition of the functions g and h where $g(x) = x^n$ and $h(x) = f(x)$. By the chain rule,

$$(g(h(x)))' = g'(h(x))h'(x) = n(f(x))^{n-1}f'(x) = nf^{n-1}(x)f'(x).$$

PROBLEMS

23.1   Differentiate each of the following functions.
  (a) x^{-7}
  (b) $x^{7/2}$
  (c) $x^{-3/2}$
  (d) $x^7 - 3x^5 + 2x^3 - 7x + 2$
  (e) $(x^6 - 7x^4 + 2x^2 + 3)(x^5 - x^3 + x)$
  (f) $(x^5 - x^3 + x)/(x^6 + x^4 + x^2 + 1)$
  (g) $[(1/x^2) - x^3]^{2/3}$
  (h) $x(x^2 + 1)^{3/2}$
  (i) $\sqrt{x}/(\sqrt{x} + 1)$
  (j) $1/(x^2 + 2x + 7)$
23.2   Find the slope of the tangent line to each of the following graphs at the indicated points.
  (a) $y = x^4 - 3x^3 + 2x^2 + 1$ at $(0, 1)$ and $(2, 1)$
  (b) $y = x(x^3 + 1)^{1/2}$ at $(1, \sqrt{2})$ and $(2, 6)$

(c) $y = (x - 1)(x + 1)^9$ at $(0, -1)$ and $(1, 0)$

(d) $y = \sqrt[3]{x}$ at $(-1, -1)$, $(0, 0)$, and $(8, 2)$

(e) $y = x^{2/3}$ at $(\sqrt{8}, 2)$ and $(0, 0)$

23.3   Find the velocity at $t = 1$ and $t = 2$ of a particle P that moves along a horizontal coordinate line with position function $s(t) = t^4 + t^3 - 3t^2 + 4t - 2$.

23.4   (a)   Prove that if f, g, and h are differentiable on some interval J, then
$(fgh)' = f'gh + fg'h + fgh'$.

  (b)   Compute the derivative of the function $F(x) = x^{3/2}(x^3 + 1)^{1/2}(x^2 - 1)^8$.

23.5   (a)   Prove by using techniques from analytic geometry that the slope of the tangent line to the graph of the circle $x^2 + y^2 = 1$ at an arbitrary point (x, y), $y \neq 0$, is $-x/y$.

  (b)   Prove the result mentioned in (a) by using the calculus.

23.6*   Let $y = ax^2 + bx + c$ be the equation of an arbitrary parabola with a vertical axis of symmetry. Prove that the tangent lines through the endpoints of the latus rectum are perpendicular and intersect at a point that is on the directrix. [See Problem 22.5(b).]

23.7   Let

$$f(x) = \begin{cases} 1 & \text{if } x \text{ is rational} \\ -1 & \text{if } x \text{ is irrational} \end{cases}$$

and let

$$g(x) = \begin{cases} -1 & \text{if } x \text{ is rational} \\ 1 & \text{if } x \text{ is irrational} \end{cases}$$

(a)   Show that f' and g' do not exist at any real number a.

(b)   Show that $(f + g)'$ exists at every real number a.

(c)   Show that $(fg)'$ exists at every real number a.

(d)   Show that $(f/g)'$ exists at every real number a.

23.8   Find the tangent of the *angle* between the tangent lines to the graphs of the parabola $y = x^2 - 1$ and the hyperbola $y^2 - x^2 = 1$ at each of their points of intersection. (See Example 12.7 and Theorem 4.1.)

HIGHER DERIVATIVES AND IMPLICIT DIFFERENTIATION

Definition 24.1 Since the derivative f' of a function f is also a function (see the discussion following Example 22.5), we may ask whether the function f' has a derivative. If so, we call it the *second derivative* of f and denote it by f''.

EXAMPLE 24.1 If $f(x) = x^3$, then $f'(x) = 3x^2$ and $f''(x) = 6x$. If $f(x) = 1/x$, then $f'(x) = -1/x^2$ and $f''(x) = 2/x^3$.

Definition 24.2 If f is a function with the property that f'' exists, then since f'' is also a function, we may ask whether f'' has a derivative. If so, we call it the *third derivative* of f and denote it by f'''. *In general*, if a function f has an nth derivative, for any positive integer $n > 3$, we denote it by $f^{(n)}$.

EXAMPLE 24.2 If $f(x) = x^3$, then $f'''(x) = 6$ and $f^{(4)}(x) = 0$. If $f(x) = 1/x$, then $f'''(x) = -6/x^4$, $f^{(4)}(x) = 24/x^5$ and $f^{(5)}(x) = -120/x^6$ (note that in this case f has infinitely many nonzero derivatives).

Theorem 24.1 Every polynomial $f(x) = a_n x^n + a_{n-1} x^{n-1} + \cdots + a_2 x^2 + a_1 x + a_0$, $a_n \neq 0$, has *exactly* n nonzero derivatives.

PROOF If we look at the term $a_n x^n$, we see that this term differentiates to $na_n x^{n-1}$, which becomes the first term of $f'(x)$. Differentiating again leads to $n(n-1)a_n x^{n-2}$ as the first term of $f''(x)$. Finally, the first term of $f^{(n-1)}(x)$

will be $n(n-1)(n-2)\cdots(3)(2)a_n x$ and then the *only* term of $f^{(n)}(x)$ will be $n!\,a_n$, which is a constant. Therefore, $f^{(n+1)}(x) = 0$, as do all the higher derivatives. ∎

EXAMPLE 24.3 Find the first and second derivatives of $f(x) = (x^3 + 2x)^5$.

SOLUTION Using the chain rule, we find that $f'(x) = 5(x^3 + 2x)^4$ $3(x^2 + 2)$; using the chain rule and the product formula, we obtain $f''(x) = 20(x^3 + 2x)^3(3x^2 + 2)^2 + 5(x^3 + 2x)^4(6x)$.

IMPLICIT DIFFERENTIATION

We mentioned in Remark 3 near the end of Section 16 that there are many equations that have graphs that are *not* graphs of functions. Some of these graphs certainly possess tangent lines at arbitrary points on the graph. The question is, how do we use the notion of the derivative to find the slopes of these tangent lines?

In part (a) of the previously mentioned Remark 3, we showed that the circle equation $x^2 + y^2 = 1$ gave rise to two distinct functions $f(x) = (1 - x^2)^{1/2}$ and $g(x) = -(1 - x^2)^{1/2}$, which have graphs that are the upper and lower semicircles, respectively. Of course, in this case, we can compute $f'(x)$ and $g'(x)$ directly and thus find the slope of a tangent line to the upper semicircle by using f' and to the lower semicircle by using g'.

The same thing happened in part (b) of Remark 3, where we showed that the hyperbola equation $x^2 - y^2 = 1$ gave rise to the functions $f(x) = (x^2 - 1)^{1/2}$ and $g(x) = -(x^2 - 1)^{1/2}$. Once again, we may use f' and g' to compute slope of a tangent line to the hyperbola, using f' at points on the hyperbola above the x axis and g' at points on the hyperbola below the x axis.

Do the following related problems as exercises.

1. In the case of the circle $x^2 + y^2 = r^2$, prove by using techniques of analytic geometry that the slope of the tangent line at any point (x, y), $y \neq 0$, on this circle is given by $-x/y$. Then compute the derivatives of $f(x) = (r^2 - x^2)^{1/2}$ and $g(x) = -(r^2 - x^2)^{1/2}$ and show that the slopes that come from f' and g' agree with the result obtained from analytic geometry (see Problem 23.5).

2. In the case of the hyperbola $x^2 - y^2 = 1$, use the derivative of $f(x) = (x^2 - 1)^{1/2}$ and $g(x) = -(x^2 - 1)^{1/2}$ to find the slope of the tangent lines to this hyperbola at the points $(2, \sqrt{3})$ and $(2, -\sqrt{3})$.

Suppose we are given an equation like $xy + y^2 - 1 = 0$ and are asked for the slope of the tangent line at some point on the graph of this equation. It is not possible to solve this equation easily for y as an expression in x; nevertheless, we know that this equation has a graph, parts of which correspond to the graph of one or more functions that are defined *implicitly* by this equation [i.e., there exists a function f such that the equation $xf(x) + (f(x))^2 - 1 = 0$ is true for every x in the domain of f]. If we assume that there is such a function $y = f(x)$, which is defined implicitly by this equation, then f has a graph that belongs to the graph of the equation $xy + y^2 - 1 = 0$ and we may rewrite the equation in the form $xf(x) + (f(x))^2 - 1 = 0$. Now, every term in the equation is a function of x, so we may differentiate this sum term by term to obtain $f(x) + xf'(x) + 2f(x)f'(x) = 0$. Solving for $f'(x)$, we obtain $f'(x) = -f(x)/(x + 2f(x)) = -y/(x + 2y)$. Therefore, we have found the derivative of f without actually knowing f in the first place. If, in addition, we wanted to know the slope of the tangent line to the graph of $xy + y^2 - 1 = 0$ at $(\frac{3}{2}, -2)$, then $x = \frac{3}{2}$, $y = -2$, and so $f'(\frac{3}{2}) = -(-2)/(\frac{3}{2} + 2(-2)) = -\frac{4}{5}$.

This technique of differentiating a function f that is defined implicitly by an equation is called *implicit differentiation*. It turns out that in the preceding case we can actually express $f(x)$ *explicitly* by using the following method.

Consider the equation $xy + y^2 - 1 = 0$ as a *quadratic* equation in y with x treated as a constant. Then $y^2 + xy - 1 = 0$ has the form $ay^2 + by + c = 0$ and, using the quadratic formula, we solve for y, obtaining

$$y = \tfrac{1}{2}[-x \pm (x^2 - 4(-1))^{1/2}].$$

This equation yields two functions $f(x) = \tfrac{1}{2}[-x + (x^2 + 4)^{1/2}]$ and $g(x) = \tfrac{1}{2}[-x - (x^2 + 4)^{1/2}]$. Now

$$f'(x) = \tfrac{1}{2}[-1 + \tfrac{1}{2}(x^2 + 4)^{-1/2}(2x)] = \tfrac{1}{2}[-1 + x(x^2 + 4)^{-1/2}].$$

Note that if we substitute $f(x) = \tfrac{1}{2}[-x + (x^2 + 4)^{1/2}]$ into the equation $f'(x) = -f(x)/(x + 2f(x))$, we obtain

$$f'(x) = \frac{-\tfrac{1}{2}[-x + (x^2 + 4)^{1/2}]}{x + 2(\tfrac{1}{2})[-x + (x^2 + 4)^{1/2}]}$$

This reduces to $f'(x) = \tfrac{1}{2}[-1 + x(x^2 + 4)^{-1/2}]$, which agrees with the result of our *explicit differentiation*. As an exercise, compute $g'(x)$ and show that it satisfies the equation $g'(x) = -g(x)/(x + 2g(x))$.

Consequently, if we have an equation that defines one or more functions implicitly, it is possible to simply set $y = f(x)$ and then find $f'(x)$ in terms of x and y without actually computing $f(x)$. This technique yields the derivative of every differentiable function defined implicitly by the given equation. In *some* cases, we can compute $f(x)$ *explicitly*, but this is not really necessary if all that we require is the *derivative* of the given function.

We conclude this section with two examples.

EXAMPLE 24.4   Find the slope of the tangent line to the graph of the equation $x^3 + xy - y^3 + 11 = 0$ at $(-1, 2)$.

SOLUTION   Let $y = f(x)$. Then rewriting the equation, we obtain $x^3 + xf(x) - (f(x))^3 + 11 = 0$. Differentiating, we obtain

$$3x^2 + f(x) + xf'(x) - 3(f(x))^2 f'(x) = 0.$$

Solving for $f'(x)$, we arrive at $f'(x) = (3x^2 + f(x))/(3(f(x))^2 - x) = (3x^2 + y)/(3y^2 - x)$. Then the required slope is $f'(-1)$

$$f'(-1) = (3(-1)^2 + 2)/(3(2)^2 - (-1)) = \tfrac{5}{13}.$$

EXAMPLE 24.5   Using implicit differentiation, prove that the derivative of $y = f(x) = x^{1/n}$ is $(1/n)x^{(1/n)-1}$ where n is any nonzero integer [see Problem 22.7(b)].

SOLUTION   If $y = x^{1/n}$, then $y^n = x$. Letting $y = f(x)$, we obtain $(f(x))^n = x$. Differentiating both sides of this equation leads to the equation $n(f(x))^{n-1}f'(x) = 1$. This implies $f'(x) = (1/n)(f(x))^{1-n} = (1/n)y^{1-n}$. But $y = x^{1/n}$; hence, $f'(x) = (1/n)x^{(1-n)/n} = (1/n)x^{(1/n)-1}$.

PROBLEMS

24.1 (a) Compute the first three derivatives of the function $f(x) = \sqrt{x}$.
      (b) Find a general formula for the nth derivative of $f(x) = \sqrt{x}$.
24.2 Find a general formula for the nth derivative of the function $f(x) = 1/x$ (see Example 24.2).
24.3 (a) Compute the first five derivatives of the function $f(x) = x^7 + x^6 + x^5 + x^4 + x^3 + x^2 + x + 1$.
      (b) Compute the first two derivatives of $f(x) = (x^2 + 1)/(x^3 + 1)$.
      (c) Compute the first three derivatives of $f(x) = (x^3 + 2x)^5$ (see Example 24.3).
24.4 (a) Compute the hundredth derivative of $f(x) = x^{100} - x^{50} + x^{25} - 1$.
      (b) Compute the thousandth derivative of $f(x) = x^{999} + x^{99} - x^9 + 1$.
24.5 (a) Compute the first five derivatives of $f(x) = 1/(1-x)$.
      (b) Find a formula for the nth derivative of $f(x) = 1/(1-x)$.
24.6 Prove that if f is a polynomial function, then $f^{(n)} = 0$ for some positive integer n.
24.7* (a) Prove by successive differentiation of $f(x) = (1+x)^4$ that the *binomial expansion* of $(1+x)^4 = 1 + 4x + 6x^2 + 4x^3 + x^4$. [*Hint:* We know that $(1+x)^4$ must be a polynomial of degree 4; therefore, we may assume that $(1+x)^4 = a_0 + a_1x + a_2x^2 + a_3x^3 + a_4x^4$. Then, $f(x) = (1+x)^4$ and $f(0) = 1 = a_0$; $f'(x) = 4(1+x)^3$ and $f'(0) = 4 = a_1$; $f''(x) = (4)(3)(1+x)^2$ and $f''(0) = 12 = 2a_2$, which implies $a_2 = 6$; and so on.]
      (b) Using the method of part (a), derive the formula for the binomial expansion of $(1+x)^n$ for any positive integer n.
      (c) Find the first four terms of the binomial expansion for $(1+x)^{-2}$. How many terms are there in this expansion?
      (d) Find the first four terms of the binomial expansion for $(1+x)^{1/2}$. How many terms are there in this expansion?

24.8 In each of the following, find the slope of the tangent to the graph of the given equation at the indicated point.

(a) $x^2 + xy + y^2 = 7$ at $(1, 2)$

(b) $\sqrt{xy} + x^4 y = 9$ at $(2, \frac{1}{2})$

(c) $2y^3 + xy - x^3 + 12 = 0$ at $(2, -1)$

(d) $x^3 y - 5x^2 y^3 + x^4 + 3 = 0$ at $(1, 1)$

24.9 In each of the parts of Problem 24.8, find the equation of the normal line to the graph of the given equation at the indicated point [see Problem 22.5(c)].

24.10 In each of the following, use implicit differentiation to find the first and second derivative of each of the functions f defined implicitly by the given equation.

(a) $x^2 + xy + y^2 = 7$ [see Problem 24.8(a)].

(b) $x^3 + xy - y^3 + 11 = 0$ (see Example 24.4.)

24.11 Find the slope of the tangent to the graph of the equation $x(4 - y^2)^{1/2} = 2$ at $(2, \sqrt{3})$ and $(2, -\sqrt{3})$ (a) by implicit differentiation; (b) by solving for y as an explicit function of x and then by differentiating directly.

24.12 Find the slope of the tangent to the graph of the equation $x^3 + y^3 = 3xy$ at the point $(-1/\sqrt[3]{2}, 1 - [1/\sqrt[3]{2}])$ (see Figure 16.7).

24.13 In each of the following, find the slope of the tangent to the graph of each of the given conics at the indicated points.

(a) $2x^2 - y^2 = 4$ at the points $(2, 2)$, $(2, -2)$, $(-2, 2)$, and $(-2, -2)$

(b) $x^2 + 2y^2 = 27$ at the points $(3, 3)$, $(3, -3)$, $(-3, 3)$, and $(-3, -3)$

(c) $x^2 + xy = 1$ at the points $(2, -\frac{3}{2})$ and $(-2, \frac{3}{2})$

MAXIMA AND MINIMA, CURVE TRACING

We begin this section with an *informal* discussion of the graph of a differentiable function from a purely geometrical viewpoint. There are initially no statements of definitions and theorems as such, but we endeavor to cover all the important information relevant to graphing a differentiable function $y = f(x)$. We also assume throughout most of this section, that f is *twice* differentiable, that is, f'' exists at each point in the domain of f. We conclude with many examples and applications of problems relating to the theory of maxima and minima.

We are now interested in applying our knowledge of the derivative of a function to the problem of plotting the graph of a function. Since the derivative of f at a gives the slope of the tangent line to the graph of $y = f(x)$ at the point $(a, f(a))$, we may determine important characteristics of the graph near $(a, f(a))$ by simply studying the slopes of the tangent lines at $(a, f(a))$ and at points nearby. If we consider the graph of $y = f(x)$ as the path of a point in the plane that moves from left to right, the slopes of the tangent lines at points on the path actually indicate the *direction* in which the point is moving.

For example, in Figure 25.1, we have a graph that rises as we move from left to right, in Figure 25.2, we have a graph that is falling as we move from left to right; in Figure 25.3, we have a graph that neither rises nor falls. Notice also that the tangent lines in Figures 25.1 and 25.2 have the same general direction at a point as the curve itself. Of course, in Figure 25.3, the tangent at any point is the line itself and is therefore horizontal. Consequently, this graph is formed by a point that moves horizontally and corresponds to a function that is constant in the interval $c \le x \le d$. In any event, the tangent line still indicates the direction of the moving point that forms the graph.

143

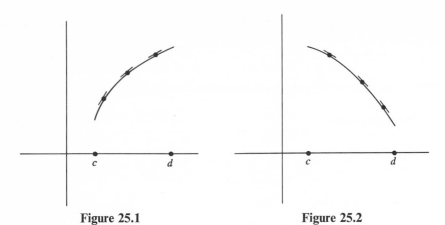

Figure 25.1   **Figure 25.2**

The graph in Figure 25.1. is an example of the graph of a function that is *increasing* at each point in the interval $c \leq x \leq d$. Note that the tangent lines all have *positive* slopes. In general, if a differentiable function is *increasing* at each point in an interval $c \leq x \leq d$, the tangent lines at points on the graph all have *positive* slopes.

The graph in Figure 25.2 is an example of a function that is *decresaing* at each point in the interval $c \leq x \leq d$. Note that the tangent lines all have *negative* slopes. In general, if a differentiable function is *decreasing* at each point in an interval $c \leq x \leq d$, the tangent lines at points on the graph will all have *negative* slopes.

Finally, in Figure 25.3, we have an example of a graph that is neither increasing nor decreasing at every point and the horizontal tangent at every point has slope equal to zero. In general, if a differentiable function is

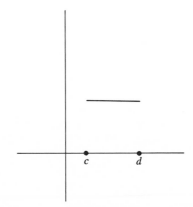

Figure 25.3

neither increasing nor decreasing at a point, the slope of the tangent to the graph at the point is zero.

Suppose we look more closely at the graphs in Figure 25.1 and 25.2. Notice that the tangent lines in both figures lie wholly *above* the graph. When this occurs at each point of an interval, we say that the graph is *concave down* over the interval. In Figures 25.4 and 25.5, we exhibit the graphs of an increasing function and a decreasing function, both of which have tangent lines that lie wholly *below* the graphs. When this occurs at each point of an interval, we say that the graph is *concave up* over the interval. Consequently, the concavity of a graph indicates whether the arc of the graph is bending upward or downward over an interval.

Observe that in Figures 25.4 and 25.5, moving from left to right, the slopes of the tangent lines to the graph of $y = f(x)$ are increasing. But this means that the first derivative f' is an *increasing* function at each point of the interval $c \leq x \leq d$. Therefore, the first derivative of the function f' (the second derivative f'' of f) must be *positive* over the interval. In a similar way, we observe that in Figures 25.1 and 25.2 the slopes of the tangent lines to the graph of $y = f(x)$ are decreasing and this implies that f' is a decreasing function at each point. Therefore, f'' must be negative over this interval.

Finally, we may summarize the discussion of concavity for twice-differentiable functions by stating that $f'' > 0$ at points in the interval where the graph is concave up and $f'' < 0$ at points in the interval where the graph is concave down.

Finally, it may happen that at some point on the graph of a twice-differentiable function $y = f(x)$ the tangent line is *not* wholly above or below the graph (see Figure 25.6). At such points, the graph is neither concave up nor concave down. As a matter of fact, the graph is actually *changing* its concavity at such points. These points are called *inflection points* of the

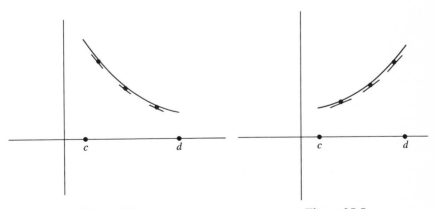

Figure 25.4   **Figure 25.5**

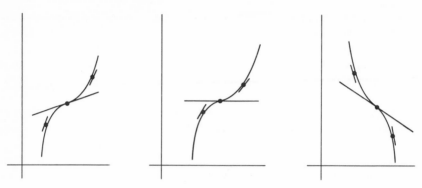

Figure 25.6

graph. Since the graph is not concave up at an inflection point, f'' is not positive there; and since the graph is not concave down, f'' is not negative. Consequently, $f'' = 0$ at such an inflection point.

We now come to a discussion of differentiable functions that are both *increasing and decreasing* in an interval $c \le x \le d$. The graphs of such functions must include *turning points* where the graph changes from increasing to decreasing, or vice versa. Once again, if we consider the graph as the path of a moving point, a turning point would be a place on the graph where the moving point changes its direction from upward to downward or from downward to upward (see Figure 25.7). If at the point $(a, f(a))$ the graph changes from increasing to decreasing, $x = a$ is called a *relative* or *local maximum* of the function f. Note that in this case $f(x) \le f(a)$ for all x near a. Conversely, if the graph changes from decreasing to increasing at $(a, f(a))$, then

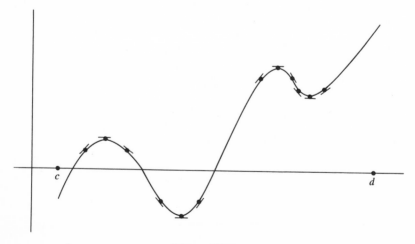

Figure 25.7

$x = a$ is called a *relative* or *local minimum* of f. In this case, notice that $f(x) \geq f(a)$ for all x near a. Observe also that just to the left of a relative maximum, the tangent lines to the graph have positive slopes; just to the right, the tangent lines have negative slopes; while at the point itself, the tangent is horizontal with slope zero. Conversely, just to the left of a relative minimum, the tangent lines to the graph have negative slopes; just to the right, the tangent lines have positive slopes; while at the point itself, the tangent is once again horizontal. This gives rise to the *first derivative test* for a relative maximum or minimum; that is, a differentiable function f has a *relative maximum* at $x = a$ if $f'(a) = 0$ and if $f'(x) > 0$ at values of x close to a but less than a, while $f'(x) < 0$ at values of x close to a but greater than a. On the other hand, a differentiable function f has a *relative minimum* at $x = a$ if $f'(a) = 0$ and if $f'(x) < 0$ just to the left of a, while $f'(x) > 0$ just to the right of a.

Also, note that at a relative maximum, the graph must be concave down, whereas at a relative minimum, the graph must be concave up. This gives rise to the *second derivative test* for a relative maximum or minimum; that is a twice-differentiable function f has a *relative maximum* at $x = a$ if $f'(a) = 0$ and $f''(a) < 0$, whereas f has a *relative minimum* at $x = a$ if $f'(a) = 0$ and $f''(a) > 0$.

Finally, if we examine the first and second derivative of a twice-differentiable function $y = f(x)$ at and near a point of inflection (see Figure 25.6), we may make the following observations. If the point $(a, f(a))$ is an inflection point of the graph of $y = f(a)$, then $f''(a) = 0$, while $f'(a)$ may be positive, negative, or zero. If $f'(a) = 0$, $(a, f(a))$ is called a *horizontal* inflection point. We also note that the concavity changes at $(a, f(a))$. Consequently, if $(a, f(a))$ is an inflection point, $f''(x)$ *changes sign* as we move from left to right along the graph through $(a, f(a))$, while $f'(x)$ keeps the *same sign* as we move from left to right through $(a, f(a))$. Hence by itself, $f''(a) = 0$ does *not* always imply the existence of an inflection point at $x = a$ [consider $f(x) = x^4$ at $x = 0$]. Finally, it is even possible for a function to have an inflection point at $x = a$ while $f''(a)$ does not exist [consider $f(a) = x^{4/3}$ at $x = 0$]. Such an inflection point is called a *vertical* inflection point (since the tangent line at the point is vertical).

If a differentiable function $y = f(x)$ has a relative maximum, minimum, or a horizontal inflection point at $(a, f(a))$, then $f'(a) = 0$ and $x = a$ is called a *critical point* of f. Also, if f is defined over a finite interval $c \leq x \leq d$, the endpoints $x = c$ and $x = d$ are included among the critical points of f. In addition, if f is continuous but *not differentiable* at some interior point a of its domain, $x = a$ is also called a critical point of f.

Finally, if there is a point $(a, f(a))$ (or more than one point) on the graph with the property that no other point on the graph is higher, $x = a$ is a critical point of f and the y coordinate $f(a)$ is called *the absolute maximum value of f.* Similarly, if there is a point $(a, f(a))$ (or more than one point) with the property that no other point on the graph is lower, $x = a$ is a critical point of f and $f(a)$ is called *the absolute minimum value of f.* Hence, a function may attain its absolute maximum or minimum at many points in its domain; however, there exists only *one* absolute maximum or minimum value of f. Notice also that an absolute maximum or minimum may occur at an endpoint of the graph (see Figure 25.7).

This concludes the geometrical discussion of the theory of maxima and minima. Armed with the machinery of the preceding part of this section, we undertake some examples, beginning, ironically, with an example of a differentiable function that has no critical points at all.

EXAMPLE 25.1 Find the critical points of $f(x) = 2x^3 + 3x - 1$ and sketch the graph.

SOLUTION $f'(x) = 6x^2 + 3$ and $f''(x) = 12x$. Setting $f'(x) = 0$, we obtain $x^2 = -\frac{1}{2}$, which is impossible for any real number x. Consequently, there are no points on the graph that have a horizontal tangent. Since $6x^2 + 3 > 0$ for all x, we have $f'(x) > 0$ for all x, indicating that the graph is increasing everywhere. Finally, setting $f''(x) = 0$, we obtain $x = 0$. This means that $(0, -1)$ is the only point of inflection (nonhorizontal). Since $f''(x) < 0$ for $x < 0$, and $f''(x) > 0$ for $x > 0$, we see that the graph is concave down for all $x < 0$ and concave up for all $x > 0$. A sketch appears in Figure 25.8.

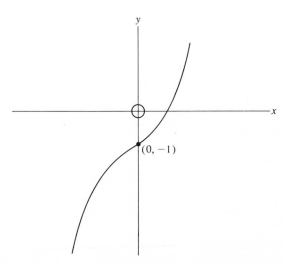

Figure 25.8

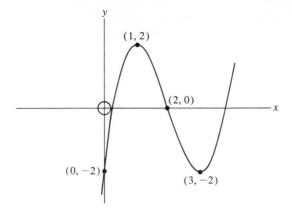

Figure 25.9

EXAMPLE 25.2 Find the critical points of $f(x) = x^3 - 6x^2 + 9x - 2$ and sketch the graph.

SOLUTION $f'(x) = 3x^2 - 12x + 9$ and $f''(x) = 6x - 12$. Setting $f'(x) = 0$, we obtain $3(x^2 - 4x + 3) = 3(x - 3)(x - 1) = 0$. Therefore, $x = 3$ and $x = 1$ are critical points and the graph of $y = f(x)$ has a horizontal tangent at the points $(3, -2)$ and $(1, 2)$. Now $f''(3) > 0$, hence, $x = 3$ is a relative minimum, whereas $f''(1) < 0$, hence, $x = 1$ is a relative maximum. Setting $f''(x) = 0$, we obtain $x = 2$; therefore, the point $(2, 0)$ is a nonhorizontal inflection point. A sketch of the graph is shown in Figure 25.9. Observe that if we graph the equations $y = f'(x)$

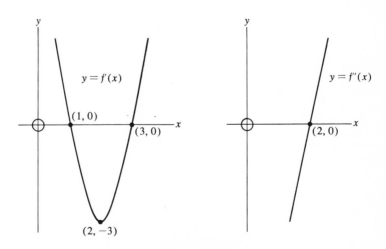

Figure 25.10

and $y = f''(x)$, we obtain the parabola and the line shown in Figure 25.10. Comparing the graphs, we see that f is increasing over the intervals $-\infty < x < 1$ and $3 < x < \infty$, while f' is positive over the same intervals. We also see that f is decreasing over the interval $1 < x < 3$, while f' is negative over this interval. Finally, f is concave down for $x < 2$ and concave up for $x > 2$, while $f'' < 0$ for $x < 2$ and $f'' > 0$ for $x > 2$.

As far as the critical points are concerned, we see that f' is changing sign at $x = 1$ and $x = 3$ and this corresponds to the turning points of f. Also, f'' changes sign at $x = 2$, which corresponds to the change in concavity that takes place at the point $(2, 0)$ on the graph of f.

EXAMPLE 25.3   Find the critical values of the function $f(x) = x^3 + 3/x$ and sketch the graph.

SOLUTION $f'(x) + 3x^2 - 3/x^2$ and $f''(x) = 6x + 6/x^3$.   Setting $f'(x) = 0$ we obtain $x^4 = 1$, which has two real roots 1 and -1. Therefore, the graph of f has a horizontal tangent at $(1, 4)$ and $(-1, -4)$. Now $f''(1) > 0$ implies $x = 1$ is a relative minimum, while $f''(-1) < 0$ implies $x = -1$ is a relative maximum.   Setting $f''(x) = 0$, we obtain $x^4 = -1$, which is impossible for any real number x. Therefore, this graph has

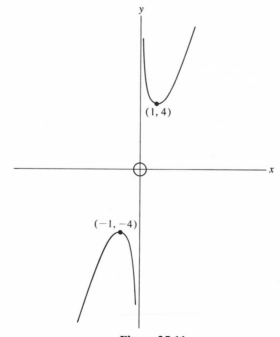

Figure 25.11

no inflection points. Consequently, it must be concave up for all $x > 0$ and concave down for all $x < 0$. A sketch appears in Figure 25.11. Notice that we did not concern ourselves with $x = 0$ since $x = 0$ is not in the domain of f. Also, note that this function has a relative minimum value that is larger than its relative maximum value. Finally, the function does not have an absolute maximum or minimum value.

Following are some examples of functions that are *not* differentiable over their entire domain but that nevertheless can be studied through the use of the preceding methods.

EXAMPLE 25.4 Find the critical points of the function $f(x) = |x^2 - 1|$ and sketch the graph.

SOLUTION Refer to Example 13.1(c) and Figure 13.6; note that this function is differentiable everywhere except at $x = -1$ and $x = 1$, where it attains its absolute minimum value 0. The function also has a horizontal tangent at $x = 0$, where it attains a relative maximum value 1. This graph could have been studied by looking at the derivatives of the functions $g(x) = 1 - x^2$ for $|x| < 1$ and $h(x) = x^2 - 1$ for $|x| > 1$. Since $g'(x) = -2x$ and $g''(x) = -2$, we obtain a relative maximum at $x = 0$, with a graph that is concave down for $|x| < 1$. Also, $h'(x) = 2x$ and $h''(x) = 2$; therefore, $h'(x) \neq 0$ for $|x| > 1$ and there are no turning points. Since $h'' > 0$, the graph must be concave up for all $|x| > 1$. Observe also that the left-hand derivative of f at $x = 1$ (see Problem 22.9) is $g'(1) = -2$, while the right-hand derivative at $x = 1$ is $h'(1) = 2$. Since these two derivatives are not equal, $f'(1)$ does not exist. By a similar argument, we can show that $f'(-1)$ does not exist. For this reason we must include among the *critical points* of a function f those points in the domain of f at which f' does not exist.

In the following example, the second derivative is more difficult to compute than in the previous examples; therefore, we use the first derivative exclusively in our analysis of the graph. There are other examples (some of which appear in the problems following this section) where it is extremely difficult and time consuming to compute second derivatives. This time could be better spent analyzing the graph with the use of the first derivative.

EXAMPLE 25.5 Find the critical values of $f(x) = (x^3 - 9x)^{1/3}$ and sketch the graph.

SOLUTION

$$f'(x) = \tfrac{1}{3}(x^3 - 9x)^{-2/3}(3x^2 - 9) = (x^3 - 9x)^{-2/3}(x^2 - 3)$$
$$= (x^2 - 3)/(x^3 - 9x)^{2/3}.$$

Setting $f'(x) = 0$, we obtain $x^2 = 3$, which implies that $x = \sqrt{3}$ and $x = -\sqrt{3}$ are critical points. The corresponding y coordinates are *approximately* -2 and 2, respectively. If we consider f' at values of x slightly less than $\sqrt{3}$, we see that $f' < 0$, whereas at values of x slightly greater than $\sqrt{3}, f' > 0$. Therefore, f has a relative minimum at $x = \sqrt{3}$. In a similar fashion, we obtain $f' > 0$ just to the left of $x = -\sqrt{3}$, whereas $f' < 0$ just to the right of $x = -\sqrt{3}$. Therefore, f has a relative maximum at $x = -\sqrt{3}$. This takes care of the critical points where $f'(x) = 0$. Now there are three more critical points of f that correspond to points where the derivative does not exist. We find these points by setting the *denominator* in the expression $(x^2 - 3)/(x^3 - 9x)^{2/3}$ equal to 0. But this means $x^3 - 9x = x(x^2 - 9) = x(x - 3)(x + 3) = 0$. Consequently, $x = 0$, $x = 3$, and $x = -3$, are the critical points where f' does not exist. The corresponding points on the graph are $(0, 0)$, $(3, 0)$, and $(-3, 0)$, and the tangent lines to the graph at all three points are vertical. If the second derivative is computed, it can be verified that all three of these points are vertical inflection points, since the concavity changes as the graph passes through each one of these points. A rough sketch appears in Figure 25.12.

ROLLE'S THEOREM AND THE MEAN VALUE THEOREM

Since both of these theorems are treated at length in regular texts we restrict ourselves to a geometrical discussion, beginning with the following *geometrical* statement of Rolle's theorem.

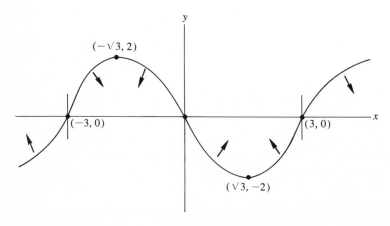

Figure 25.12

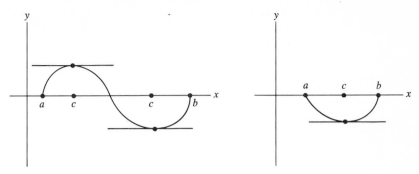

Figure 25.13

Rolle's Theorem   Let $f(x)$ be defined and continuous on an interval $a \leq x \leq b$ such that the graph of $y = f(x)$ intersects the x axis at $x = a$ and $x = b$ [i.e., $f(a) = f(b) = 0$].   If the graph of $y = f(x)$ has a tangent line at every point between $(a, 0)$ and $(b, 0)$, there exists at least one point $(c, f(c))$ on the graph between $(a, 0)$ and $(b, 0)$ that possesses a horizontal tangent [i.e., $f'(c) = 0$]. Diagrams appear in Figure 25.13.

> **EXAMPLE 25.6**   Given the function $f(x) = 3(x^2 - 4x + 3)$, $1 \leq x \leq 3$ (see Figure 25.10), find the point that may be used for c in Rolle's Theorem.
>
> SOLUTION   In the terminology of Rolle's theorem, $a = 1, b = 3$, and since there is a horizontal tangent at $x = 2$, we have $c = 2$.

The following is a *geometrical* statement of the *mean value theorem*.

Mean Value Theorem   Let f be defined and continuous on an interval $a \leq x \leq b$.   If the graph of $y = f(x)$ has a tangent line at every point between

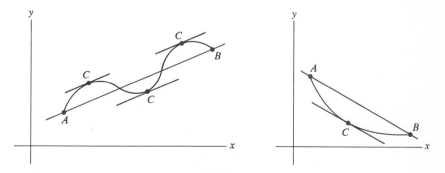

Figure 25.14

$A = (a, f(a))$ and $B = (b, f(b))$, there exists at least one point $C = (c, f(c))$ between A and B with the property that the tangent line to the graph at C is *parallel* to the secant line joining A to B (see Figure 25.14).   Observe that this result includes Rolle's theorem as a special case with $f(a) = f(b) = 0$.

EXAMPLE 25.7   Given the function

$$f(x) = x^3 - 6x^2 + 9x - 2, \quad 0 \le x \le 2$$

(see Figure 25.9), find the point that may be used for c in the mean value theorem.

SOLUTION   In the terminology of the mean value theorem, $A = (0, -2)$, $B = (2, 0)$, and the slope of the secant line from A to B is 1.   Hence, we must find a number c, $0 < c < 2$, such that $f'(c) = 1$.   But $f'(c) = 3c^2 - 12c + 9$ (see Example 25.2), and setting this equal to 1, we solve for c, obtaining the two values $2 + \frac{2}{3}\sqrt{3}$ and $2 - \frac{2}{3}\sqrt{3}$.   Since $0 < 2 - \frac{2}{3}\sqrt{3} < 2$ we conclude that the required value of c is $2 - \frac{2}{3}\sqrt{3}$.

PROBLEMS

25.1   In each of the following, find all the critical values (including *all* the inflection points) of the given function and sketch the graph.
  (a) $f(x) = x^3 + 2x^2 + x + 1$
  (b) $f(x) = x^3 + x^2 - x - 4, \quad -2 \le x \le 2$
  (c) $f(x) = x^3 - 3x^2 + 1$
  (d) $f(x) = x^3 - 3x^2 - 9x + 1, \quad -4 \le x \le 6$
  (e) $f(x) = x^4 - 2x^2$
  (f) $f(x) = x^4 - x^3, \quad -1 \le x \le 2$
  (g) $f(x) = 3x^4 - 4x^3 - 36x^2 + 24$
  (h) $f(x) = x^5 - 5x$
25.2   In each of the following, plot the graphs of the first and second derivatives of each of the given functions and use these graphs to plot the graph of the given function (see Example 25.2).
  (a) $f(x) = x^3 - 2x^2 + x - 1$
  (b) $f(x) = x^4 - 4x^3 + 10$
25.3   In each of the following, find all the critical values of the given function and sketch the graph.
  (a) $f(x) = x + (1/x)$
  (b) $f(x) = x/(x^2 + 1)$
  (c) $f(x) = x^2(x + 2)^3$
  (d) $f(x) = |x + 1|$
  (e) $f(x) = |x^2 - 2x|$
25.4   In each of the following, find all the critical values of the given function and sketch the graph.
  (a) $f(x) = x^{1/3}$
  (b) $f(x) = x^{2/3}$
  (c) $f(x) = x^{2/3}$
  (d) $f(x) = x(x + 1)^{1/2}$
  (e) $f(x) = x^{2/3}(x + 1)$

(f) $f(x) = x - [x]$ $x \geq 0$

(g) $f(x) = (x - [x])^{1/2}$ $x \geq 0$

25.5 (a) Find a value of k for which the function $f(x) = x + (k/x)$ has a relative minimum at $x = 3$.

 (b) Find a value of k for which the function $f(x) = k[x + (1/x)]$ has a relative minimum at $x = -1$.

25.6 Let a, b, c be three arbitrary real numbers. Determine x such that the sum $(a - x)^2 + (b - x)^2 + (c - x)^2$ is a minimum.

25.7 (a) Construct a polynomial function that has a relative minimum at $x = 1$, a relative maximum at $x = -1$, a nonhorizontal inflection point at $x = 0$, and y intercept $(0, 2)$.

 (b) Construct a polynomial function that has a relative minimum at $x = -2$, a relative maximum at $x = 2$, a nonhorizontal inflection point at $x = 0$, and y intercept $(0, 5)$.

25.8* Prove that every cubic polynomial $f(x) = ax^3 + bx^2 + cx + d$ has exactly one inflection point P and that its graph is symmetrical with respect to the point P (see Figures 25.8 and 25.9).

25.9 Give *two* examples of each of the following.

 (a) a function that is defined at $x = a$ but that is not continuous at $x = a$

 (b) a function that has a tangent to its graph at $(a, f(a))$ but is not differentiable at $x = a$

 (c) a function that is defined at $x = a$ but that has neither a tangent to its graph at $(a, f(a))$ nor a derivative at $x = a$

 (d) a function that has a left-hand and a right-hand derivative at $x = a$ that are unequal

 (e)* a nonconstant function that has a zero derivative at $x = a$ but does not have a relative maximum, a relative minimum, nor an inflection point at $x = a$. (*Hint:* Such a function must have an undefined *second* derivative at $x = a$.)

25.10 Given the following functions and intervals, find the point or points that may be used for c in Rolle's theorem.

 (a) $f(x) = x^3 - 6x^2 + 9x - 2$, $2 - \sqrt{3} \leq x \leq 2 + \sqrt{3}$ (See Figure 25.9.)

 (b) $f(x) = (x^3 - 9x)^{1/3}$, $0 \leq x \leq 3$ (See Figure 25.12.)

25.11 Use Rolle's theorem to prove the following result. If f'' exists for each x in an interval $a \leq x \leq b$, and the graph of $y = f(x)$ intersects the x axis at three distinct points in this interval, then f has at least one inflection point in the interval.

25.12 Given the following functions and intervals, find the point or points that may be used for c in the mean value theorem.

 (a) $f(x) = x^{3/2}$, $0 \leq x \leq 1$

 (b) $f(x) = 2x^3 + 3x - 1$, $0 \leq x \leq 2$ (See Example 25.1.)

 (c) $f(x) = 2x^3 + 3x - 1$, $-1 \leq x \leq 1$ (See Example 25.1)

 (d) $f(x) = x^3 - 6x^2 + 9x - 2$, $0 \leq x \leq 1$ (See Example 25.2.)

 (e) $f(x) = x^3 - 6x^2 + 9x - 2$, $1 \leq x \leq 3$ (See Example 25.2.)

 (f) $f(x) = x + (k/x)$, $1 \leq x \leq 2$, k any positive constant

25.13* (a) Using the mean value theorem, prove the following result. If $f'(x) = 0$ for all x in an interval $a \leq x \leq b$, $f(x)$ is constant over the interval.

 (b) If the functions f and g have equal derivatives over an interval J, there exists a constant K with the property that $f - g = K$ on the interval J.

25.14 Prove the following results. If $y = f(x) = a_2 x^2 + a_1 x + a_0$, $a_2 \neq 0$, is defined on any interval $a \leq x \leq b$, then $c = \frac{1}{2}(a + b)$ is the value that satisfies the mean value theorem.

25.15* Let f be a function having a positive *third* derivative on the interval $a \le x \le b$. Prove that

$$\frac{f(b) - f(a)}{b - a} > f'\left(\frac{a + b}{2}\right).$$

What is the geometrical significance of this result?

25.16* Let f be a continuous function for all real numbers x and let f be differentiable for all $x \ne a$ where a is some real number. Show that if $\lim\limits_{x \to a} f'(x) = K$, K a constant, then $f'(a) = K$. What is the geometrical significance of this result?

Section 26

OTHER APPLICATIONS
OF THE THEORY OF
MAXIMA AND MINIMA

We now give some examples of practical problems that are solvable by the methods of the preceding section. The first example belongs to the category of problems that ask for the minimum distance of a curve to a point not on the curve. If the curve has equation $y = f(x)$ and the given point has coordinates (a, b), we simply find the distance between (a, b) and an arbitrary point $(x, f(x))$ on the graph. This gives the distance as a function of x and we then use the first derivative to find the minimum of the function.

EXAMPLE 26.1 Find the point on the graph of the equation $y = 2\sqrt{x}$ that is closest to the point $(2, 1)$.

SOLUTION An arbitrary point on the graph has coordinates $(x, 2\sqrt{x})$. Therefore, the *square* of the distance from $(2, 1)$ to an arbitrary point on the graph is

$$D(x) = (2 - x)^2 + (1 - 2\sqrt{x})^2 = 4 - 4x + x^2 + 1 - 4\sqrt{x} + 4x$$
$$= x^2 - 4\sqrt{x} + 5.$$

To find the minimum distance, it is enough to minimize the square of the distance $D(x) = x^2 - 4\sqrt{x} + 5$. Hence, we compute $D'(x) = 2x - 2/\sqrt{x}$ and set this equal to zero, obtaining $2x = 2/\sqrt{x}$. This implies $x^{3/2} = 1$ and finally $x = 1$. Therefore, the closest point to $(2, 1)$ on the graph of $y = 2\sqrt{x}$ is $(1, 2)$. [If we plot the graph and the point $(2, 1)$, it is clear that $D(x)$ has no maximum value.]

157

Another type of practical problem concerns itself with profit and loss. The concrete situation is put into the form of an algebraic relation that results in a function of x. We then look for either the maximum or minimum value of this function.

EXAMPLE 26.2   If 500 people will attend a movie when the admission price is \$1.00, and if attendance decreases by 25 for each 10 cents added to the price, which price of admission will yield the greatest gross receipts?

SOLUTION   Let $x =$ the number of 10 cent increases. Then the new admission price is $100 + 10x$ cents, while the attendance becomes $500 - 25x$. Therefore, the equation for the gross receipts is

$$g(x) = (100 + 10x)(500 - 25x) = 50{,}000 + 2500x - 250x^2$$
$$= 250(200 + 10x - x^2).$$

Differentiating, we obtain $g'(x) = 250(-2x + 10)$. Setting $g'(x) = 0$, we obtain $x = 5$ and, since $g''(x) = -2$, $x = 5$ must be a maximum. Therefore, the greatest gross receipts will come when the admission price is raised to \$1.50. Observe also that $g(x)$ has a graph that is a parabola, with a vertical axis of symmetry, opening in a downward direction. Consequently, we merely found the x coordinate of the vertex that must be the highest point on the graph and hence yields the required maximum.

Another category of practical problems involves the motion of an object or particle on a straight line. In Problem II, which follows Example 22.7, we introduced the position function $s(t)$ of a particle P that moves along a horizontal coordinate line. We explained that $s'(t)$ gives the velocity, denoted by $v(t)$, of the particle at any time t. Since then, we have introduced the *second derivative* of a function. If we compute $v'(t)$, which is really $s''(t)$, we see that this represents the *instantaneous rate of change of the velocity with respect to time*, but this is just the *acceleration* of the particle, denoted by $a(t)$, at any time t. The acceleration is measured in units of velocity per unit of time and usually in feet per second per second, abbreviated as ft/sec/sec.

Connected with the preceding type of problem, we have another that involves an object or particle P being propelled straight upward or downward (see Example 22.8). In this case P can be thought of as moving up and down a vertical coordinate line, and its position function is usually denoted by $h(t)$, denoting the height of the object at any time t in the former case, and $s(t)$, denoting the distance traveled by the falling object after t seconds in the latter case.

In any event, whether the motion of the object or particle is on a horizontal or vertical coordinate line, we can give some general rules describing this motion. If $f(t)$ is the *position* function, $f'(t) = v(t)$ is the *velocity* at any time t, and $f''(t) = v'(t) = a(t)$ is the *acceleration* at any time t. If $v(t) > 0$ at any time t, the object is moving either to the *right* or in an *upward* direction. If $v(t) < 0$ at time t, the object is moving either to the *left* or in a *downward* direction. Consequently, the sign of $v(t)$ will indicate the *direction* in which the object is moving at time t. However, if only the speed is required, simply compute $|v(t)|$. Finally, if $t \neq 0$, $v(t) = 0$, and the object is still in motion, the object is *reversing* its direction. This means that on a horizontal line it is changing its direction of motion from left to right or from right to left. In the case of a vertical line, the object is changing its direction from up to down or from down to up; therefore, it must be at its maximum height or depth at the instant $v(t) = 0$. Hence, in both of the preceding cases $v(t)$ must change sign in an ε neighborhood of the number t for which $v(t) = 0$.

With regard to the sign of the acceleration $a(t)$ at any time t, the general rule is that the object is *speeding up* (accelerating) over any time interval where $v(t)$ and $a(t)$ have the *same sign* and the object is *slowing down* (decelerating) over any time interval where $v(t)$ and $a(t)$ have *opposite sign*.

We now give an example of each one of these types of problems.

EXAMPLE 26.3   The position function of a particle P moving along a horizontal coordinate line is given by $s(t) = t^3 - 6t^2 + 9t + 1$.
(a) At $t = 2$, find the position, velocity, and acceleration of the particle.
(b) Find the times and the positions at which the particle reverses its direction.
(c) Give the time intervals over which the particle is speeding up and the time intervals over which the particle is slowing down.

SOLUTION

(a) $v(t) = s'(t) = 3t^2 - 12t + 9 = 3(t^2 - 4t + 3) = 3(t - 1)(t - 3)$

and $a(t) = v'(t) = s''(t) = 6t - 12 = 6(t - 2)$. The position of P at $t = 2$ is $s(2) = 3$; that is, P is 3 units to the right of the origin. The velocity of P at $t = 2$ is $v(2) = -3$ ft/sec, which indicates that P is moving to the left at a speed of 3 ft/sec. The acceleration is $a(2) = 0$ ft/sec/sec, which indicates that P is neither speeding up nor slowing down at $t = 2$.
(b)   The particle reverses direction at $t = 1$ and $t = 3$ since $v(t) = 0$ for these values of t and $v(t)$ changes sign in a neighborhood of each of these points. The position of P at these times is $s(1) = 5$ and $s(3) = 1$, respectively.
(c)   Over the interval $0 < t < 1$, $v(t) > 0$ and $a(t) < 0$; hence, P is slowing down. Over $1 < t < 2$, $v(t) < 0$ and $a(t) < 0$, hence, P is speeding up. Over $2 < t < 3$, $v(t) < 0$ and $a(t) > 0$; hence, P is slowing down. Finally, for $t > 3$, $v(t) > 0$ and $a(t) > 0$; therefore, P is speeding up. In Figure 26.1, we show a diagram describing the motion of the particle P.

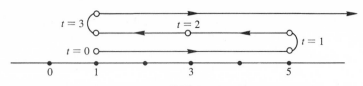

Figure 26.1

EXAMPLE 26.4   If an object is propelled straight up from an initial height of h_0 ft with an initial velocity of v_0 ft/sec, its height h measured after t sec is given by the function of t, $h = h(t) = h_0 + v_0 t - 16t^2$ We would like to answer each of the following questions.
  (a) What is the maximum height attained by the object?
  (b) How much time does it take for the object to hit the ground?
  (c) How fast is the object going when it hits the ground?

SOLUTION (a)   Since $h(t)$ gives the height at any time t, we need only find the maximum value of the function.   We can do this by simply computing $h'(t) = v_0 - 32t$.   Setting $h'(t) = 0$, we obtain $t = v_0/32$ and since $h''(t) = -32 < 0$, $t = v_0/32$ must be a maximum.   Observe that $h'(t) = v(t) = 0$ means that the object is reversing its direction; consequently, it is at its maximum height.   Hence, the maximum height is $h(v_0/32) = h_0 + v_0^2/32 - 16(v_0/32)^2 = h_0 + (v_0^2/32)(1 - \frac{1}{2}) = h_0 + v_0^2/64$.
Note also that $h''(t) = a(t) = -32 < 0$ for all t.   Therefore, the object is *slowing down* on its way *up* since $h'(t) = v(t) > 0$ for $0 < t < v_0/32$, and the object is *speeding up* on its way *down* since $v(t) < 0$ for $t > v_0/32$.
(b)   The object hits the ground when $h(t) = 0$.   Setting $h(t) = 0$, we obtain $-16t^2 + v_0 t + h_0 = 0$.   Solving for t, we have

$$t = \frac{-v_0 \pm (v_0^2 - 4(-16)h_0)^{1/2}}{2(-16)} = \left(\frac{1}{32}\right)[v_0 \pm (v_0^2 + 64h_0)^{1/2}]$$

Now, only one of the preceding roots is positive and therefore meaningful. Consequently, the object hits the ground at

$$t_0 = (1/32)[v_0 + (v_0^2 + 64h_0)^{1/2}].$$

(Note that if $h_0 = 0$, then $t_0 = v_0/16$).
(c)   The speed of the object as it hits the ground is simply $|h'(t_0)|$ since $h'(t) = v(t)$ is the velocity at any time t and $|v(t)|$ is the speed at any time t.

The following example is slightly different from the preceding ones; however, it still involves the rate of change of distance with respect to time.

EXAMPLE 26.5   A ladder 6 ft long is leaning against a wall, with the bottom of the ladder 2 ft from the wall.   If the lower end is pulled away from the wall at the rate of 1 ft/sec, find the rate at which the upper end comes down along the wall.   What is this rate at the end of 2 sec?

SOLUTION   In Figure 26.2, we show on the left the position of the ladder at the start, and on the right the position of the ladder after t sec where the bottom of the ladder has been pulled away t ft from its starting position.   If we let $d(t)$ indicate the distance of the top of the ladder from the ground at any time t, the diagram on the right yields the equation $(d(t))^2 = 36 - (2 + t)^2 = 36 - 4 - 4t - t^2 = 32 - 4t - t^2$.   Differentiating, we obtain $2d(t)d'(t) = -4 - 2t$, which means that $d'(t) = -(2 + t)/d(t)$ = rate of descent of the top of the ladder along the wall. At $t = 2$, $d'(2) = -4/d(2)$, where $d(2) = (36 - 16)^{1/2} = 2\sqrt{5}$.   Therefore, $d'(2) = -4/2\sqrt{5} = -2/\sqrt{5}$ ft/sec, which indicates that the top of the ladder is moving *down* the wall at the speed of $2/\sqrt{5}$ ft/sec.

We conclude this section with two examples that involve more than one rate of change with respect to time and the *relation* between these rates.

EXAMPLE 26.6   A spherical balloon is blown up in such a way that its *radius* increases at the constant rate of 2 in./sec.   Find the rate of increase of the *volume* enclosed by the balloon at any time t after inflation begins.   What is this rate at $t = 4$?

SOLUTION   If the radius increases at the constant rate of 2 in./sec, we can consider the radius r as a function of t, that is, $r = r(t)$, with the property that $r'(t) = 2$.   Therefore, $r(t) = 2t$ (since $2t$ has derivative 2) and the volume V, as a function of t, can be expressed $V(t) = \frac{4}{3}\pi(r(t))^3 = \frac{4}{3}\pi 8t^3$.   The rate of increase of the volume with respect to time is then given by $V'(t) = 32\pi t^2$ and the rate at $t = 4$ is $V'(4) = 32\pi 16 = 512\pi$ cubic in./sec.

EXAMPLE 26.7   A cube is shrinking in such a way that at the instant its edge is 4 in., its volume is decreasing at the rate of 2 cubic in./per sec. How fast is the edge changing at that instant?

SOLUTION   If the cube has edge E inches at time t, we can write E as a function of t, $E = E(t)$.   Then this problem requires that we find the value of $E'(t)$ when $E(t) = 4$.   Let t_0 be the time at which the edge is

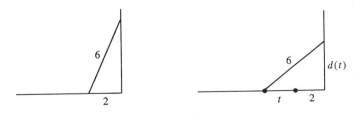

Figure 26.2

4 in. long, that is, $E(t_0) = 4$; then we are looking for $E'(t_0)$.   The volume of the cube at any time t is given by $V(t) = (E(t))^3$; therefore, the rate of change of the volume with respect to time is simply $V'(t) = 3(E(t))^2 E'(t)$. But we are given that the volume is *decreasing* at the rate of 2 cubic in./sec at the instant its edge is 4 in. long, namely, at $t = t_0$.   Therefore, $V'(t_0)$ $= 3(E(t_0))^2 E'(t_0) = -2$. This implies that $E'(t_0) = -2/(3(E(t_0))^2) = -2/(3)(16) = -1/24$; hence, the edge is *decreasing* at the rate of $1/24$ in./sec.   Therefore, we were able to obtain the required answer *without ever knowing the edge as a function of t*.   In Example 26.6, we were able to express the volume as a function of t and then, by direct differentiation, we obtained the required rate of change.   This example, however, shows that if we are given enough information, we can still determine a rate of change without actually expressing the required quantity as a function of t and differentiating it directly.

PROBLEMS

26.1 (a) Find the point on the graph of $y = 2/x$ that is closest to the point $(-1, 1)$.
   (b) Find the point on the graph of $y = -x^3 + x^2 - 2x + 1$ where the tangent line to the graph has maximum slope.
   (c) Find the point on the graph of $y = 2x^3 + 3x^2 + 3x - 1$ where the tangent line to the graph has minimum slope.
26.2 Let $f(x) = x^3 - 4x$ and let $g(x) = x^3 - x^2 - 5x - 1$.   Plot the graphs of these functions and find the shortest distance between the graphs.
26.3 (a) Prove that among all triangles having a given area, the equilateral triangle has the smallest perimeter.
   (b) Prove that among all triangles having a given perimeter, the equilateral triangle has the greatest area.
26.4 (a) Prove that among all the rectangles having a given perimeter, the square has the greatest area.
   (b) Prove that among all closed rectangular boxes having a square base and specified volume V, the cube has the least surface area.
26.5 Suppose a 12-in. piece of wire is cut into *two* pieces and suppose the pieces are bent into the shape of (a) a circle and a square; (b) a square and an equilateral triangle; (c) an equilateral triangle and an isosceles right triangle; (d) an isosceles right triangle and a circle.   In each case, determine how the wire should be cut if the sum of the areas enclosed by the two figures is to be a minimum?   A maximum?
26.6* Suppose an 18-in. piece of wire is cut into *three* pieces where one of the three pieces is 6 in. long.   Suppose the three pieces are bent into the shape of a circle, a square, and an equilateral triangle.   How should the wire be cut if the sum of the areas enclosed by the three figures is to be a minimum?   A maximum?
26.7* (a) Prove that a tin can enclosing a specified volume V will be made of the least amount of metal if its height equals the diameter of its base.
   (b) Prove that a tin can having S square inches of surface area will enclose the greatest volume if its height equals the diameter of its base.

26.8   A car rental agency has 80 automobiles to rent.   If it charges $60 per month, all the cars are rented.   However, it decides to raise its rates and discovers that for each $2 increase in monthly rates, *one* less car is rented.   If each rented car costs the agency $6 a month in maintenance costs, which rental charge yields the most profit?

26.9   Suppose the position function of a particle moving along a horizontal coordinate line is given by the equation $s(t) = t + 4/t^2$, $t \geq 1/4$.

(a)   At $t = \frac{1}{2}$, find the position, velocity, and acceleration of the particle.

(b)   Find the times and the positions at which the particle reverses its direction.

(c)   Give the time intervals over which the particle is speeding up and the time intervals over which the particle is slowing down.

(d)   As $t \to \infty$, what does the velocity approach?

(e)   As $t \to \infty$, what does the acceleration approach?

26.10   A ball is thrown straight up from a roof 24 ft high with an initial velocity of 40 ft/sec.   Find the maximum height attained by the ball; the time it takes for the ball to come down to the ground; and how fast the ball is going when it hits the ground.

26.11   A rocket is fired from the ground at an angle of 30° with an initial velocity of 320 ft/sec.   Let $x = (\cos 30°)320t$ be the horizontal distance traveled after t sec, and let $y = (\sin 30°)320t - 16t^2$ be the height of the rocket at any time t.

(a)   Find the horizontal distance traveled by the rocket before it hits the ground.

(b)   Find the time at which the rocket reaches its maximum height.

(c)   Find the maximum height attained by the rocket.

26.12*   Suppose a particle P is at the point $(0, 0)$ in the plane and a particle Q is at the point $(0, -6)$.   If P starts to move along the x axis to the right at the rate of 2 units/sec at the same instant that Q starts to move up the y axis at the rate of 1.5 units/sec, find a function $d(t)$ that gives the distance between the particles at any time t sec after they start to move.   Also, find the rate of change of the distance between the particles at any time t.   At what time are the particles closest together?

26.13   A cube is expanding in such a way that its edge is increasing at the rate of 2 in./sec.   How fast is the volume increasing at the instant the edge is 10 in. long?

26.14   (a)   A spherical balloon is blown up in such a way that its volume increases at the constant rate of 36 cubic in./min.   Find the rate of increase of the radius of the balloon at any time t min after inflation begins.   What is the rate of increase at $t = 8$?

(b)   A spherical balloon is being *deflated* in such a way that at the instant its radius is 2 in., its volume is decreasing at the rate of 4 cubic in./min.   How fast is the radius changing at that instant?

26.15*   Two corridors 1 yard wide and 3 yards wide intersect at right angles.   What is the length of the largest telephone pole that can be carried horizontally from one corridor to the other?

26.16   (a)   Suppose on the circumference of a circle with center at the point $(0, 0)$ and radius 1 ft, a particle P in the first quadrant is moving in a counterclockwise direction away from the point $A = (1, 0)$ at the constant angular rate of 1 radian per second.   Find how fast (in feet per second) the distance AP is changing when the angle AOP is a right angle. [*Hint:* use the law of cosines (see Appendix, Problem 9) to find the length of AP.]

(b)   Repeat part (a), using a constant angular rate of 2 rad/sec with angle *AOP* equal to 60°.

26.17   A delivery truck can travel at a maximum speed of 75 mph.   When it is going at the rate of x mph, it uses gasoline at the rate of $(1/x + x/25)$ gal/mile with the gasoline costing 30 cents per gallon. The truck must make a 400-mile trip and its driver must be paid at the rate of \$2.50 per hour.   Find the most economical speed at which the truck should travel.

TRIGONOMETRIC FUNCTIONS AND THEIR INVERSES

In this section, we discuss the trigonometric functions, their inverse functions, and the various derivatives involved. We then attempt to solve the same kinds of problems for these functions as we solved previously for algebraic functions; for example, problems that involve limits, derivatives, implicit differentiation, maxima and minima, and curve tracing.

THE TRIGONOMETRIC FUNCTIONS

Before beginning the treatment of the trigonometric functions, we recall some basic results from high school trigonometry (see the Appendix) with the following three tables. Table 1 gives the *sign* of the three functions listed in each of the four quadrants. Table 2 gives the values of the listed functions at the angles of $30° = \pi/6$ rad, $45° = \pi/4$ rad, and $60° = \pi/3$ rad. Note that the $\sin \theta$ row (in Table 2) can be best remembered as $\frac{1}{2}\sqrt{1}, \frac{1}{2}\sqrt{2}, \frac{1}{2}\sqrt{3}$; the

Table 1

	1	2	3	4
$\sin \theta$	+	+	−	−
$\cos \theta$	+	−	−	+
$\tan \theta$	+	−	+	−

Table 2

θ	30°	45°	60°
$\sin\theta$	$\frac{1}{2}$	$\frac{1}{2}\sqrt{2}$	$\frac{1}{2}\sqrt{3}$
$\cos\theta$	$\frac{1}{2}\sqrt{3}$	$\frac{1}{2}\sqrt{2}$	$\frac{1}{2}$
$\tan\theta$	$1/\sqrt{3}$	1	$\sqrt{3}$

$\cos\theta$ row is just the reverse of the $\sin\theta$ row; and finally, each of the terms in the $\tan\theta$ row is the quotient of the two terms immediately above it.

Table 3, in the form of coordinate axes, gives the values of the listed functions at the quadrant angles of $0° = 0$ rad, $90° = \pi/2$ rad, $180° = \pi$ rad, $270° = 3\pi/2$ rad and $360° = 2\pi$ rad. The symbol ∞ at $90°$ indicates that the $\tan 90°$ is *not defined*, but that as the angle approaches $90°$, say, through values $89.1°$, $89.2°$, $89.3°$, …, the values of the tangent function become large and positive without bound. The $-\infty$ at $270°$, in a similar way, indicates that as the angle approaches $270°$, the values of the tangent function become large and negative without bound.

Table 3

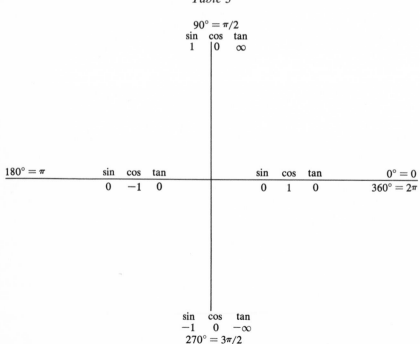

We assume you are familiar with the *definitions* and the *graphs* of the trigonometric functions sin x, cos x, and tan x (see the Appendix).   We further assume that you know that $\lim_{h\to 0}[(\sin h)/h] = 1$ and $\lim_{h\to 0}[(1 - \cos h)/h] = 0$ (see Problem 21.6).   Finally, we assume that you know that the derivative of sin x is cos x and the derivative of cos x is $-\sin x$.

EXAMPLE 27.1   If $f(x) = \tan x$, show that $f'(x) = \sec^2 x$.

SOLUTION   Since tan $x = \sin x/\cos x$, we may use the quotient formula for the derivative.   Hence

$$(\tan x)' = \frac{\cos x(\cos x) - \sin x(-\sin x)}{\cos^2 x} = \frac{1}{\cos^2 x} = \sec^2 x$$

EXAMPLE 27.2   Show that $\lim_{h\to 0}[(\tan h)/h] = 1$.

SOLUTION   tan $h = \sin h/\cos h$; therefore

$$\frac{\tan h}{h} = \frac{\sin h}{h \cos h} = \frac{\sin h}{h} \frac{1}{\cos h}$$

Now,

$$\lim_{h\to 0}\left(\frac{\sin h}{h} \frac{1}{\cos h}\right) = \lim_{h\to 0}\frac{\sin h}{h} \lim_{h\to 0}\frac{1}{\cos h} = 1 \cdot 1 = 1$$

Note that $\lim_{h\to 0}[(\tan h)/h]$ is, by definition, the derivative of tan x at $x = 0$. But, from the preceding example, $(\tan x)' = \sec^2 x$; hence, $\sec^2 0 = 1$ and we have another solution.

In the following two examples, we make use of the chain rule formulas $(\sin f(x))' = (\cos f(x))f'(x)$ and $(\cos f(x))' = (-\sin f(x))f'(x)$.

EXAMPLE 27.3   If $f(x) = x^2 \cos 3x$, find $f'(x)$.

SOLUTION   By the product formula for the derivative, $f'(x) = 2x \cos 3x + x^2(\cos 3x)'$.   By the chain rule, $(\cos 3x)' = (-\sin 3x)(3)$.   Therefore, $f'(x) = 2x \cos 3x - 3x^2 \sin 3x$.

EXAMPLE 27.4   Find the slope of the tangent line to the graph of the equation $x \sin y = 2x^2 - 3/2$ at the point $(1, \pi/6)$.

SOLUTION   We let $y = f(x)$ and then we use the technique of *implicit differentiation* (see Section 24).   Differentiating both sides of the equation, we obtain $(x \sin f(x))' = \sin f(x) + x(\cos f(x))f'(x) = 4x$.   Hence,

$$f'(x) = \frac{4x - \sin f(x)}{x \cos f(x)} = \frac{4x - \sin y}{x \cos y}$$

Now, substituting $x = 1$, $y = \pi/6$, we obtain $(4 - \frac{1}{2})/\frac{1}{2}\sqrt{3} = 7/\sqrt{3}$, which is the required slope.

EXAMPLE 27.5 Find the critical values of $f(x) = \sin x + \cos x$, $0 \le x \le 2\pi$, and sketch the graph.

SOLUTION $f'(x) = \cos x - \sin x$ and $f''(x) = -\sin x - \cos x$. Setting $f'(x) = 0$, we obtain $\cos x = \sin x$, which occurs in the interval $0 \le x \le 2\pi$ only at $x = \pi/4$ and $x = 5\pi/4$. The corresponding y coordinates are $\sqrt{2}$ and $-\sqrt{2}$, respectively. Since $f''(\pi/4) = -\sqrt{2} < 0$, $x = \pi/4$ is a relative maximum and $f''(5\pi/4) = \sqrt{2} > 0$ implies $x = 5\pi/4$ is a relative minimum. Setting $f''(x) = 0$, we obtain $\sin x = -\cos x$, which occurs in the interval $0 \le x \le 2\pi$ only at $x = 3\pi/4$ and $x = 7\pi/4$. The corresponding y coordinate in each case is zero. Hence, the graph has inflection points at $(3\pi/4, 0)$ and $(7\pi/4, 0)$. A sketch appears in Figure 27.1.

EXAMPLE 27.6 Discuss the continuity and the differentiability of the function $f(x) = x \sin(1/x)$, $x \ne 0$ and $f(0) = 0$ at the points $x = 1/\pi$ and $x = 0$.

SOLUTION

$$f'(x) = \sin(1/x) + x(\cos(1/x))(-1/x^2) = \sin(1/x) - (\cos(1/x))/x.$$

Therefore, $f'(1/\pi) = \sin \pi - \pi \cos \pi = 0 - (-\pi) = \pi$ and f is differentiable, and hence continuous, at $x = 1/\pi$ [see Problem 22.6(a)].

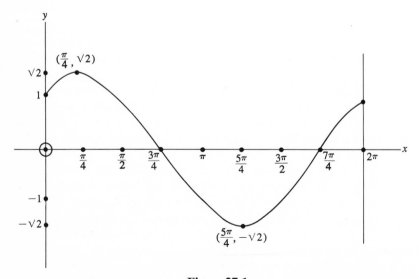

Figure 27.1

Since the function $f'(x)$ is not defined at $x = 0$, we must use the limit definition for the derivative of $f(x)$ at $x = 0$; that is, we must compute

$$\lim_{h \to 0} \frac{f(0 + h) - f(0)}{h} = \lim_{h \to 0} \frac{h \sin(1/h)}{h} = \lim_{h \to 0} \sin(1/h)$$

Now, if we choose the sequence $\{h_n\} = \{2/n\pi\}$, which approaches 0, we see that the corresponding sequence $\{\sin(1/h_n)\} = 1, 0, -1, 0, 1, 0, -1, 0, \ldots$, does not have a limit. Therefore, $\lim_{h \to 0} \sin(1/h)$ does not exist and f is *not* differentiable at $x = 0$. However, from Problem 20.8, f is *continuous* at $x = 0$ since $\lim_{h \to 0} f(0 + h) = \lim_{h \to 0} h \sin(1/h) = 0 = f(0)$ [as $h \to 0$ through any sequence of values, $|\sin(1/h)| \le 1$ and consequently, $|h| \, |\sin(1/h)| \le |h| \to 0$]. As an exercise, sketch the graph of this function in a neighborhood of the origin.

THE INVERSE TRIGONOMETRIC FUNCTIONS

At this point, you should review the essential results concerning the *inverse function* f^* of a given function f (these results both precede and follow Theorem 17.1), paying particular attention to the material presented on the *graph* of f^* (Figure 17.2) and to Examples 17.3 and 17.4. The results of Problems 17.17 and 17.19 should also be reviewed.

According to Theorem 17.1, a function f has an inverse f^* if and only if f is a one-to-one function. Now that we have studied the derivative and the applications to curve tracing, we know that any function f that has $f' > 0$ over an interval J must be an increasing function and hence one-to-one over J. On the other hand, if $f' < 0$ over an interval J, f must be a decreasing function and hence it is one-to-one over J also. Observe that it is possible for f to have horizontal inflection points and still be either increasing or decreasing over an interval; however, f cannot have a turning point in the interval J, for the graph would then double back on itself and the function would *not* be one-to-one. In Figure 27.2, we have sketched the graph of a function with a turning point in the interval J: $a \le x \le c$. If we restrict the domain of this function to the subinterval $a \le x \le b$, $f(x)$ has no turning points and it is a one-to-one function over the interval J_1: $a \le x \le b$. Therefore, f has an inverse f^* over this interval. In fact, if we restrict the domain of f to the interval J_2 : $b \le x \le c$, it is also one-to-one and has an inverse F^* over this interval. Observe that $f^* \ne F^*$ since f^* has range J_1 and F^* has range J_2 ; thus, they have *different* ranges. [In Figure 27.2, we see that $f^*(1) = x_1$, whereas $F^*(1) = x_2$.] Consequently, when dealing with inverse functions, we must be careful to define the domain and the range of both f and f^*.

In the following definition, we restrict the domains of the functions involved in such a way that the functions are one-to-one and hence have their inverses

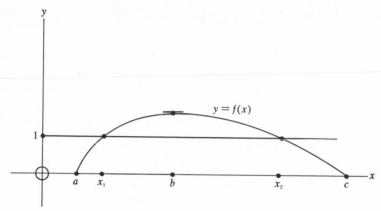

Figure 27.2

defined. We also include, after the definition, a graph consisting of each function with its inverse superimposed. We use the technique explained in the discussion associated with Figure 17.2 to obtain the graph of the inverse function from the graph of the original function.

Definition 27.1 (a) Let $f(x) = \sin x$ be defined on the interval $-\pi/2 \le x \le \pi/2$. Then we call the inverse function of sine by the name arcsine. Thus,

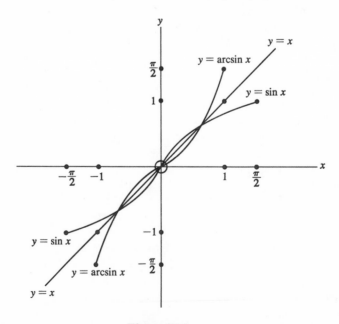

Figure 27.3

if $y = \sin x$, $f^*(y) = \arcsin y = x$.   The domain of arcsine is $-1 \leq y \leq 1$, while the range is $-\pi/2 \leq x \leq \pi/2$.   The graphs appear in Figure 27.3.

(b)   Let $f(x) = \tan x$ be defined on the interval $-\pi/2 < x < \pi/2$.   Then we call the inverse function of tangent by the name arctangent.   Thus, if $y = \tan x$, $f^*(y) = \arctan y = x$.   The domain of arctangent is the set of all real numbers and the range is $-\pi/2 < x < \pi/2$.   The graphs appear in Figure 27.4.

As an exercise consider $f(x) = \sin x$ over the interval $\pi/2 \leq x \leq 3\pi/2$ and then define the inverse function, giving its domain and range. Also, consider $f(x) = \tan x$ over $\pi/2 < x < 3\pi/2$ and define the inverse function, giving its domain and range.

We now prove an important general result on the derivative of an inverse function f^*.

Theorem 27.1   If f^* is the inverse function of f and $f(x) = y$ where x is a number in the domain of f, then $f^*(y) = x$ and $f^{*\prime}(y) = 1/f'(x)$ provided $f'(x)$ exists and is not zero.   Therefore, whenever the derivative of the inverse function f^* exists at a number y in its domain it is given by the reciprocal of the derivative of the function f at the number x where $f(x) = y$.

PROOF   Since $y = f(x)$, $f^*(y) = f^*(f(x))$.   But by the definition of the inverse function, $f^*(f(x)) = x$.   Differentiating both sides of this equation,

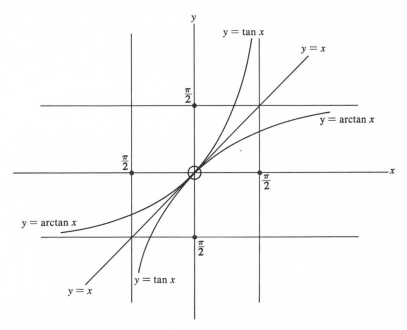

Figure 27.4

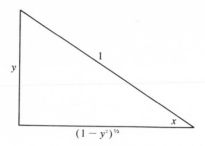

Figure 27.5

using the chain rule on $f*(f(x))$, we obtain $f*'(f(x))f'(x) = 1$. Therefore, $f*'(f(x)) = 1/f'(x)$. But $y = f(x)$; hence, $f*'(y) = 1/f'(x)$ and the proof is completed. ∎

EXAMPLE 27.7 Use Theorem 27.1 to compute the derivative of (a) the arcsine function and (b) the arctangent function.

SOLUTION (a)   Let $f(x) = \sin x = y$; then $f*(y) = \arcsin y = x$ and $f'(x) = \cos x$. Therefore, if y is any number in the domain of arcsine, $f*'(y) = 1/f'(x)$ means $(\arcsin y)' = 1/\cos x$. But $x = \arcsin y$, so $\cos x = \cos(\arcsin y) = (1 - y^2)^{1/2}$. This follows from the right triangle relation depicted in Figure 27.5. Since $x = \arcsin y$ means that x is the angle whose sine is equal to y and $\cos(\arcsin y)$ means the cosine of that angle, from the diagram we see that the $\cos x = (1 - y^2)^{1/2}$. Therefore, if y is any real number in the domain of arcsine, $(\arcsin y)' = 1/(1 - y^2)^{1/2}$ If we finally rewrite this formula as a function of x, we obtain $(\arcsin x)' = 1/(1 - x^2)^{1/2}$.

SOLUTION (b)   Let $f(x) = \tan x = y$; then $f*(y) = \arctan y = x$ and $f'(x) = \sec^2 x$. Therefore, if y is any number in the domain of arctangent, $(\arctan y)' = 1/\sec^2 x = \cos^2 x = (\cos(\arctan y))^2$. But if x is the angle whose tangent is y, $\cos x = 1/(1 + y^2)^{1/2}$ (see Figure 27.6) and

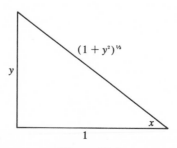

Figure 27.6

$\cos^2 x = 1/(1 + y^2)$.   Therefore, $(\arctan y)' = 1/(1 + y^2)$ for any real number y in the domain of arctangent and, rewriting this formula as a function of x, we obtain $(\arctan x)' = 1/(1 + x^2)$.

Finally, if we use the chain rule, we can derive the general derivative formulas $(\arcsin f(x))' = f'(x)/(1 - f^2(x))^{1/2}$ and $(\arctan f(x))' = f'(x)/(1 + f^2(x))$.

EXAMPLE 27.8   Find the derivative of each of the functions (a) $(\arctan x^2)^{1/2}$; (b) $x \arcsin x$; and (c) $\arcsin(\cos x)$.

SOLUTION (a) $((\arctan x^2)^{1/2})' = \frac{1}{2}(\arctan x^2)^{-1/2}[1/(1 + x^4)](2x)$ (by the chain rule).

SOLUTION (b) $(x \arcsin x)' = \arcsin x + x[1/(1 - x^2)^{1/2}] = \arcsin x + x/(1 - x^2)^{1/2}$ (by the product formula and the chain rule).

SOLUTION (c) $(\arcsin(\cos x))' = (1 - \cos^2 x)^{-1/2}(-\sin x) =$

$$-\sin x/|\sin x| = \begin{cases} -1 & \text{for } 0 < x < \pi \\ +1 & \text{for } -\pi < x < 0 \end{cases}$$

(by the chain rule and the fact that $\sqrt{r^2} = |r|$ for every real number r).

PROBLEMS

27.1   Evaluate each of the following limits.

(a) $\lim\limits_{h \to 0} \dfrac{\sin(2h^2)}{3h}$

(b) $\lim\limits_{h \to 0} \dfrac{h \sin h}{\sin(2h^2)}$

(c) $\lim\limits_{x \to \pi/2} \dfrac{\sin x}{1 + x}$

27.2   Differentiate each of the following functions:
 (a) $(\sin(x^2 + 1)^{1/2})^3$
 (b) $(\cos(x^2 + 2x))^{1/2}$

27.3   Find the first and second derivatives of each of the following functions.
 (a) $x \sin x$, $-\pi < 0 < \pi$
 (b) $(\cos x)/x$, $0 < x < 3\pi/2$
 (c) $\sin^2 x$
 (d) $\sin x^2$, $-\sqrt{\pi} < x < \sqrt{\pi}$
 (e) $|\sin x|$, $0 < x < 2\pi$
 (f) $|\tan x|$, $-\pi/2 < x < \pi/2$
 (g) $1 + x - 2 \sin x$, $-\pi \leq x \leq \pi$
 (h) $x + \cos x$
 (i) $\cos 4x + 2 \cos 2x$, $\pi/2 < x < \pi/2$

27.4   Find the critical values and sketch the graph of each of the functions in the preceding problem.

27.5 Find the slope of the tangent line to the graph of each of the given equations at the point indicated. (Use implicit differentiation; see Section 24.)
(a) $x + \cos y = 1$, $\quad$ $(1, \pi/2)$
(b) $y^2 - x^2 + \sin y - \pi^2/4 = 0$, $\quad$ $(1, \pi/2)$
(c) $x \sin y = 2x^2 - 3$, $\quad$ $(3/2, \pi/2)$

27.6* Discuss the continuity and the differentiability of each of the following functions at the point $x = 1/\pi$ and $x = 0$. Sketch the graphs over $0 \le x \le 2\pi$.
(a) $f(x) = \sin(1/x)$, $x \ne 0$ and $f(0) = 0$
(b) $f(x) = x \sin(1/x)$, $x \ne 0$ and $f(0) = 0$ (See Example 27.6.)
(c) $f(x) = x^2 \sin(1/x)$, $x \ne 0$ and $f(0) = 0$
(d) $f(x) = x^3 \sin(1/x)$, $x \ne 0$ and $f(0) = 0$
(e) $f(x) = x^4 \sin(1/x)$, $x \ne 0$ and $f(0) = 0$

27.7* Repeat the preceding problem using the derivatives of the given functions.

27.8 Differentiate each of the following functions and then find the slope (if possible) of the tangent line to the graph of the function at each of the given points.
(a) $f(x) = \arcsin(1 - x^2)^{1/2}$ at the points $(\frac{1}{2}, \pi/3)$, $(-\frac{1}{2}, \pi/3)$, and $(0, \pi/2)$
(b) $f(x) = \sin(\arctan x^2)^{1/2}$ at $x = 1$
(c) $f(x) = \log(\arctan(e^{\sin x}))$ at $x = 0$

27.9* Let $f(x) = \arcsin(\cos x)$.
(a) Prove that $f(x) = \pi/2 - x$ for $0 < x < \pi$ and $f(x) = \pi/2 + x$ for $-\pi < x < 0$.
(b) Compute $f'(x)$ and compare the results with part (c) of Example 27.8.
(c) Sketch the graph of $f(x)$ for $-\pi < x < \pi$.

27.10 (a) Let $f(x)$ be implicitly defined by the equation $x - \arctan y = 0$. Find $f'(x)$.
(b) Evaluate f' at $x = \pi/4$.

27.11 (a) State a specific domain and range for the arccos x.
(b) Use Theorem 27.1 to compute the derivative of arccos x.

27.12 (a) Graph the function $f(x) = \sec x$, $0 \le x < \pi/2$.
(b) Find the derivative of $f(x) = \sec x$ at each point of the interval $0 \le x < \pi/2$.
(c) Define, if possible, a specific domain and range for the inverse function of sec x over $0 \le x < \pi/2$.
(d) Using Theorem 27.1 and the results of parts (b) and (c), find the derivative of this inverse function.

27.13* Repeat the preceding problem with the function $f(x) = \sec x$ over the interval $\pi < x < 3\pi/2$. Are the results the same?

27.14 Given the following functions and intervals, find the point or points that may be used for c in the mean value theorem (see Example 25.7).
(a) $f(x) = \sin x$, $\quad$ $0 \le x \le \pi/2$ (See Figure 27.3.)
(b) $f(x) = \cos x$, $\quad$ $0 \le x \le \pi$
(c) $f(x) = \arcsin x$, $\quad$ $0 \le x \le 1$ (See Figure 27.3.)
(d) $f(x) = \arctan x$, $\quad$ $0 \le x \le 1$ (See Figure 27.4.)

Section 28

THE EXPONENTIAL
AND LOGARITHMIC
FUNCTIONS

We now discuss the exponential functions and their inverse functions, namely, the logarithmic functions. We compute the various derivatives involved and then solve some problems dealing with derivatives, implicit differentiation, and curve tracing.

Let a be any positive real number. At present there is a meaning for a^x when x is any positive or negative rational number. For example, if x is a positive integer n, then

$$a^n = \overbrace{a \cdot a \cdot a \cdots a}^{n \text{ factors}}$$

while $a^{-n} = 1/a^n$; if x is a positive rational number p/q, then $a^{p/q} = \sqrt[q]{a^p}$ while $a^{-(p/q)} = 1/a^{p/q}$. However, we have not yet assigned a meaning to a^x when x is an irrational number. For example, $2^{\sqrt{2}}$, $2^{\sqrt{3}}$, $3^{\sqrt{5}}$, $(\sqrt{2})^{\sqrt{3}}$, $\pi^{\sqrt{2}}$, and 2^{π} have not been defined. We shall now define such numbers, using limits.

Definition 28.1 Let a be any positive real number. Let x be any irrational number and let $\{x_n\}$ be any sequence of *rational* numbers approaching x. Then we define a^x to be the limit of the sequence of numbers (a^{x_n}) (it turns out that the limit of such a sequence always exists).

For example, $2^{\sqrt{3}}$ may be defined as the limit of the sequence 2, $2^{1.7}$, $2^{1.73}$, $2^{1.732}$, $2^{1.73205}$, $2^{1.732051}$, ... (the limit is the irrational number $3.3201\cdots$) and 2^{π} may be defined as the limit of the sequence 2^3, $2^{3.1}$, $2^{3.14}$, $2^{3.141}$, $2^{3.1415}$, $2^{3.14159}$, ... (the limit is the irrational number $8.8199\cdots$).

175

Therefore, the meaning of a^x has now been extended to include irrational numbers x. Observe that if $a = 1$, the limit definition gives 1^x as the limit of the sequence $\{1^{x_n}\} = 1, 1, 1, \ldots$, which is, of course, 1 (since $1^{x_n} = 1$ for any rational number x_n). It can also be shown that the ordinary rules for exponents, such as $a^x a^z = a^{x+z}$ and $(a^x)^z = a^{xz}$, hold for both rational and irrational numbers x and z. As an exercise, using the limit definition of a^x, prove that $a^x a^z = a^{x+z}$ in the case when x is irrational and z is rational and in the case when *both* x and z are irrational.

Suppose we now consider the general exponential function $f(x) = a^x$ for any positive real number a. It can be shown that $\lim_{h \to 0} a^h = 1$, so this function is continuous at $x = 0$. But $\lim_{h \to 0} a^{x+h} = \lim_{h \to 0} (a^x a^h) = a^x \lim_{h \to 0} a^h = a^x \cdot 1 = a^x$; therefore, the function $f(x) = a^x$ is also continuous for all real x (see Problem 20.9). The graph of such a function for $a > 1$ looks like the one in Figure 28.1. This function is clearly increasing over its entire domain $-\infty < x < \infty$; consequently, it is a one-to-one function. Therefore, it has an inverse function f^*, which we will denote by $\log_a$. Then $\log_a(y) = x$ if and only if $y = a^x$. Since the range of $f(x) = a^x$ is the set of all positive real numbers, the domain of the function $\log_a$ is the set of all positive real numbers, while the range of $\log_a$ is the set of all real numbers. Using the method explained

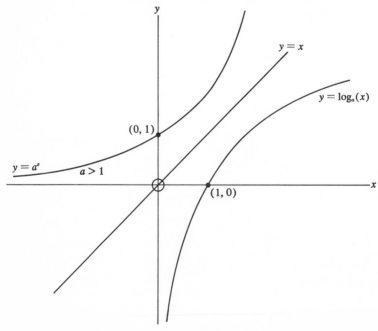

Figure 28.1

in Figure 17.2, we obtain the graphs of both functions in Figure 28.1.   As an exercise, graph the function $f(x) = a^x$ for $0 < a < 1$ and verify that it is a one-to-one function that is decreasing over its entire domain $-\infty < x < \infty$.

The graph of $f(x) = a^x$ for $a > 1$ seems to have a tangent line with a positive slope at each point and, in particular, at $(0, 1)$ [note that for any positive value of a, the point $(0, 1)$ is on the graph since $a^0 = 1$].   In fact, as a increases, the tangent lines at $(0, 1)$ are steeper and their slopes become larger.   For example, compare the tangent lines to the graphs of $y = 2^x$ and $y = 10^x$ at $(0, 1)$ (see Figure 13.10).   Hence, this function apparently has a positive derivative for each value of x including $x = 0$.   Suppose we let K represent the value of the derivative of $f(x) = a^x$ at $x = 0$.   Then, $K = \lim_{h \to 0} [(a^h - 1)/h]$, and for arbitrary x,

$$f'(x) = \lim_{h \to 0} \left(\frac{a^{x+h} - a^x}{h} \right) = a^x \lim_{h \to 0} \left(\frac{a^h - 1}{h} \right) = a^x \cdot K.$$

Consequently, the derivative of the function $f(x) = a^x$ is just a constant K times the function itself where $K = f'(0)$.   Now, if we can find a value of a for which $K = 1$, we will have found a *particular* exponential function having its derivative *equal* to itself for all values of x.

Letting $a = 2$, we have $K = \lim_{h \to 0}[(2^h - 1)/h]$, and letting h approach zero through the values $\frac{1}{2}, \frac{1}{4}, \frac{1}{8}, \frac{1}{16}, \ldots, \dfrac{1}{2n}, \ldots$, we obtain K as the limit of the sequence

$$\frac{2^{(1/2)} - 1}{(1/2)} \simeq 0.83, \quad \frac{2^{(1/4)} - 1}{(1/4)} \simeq 0.76, \quad \frac{2^{(1/8)} - 1}{(1/8)} \simeq 0.72, \quad \frac{2^{(1/16)} - 1}{(1/16)} \simeq 0.70, \ldots$$

The limit of this sequence is the irrational number $0.69315\cdots$, so K is *less* than 1 for $a = 2$.

Letting $a = 3$, we have $K = \lim_{h \to 0}[(3^h - 1)/h]$ and letting h again approach zero through the values $\frac{1}{2}, \frac{1}{4}, \frac{1}{8}, \frac{1}{16}, \ldots, \dfrac{1}{2n}, \ldots$, we obtain K as the limit of the sequence

$$\frac{3^{(1/2)} - 1}{(1/2)} \simeq 1.46, \quad \frac{3^{(1/4)} - 1}{(1/4)} \simeq 1.26, \quad \frac{3^{(1/8)} - 1}{(1/8)} \simeq 1.16, \quad \frac{3^{(1/16)} - 1}{(1/16)} \simeq 1.12, \ldots$$

The limit of this sequence is the irrational number $1.09861\cdots$; thus, K is *greater* than 1 for $a = 3$.   Now it can be shown that if a varies continuously from 2 to 3, K will vary continuously from $0.69315\cdots$ to $1.09861\cdots$ and there is a value of a for which $K = 1$.   If we denote this value by e, then $2 < e < 3$ and $\lim_{h \to 0}[(e^h - 1)/h] = 1$.   Hence, for small values of h, $(e^h - 1)/h \simeq 1$.

Therefore, $e^h - 1 \simeq h$, $e^h \simeq 1 + h$, and finally, $e \simeq (1 + h)^{1/h}$. Letting h approach zero through the values $1/n$ for $n = 1, 2, 3, 4, \ldots$, we obtain

$$e = \lim_{n \to \infty} \left(1 + \frac{1}{n}\right)^n$$

which is the usual definition of e. The first few terms of this sequence are 2, $(\frac{3}{2})^2 = 2.25$, $(\frac{4}{3})^3 \simeq 2.37$, $(\frac{5}{4})^4 \simeq 2.44$, $\ldots$ and it can be shown that the limit of this sequence is the irrational number $2.71828183 \cdots$.

As a result of the preceding discussion, we have the following Theorem.

Theorem 28.1   If $f(x) = e^x$ where e is the value of a for which $\lim_{h \to 0}[(a^h - 1)/h]$ $= 1$, then $f'(x) = e^x$; that is, the function $f(x) = e^x$ has its derivative equal to itself for all values of x.

A GEOMETRICAL APPROACH TO THE DEFINITION OF e

The real number e is *the* value of a in that exponential function $f(x) = a^x$ that yields a graph with the property that the slope of the tangent line to the graph at $(0, 1)$ is precisely equal to the number 1. Observe that if we choose $a = 2$, the slope of the tangent line to the graph of $y = 2^x$ at $(0, 1)$ is approximately equal to $(2^h - 1)/h$ for small h and, from the previous discussion, this number is close to 0.69, which is less than 1. If we choose $a = 3$, the slope of the tangent line to the graph of $y = 3^x$ at $(0, 1)$ is approximately equal to $(3^h - 1)/h$ for small h and, from the previous discussion, this number is close to 1.09, which is greater than 1. If we begin with the graph of $y = a^x$ for $a = 2 + \varepsilon$ where ε is a real number with $0 < \varepsilon < 1$, the tangent line at $(0, 1)$ will have slope m where $0.69 < m < 1.09$. Then if we let ε vary continuously from 0 to 1, it can be shown that the tangent lines to the graphs of $y = (2 + \varepsilon)^x$ at $(0, 1)$ have slopes m that vary continuously between 0.69 and 1.09. Hence, there must be a value of ε, say, $\varepsilon = \varepsilon_0$, that yields a slope m equal to 1 (the required value of ε is the irrational number $\varepsilon_0 = 0.71828183 \cdots$). We then define e to be $2 + \varepsilon_0$ and the exponential function $f(x) = e^x$ has the property that the slope of the tangent line to its graph at $(0, 1)$ is precisely 1.

We can now show that if $f(x) = e^x$, then $f'(x) = e^x$ (Theorem 28.1). Since the slope of the tangent to the graph of $f(x) = e^x$ at $(0, 1)$ is 1, the derivative of $f(x) = e^x$ at $x = 0$ must be 1. But by the definition of the derivative, $f'(0) = \lim_{h \to 0}[(e^h - 1)/h]$. Therefore, $\lim_{h \to 0}[(e^h - 1)/h] = 1$. Now, to compute $f'(x)$, we look at

$$\lim_{h \to 0} \frac{e^{x+h} - e^x}{h} = \lim_{h \to 0} e^x \frac{(e^h - 1)}{h} = e^x \lim_{h \to 0} \frac{e^h - 1}{h} = e^x \cdot 1 = e^x$$

and we are done.

Definition 28.2 The particular inverse function $\log_e(x)$ of the exponential function $f(x) = e^x$ is usually denoted by $\ln x$ or $\log x$ *without* a subscript, and it is called the *natural logarithm* function. Hence log without a subscript always means the *inverse* function of $f(x) = e^x$; when we discuss the inverse function of an *arbitrary* exponential function a^x in this book we always write $\log_a(x)$.

We assume that you are familiar with the elementary properties of the log function, such as $\log(ab) = \log a + \log b$ and $\log(a/b) = \log a - \log b$ for all positive real numbers a and b, while $\log(a^b) = b \log a$ for any real number b and any positive real number a. Also, since the exponential and log functions are inverses, for any positive real number r we have $r = e^{\log r}$, and for any real number s we have $\log e^s = s$.

Theorem 28.2 The derivative of the log function at any point y in its domain is simply $1/y$. $\quad$ *inverse*

PROOF If $f(x) = e^x = y$, then $f^*(y) = \log y = x$. Since $f'(x) = e^x$, Theorem 27.1 implies that $(\log y)' = 1/f'(x) = 1/e^x = 1/y$. Rewriting this relation in terms of x results in the formula $(\log x)' = 1/x$. ∎

REMARK By using the chain rule, we may extend the results of Theorems 28.1 and 28.2 to functions of the type $F(x) = e^{f(x)}$ and $G(x) = \log f(x)$, obtaining the derivative formulas $(e^{f(x)})' = e^{f(x)}f'(x)$ and $(\log f(x))' = f'(x)/f(x)$.

EXAMPLE 28.1 Compute the derivative of each of the functions (a) e^{x^2}, (b) $e^{\cos x}$, (c) $\log(\sin x)$, (d) $\log(\sin e^{2x})$.

SOLUTION (a) $(e^{x^2})' = e^{x^2}(x^2)' = e^{x^2}(2x)$ (b) $(e^{\cos x})' = (e^{\cos x})(\cos x)' = e^{\cos x}(-\sin x)$ (c) $(\log(\sin x))' = (1/\sin x)(\sin x)' = \cos x/\sin x = \cot x$ (d) $(\log(\sin e^{2x}))' = (1/\sin e^{2x})(\sin e^{2x})' = (1/\sin e^{2x})(\cos e^{2x})(e^{2x})' = (1/\sin e^{2x})(\cos e^{2x})(e^{2x})(2)$.

We now prove a theorem that has very wide application in the computation of derivatives of functions that are composed of exponential and other functions.

Theorem 28.3 Suppose $h(x) = (f(x))^{g(x)}$ where $f(x)$ is a differentiable function with $f(x) > 0$ for all x and where $g(x)$ is any differentiable function of x. Then $h'(x) = h(x)[g'(x)\log f(x) + g(x)f'(x)/f(x)]$.

PROOF For simplicity we write f, g, and h for $f(x)$, $g(x)$, and $h(x)$. Since $f > 0$, $f^g = h$ must also be greater than zero for all x. Taking the log of both sides, we obtain $\log h = \log f^g = g \log f$. Differentiating both sides, we have $h'/h = g' \log f + gf'/f$, which leads to $h' = h(g' \log f + gf'/f)$, and the proof is completed. ∎

EXAMPLE 28.2   Prove that the derivative of the function $h(x) = x^r$ is rx^{r-1} for *every real number r.*   (Note that we may now choose r irrational.)

SOLUTION   Apply the preceding theorem with $f(x) = x$, $g(x) = r$, $f'(x) = 1$, and $g'(x) = 0$.

EXAMPLE 28.3   Compute the derivative of each of the following functions.

    (a) $h(x) = a^x$ a any positive real number
    (b) $h(x) = 2^{x^2}$
    (c) $h(x) = x^x$ $x > 0$
    (d) $h(x) = (\cos x)^{\sin x}$, $0 \le x \le \pi/2$

SOLUTION (a)   Apply the preceding theorem with $f(x) = a$, $g(x) = x$, $f'(x) = 0$, and $g'(x) = 1$.   We obtain $(a^x)' = a^x[1(\log a) + (x)(0)] = a^x \log a$.

(b)   Apply the preceding theorem with $f(x) = 2$, $g(x) = x^2$, $f'(x) = 0$, and $g'(x) = 2x$.   We obtain $(2^{x^2})' = 2^{x^2}[2x \log 2 + (x^2)(0)] = (2 \log 2) \times (x)(2^{x^2})$.

(c)   Let $f(x) = x$, $g(x) = x$, $f'(x) = 1$, and $g'(x) = 1$; then $(x^x)' = x^x[(1)(\log x) + (x)(1/x)] = x^x(\log x + 1)$.

(d)   Let $f(x) = \cos x$, $g(x) = \sin x$, $f'(x) = -\sin x$, and $g'(x) = \cos x$; then $(\cos x)^{\sin x} = (\cos x)^{\sin x}[(\cos x)(\log \cos x) + (\sin x)(-\sin x/\cos x)]$.

EXAMPLE 28.4   Find the derivative of the function $y = f(x)$ defined *implicitly* by the equation $e^y + x \sin y = 0$.

SOLUTION   Substituting $f(x)$ for y yields the equation $e^{f(x)} + x \sin f(x) = 0$.   Differentiating, we obtain $e^{f(x)}f'(x) + \sin f(x) + x[\cos f(x)]f'(x) = 0$. Solving for $f'(x)$, we have

$$f'(x) = -\frac{\sin f(x)}{e^{f(x)} + x \cos f(x)} = -\frac{\sin y}{e^y + x \cos y}$$

EXAMPLE 28.5   Find the critical values of $f(x) = e^{-x^2}$ and sketch the graph.

SOLUTION $f'(x) = -2xe^{-x^2}$ and

$$f''(x) = -2e^{-x^2} + 4x^2 e^{-x^2} = 2e^{-x^2}(2x^2 - 1).$$

Setting $f'(x) = 0$, we obtain $x = 0$ as the only critical value, and since $f''(0) < 0$, the graph has a relative maximum at $(0, 1)$. Setting $f''(x) = 0$, we obtain the two inflection points $(1/\sqrt{2}, 1/\sqrt{e})$ and

$(-1/\sqrt{2}, 1/\sqrt{e})$. Since there are no other inflection points, the graph must be concave down between $x = -1/\sqrt{2}$ and $x = 1/\sqrt{2}$ and concave up at all points outside this interval. A sketch appears in Figure 28.2.

EXAMPLE 28.6 Find the critical values of the function $f(x) = e^{-1/x}$, $x \neq 0$, and $f(0) = 0$. Sketch the graph.

SOLUTION $f'(x) = (1/x^2)e^{-1/x}$, which is never equal to zero. Therefore, f has no points on its graph that are *relative* maximum or minimum points. Now, $f''(x) = (1/x^4)e^{-1/x} - (2/x^3)e^{-1/x} = e^{-1/x}(1/x^4 - 2/x^3)$. Setting $f''(x) = 0$, we obtain $1/x^4 = 2/x^3$ and this implies that $1/x = 2$ and finally, $x = \frac{1}{2}$. Therefore, the graph of f has an inflection point at $(\frac{1}{2}, 1/e^2)$. For all nonzero values of $x < \frac{1}{2}$, $f''(x) > 0$ and so the graph is concave up. For $x > \frac{1}{2}$, $f''(x) < 0$ and the graph is concave down. Observe also that as $x \to 0$ through positive values, $f(x) \to 0$; however, as $x \to 0$ through negative values, $f(x) \to +\infty$. Consequently, f is *not* continuous at $x = 0$ and hence cannot be differentiable there. As $x \to \infty$, $-1/x \to 0$ through negative values and therefore $f(x) \to e^0 = 1$ through values less than 1. As $x \to -\infty$, $-1/x \to 0$ through positive values, and therefore $f(x) \to e^0 = 1$ through values greater than 1. A sketch appears in Figure 28.3.

As a final remark regarding the function $f(x) = e^{-1/x}$, we note that as $x \to 0$ through positive values, the tangent lines to the graph become more and more horizontal. In fact, it can be shown by analytic methods that $f(x)$ has *infinitely many right-hand derivatives* at $x = 0$ and they are all equal to zero. [This follows from the fact that every derivative of $f(x)$ is composed of terms of the type $e^{-1/x}/x^n$ where n is a positive integer (consider f' and f'' above).

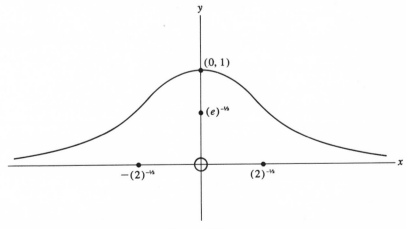

Figure 28.2

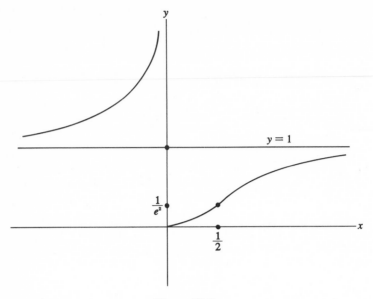

Figure 28.3

Hence, as $x \to 0$ through positive values, the numerator goes to zero much faster than the denominator; thus the limit of the quotient will be zero.] As $x \to 0$ through negative values, we see that the tangent lines are becoming more and more vertical.   Therefore, $f(x)$ does not possess even *one* left-hand derivative at $x = 0$.

EXAMPLE 28.7   Find the critical values of $f(x) = x^{1/2} \log x$ for $x > 0$, $f(0) = 0$.   Sketch the graph.

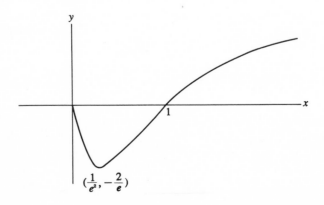

Figure 28.4

SOLUTION $f'(x) = \frac{1}{2}x^{-1/2} \log x + x^{1/2}(1/x) = \frac{1}{2}x^{-1/2} \log x + x^{-1/2} = x^{-1/2}[\frac{1}{2}\log x + 1]$. Now, using the product formula, we obtain

$$f''(x) = -\frac{1}{4}x^{-3/2}[\frac{1}{2} \log x + 1] + x^{-1/2}[\frac{1}{2}(1/x)] = -\frac{1}{4}x^{-3/2}[\frac{1}{2} \log x].$$

Setting $f'(x) = 0$, we obtain $\frac{1}{2} \log x + 1 = 0$, which implies $\log x = -2$ and therefore, $x = e^{-2}$. The corresponding y coordinate is $y = -2e^{-1}$. Since $f''(e^{-2}) > 0$, f has a relative minimum at $(e^{-2}, -2e^{-1})$. Setting $f''(x) = 0$, we obtain $\log x = 0$, which implies $x = 1$. The corresponding y coordinate is 0; therefore, the graph has an inflection point at $(1, 0)$. Since $f''(x) < 0$ for $x > 1$, the graph is concave down for $x > 1$ and concave up for x between 0 and 1. A sketch appears in Figure 28.4. Observe that f has a vertical tangent at $(0, 0)$ and so is not differentiable at $x = 0$. Also, f is continuous at $x = 0$, since for any sequence of points $\{x_n\}$ approaching the origin through positive values, the corresponding $\{y_n\}$ sequence approaches 0, the functional value at $x = 0$. Since f is not defined for $x < 0$, we need not concern ourselves with points to the left of the origin.

We now conclude this section with an example that involves the product of an exponential and a trigonometric function.

EXAMPLE 28.8 Find the critical values and sketch the graph of the function $f(x) = e^{-x} \sin x$, $0 \le x \le 2\pi$.

SOLUTION $f'(x) = -e^{-x} \sin x + e^{-x} \cos x = e^{-x}(\cos x - \sin x)$ and

$$f''(x) = -e^{-x}(\cos x - \sin x) + e^{-x}(-\sin x - \cos x) = -e^{-x}(2 \cos x).$$

Setting $f'(x) = 0$, we obtain $\cos x = \sin x$, which occurs only at $x = \pi/4$ and $x = 5\pi/4$. Since $f''(\pi/4) < 0$, f has a relative maximum at $x = \pi/4$ and since $f''(5\pi/4) > 0$, f has a relative minimum at $x = 5\pi/4$. Setting $f''(x) = 0$, we obtain $\cos x = 0$, which occurs at $x = \pi/2$ and $x = 3\pi/2$. Therefore, the graph has inflection points at $x = \pi/2$ and $x = 3\pi/2$. A sketch appears in Figure 28.5.

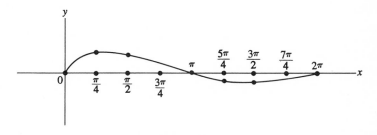

Figure 28.5

PROBLEMS

28.1   Find all the critical values and sketch the graphs of each of the following functions.
(a) xe^{-x}
(b) xe^{-x^2}
(c) $x^2 e^{-x}$
(d) $xe^{-1/x}$
(e) e^{x^2}
(f) $e^{-x^2/2}$
(g) $x - e^x$

28.2*   Let $f(x) = e^{-1/x^2}$ for $x \neq 0$, $f(0) = 0$ (see Example 28.6).
(a) Prove that f has infinitely many derivatives at $x = 0$ and that they are all equal to zero.
(b) Find all the inflection points.
(c) Sketch the graph.

28.3*   (a) Prove that the equation $e^x = kx$ has at least one solution for every real number $k \geq e$.
(b) Show that the equation $e^x - 1 = (e - 1)x$ has two and only two real roots.

28.4   Let a be any real number greater than 1.   Find the critical values of the function $f(x) = x^2/a^x$.

28.5*   Let $f(x)$ be a differentiable function over some interval J with the property that $f'(x) = Af(x)$ for some constant A.   Show that $f(x) = Be^{Ax}$ for some constant B.   *Note* that this means there is essentially only one type of function that has a derivative equal to a constant times itself, namely, an exponential function.

28.6   Let $\sinh x = (e^x - e^{-x})/2$ and $\cosh x = (e^x + e^{-x})/2$.   These functions are called the *hyperbolic sine* and the *hyperbolic cosine* functions, respectively.
(a) Prove that $(\sinh x)' = \cosh x$ and $(\cosh x)' = \sinh x$.
(b) Prove that $\cosh^2 x - \sinh^2 x = 1$.
(c) Sketch the graphs of $\sinh x$ and $\cosh x$.

28.7*   Let $\tanh x = \sinh x/\cosh x$; this function is called the *hyperbolic tangent* function.
(a) Find $(\tanh x)'$.
(b) Sketch the graph of $f(x) = \tanh x$.
(c) Sketch the graph of $f(x) = \tanh(1/x)$.
(d) Sketch the graph of $f(x) = x \tanh(1/x)$.

28.8   Find all the critical values and sketch the graph of each of the following functions.
(a) $x - \log x$
(b) $e^x - \sin x$
(c) $e^{-x} \cos x$
(d) $e^x + \cos x$

28.9   Find the slope of the tangent line to the graph of each of the given equations at the point indicated (use implicit differentiation; see Section 24).
(a) $xe^y + \cos(xy) = 2$ $(1, 0)$
(b) $e^y + x \cos y = 0$ $(-1, 0)$
(c) $x \log y + ye^x = 1$ $(0, 1)$
(d) $ye^x - x^2 \sin y = \pi$ $(0, \pi)$

28.10   Differentiate each of the following functions.
  (a) $\log x/\sin x$
  (b) $x^2 \sin e^{2x}$
  (c) $e^x \sin(5e^x)$
  (d) $\log(\cos x)$
  (e) $\log(\cos e^x)$
  (f) $e^x \cos x$
  (g) $\arcsin (1 - x^2)^{1/2}$
  (h) $\log(\sin(\arctan e^{x^2})^{1/2})$
  (i) $\sin(\arctan x^2)^{1/2}$
  (j) $e^{x \cos x}$
  (k) $e^{\sin(\log x)}$
  (l) $(\sin x)^{\cos x}$
  (m) $(\sin x)^{e^x}$
  (n) $(x^2 + x)^{x^2}$
  (o) x^{x^2}
  (p)* x^{x^x}
  (q)* $x^{x^{x^2}}$

28.11*   Find the slope of the tangent line to the graph of each of the following functions at the points indicated.
  (a) $f(x) = x^{x^2}$ at the points $(1, 1)$ and $(2, 16)$
  (b) $f(x) = \sin(\log(\arctan e^{x^2}))$ at $x = 0$

28.12*   (a)   Let $f(x)$ be implicitly defined by the equation $xe^y y^2 - 1 = 0$. Find $f'(x)$.
  (b)   Evaluate f' at $x = 1/e$.

28.13   Given the following functions and intervals, find the point or points that may be used for c in the mean value theorem (see Example 25.7).
  (a) $f(x) = e^x$, $0 \leq x \leq 1$
  (b) $f(x) = \log x$, $1 \leq x \leq e$
  (c) $f(x) = 10^x$, $0 \leq x \leq 1$

28.14*   (a)   Find all the critical points and plot the graph of the function $f(x) = x/(1 + e^{1/x})$, $x \neq 0$ and $f(0) = 0$
  (b)   Discuss the continuity and differentiability of the function of part (a) at the point $(0, 0)$.

28.15*   Find a real number a such that the graph of $f(x) = a^x$ is tangent to the line $y = x$ at some point.

28.16*   Prove that the limit of the sequence $\{(n!/n^n)^{1/n}\}$ is $1/e$.

L'HOSPITAL'S RULE

In many cases, we are interested in the limit as x approaches a number a of the quotient of two functions $f(x)$ and $g(x)$. However, we cannot always apply the theorem that states that the limit of a quotient of two functions equals the quotient of the individual limits. For example, we have seen that $\lim_{x \to 0}((\sin x)/x) = 1$, yet the quotient of the individual limits yields the indeterminant form $0/0$. In situations where x approaches infinity, we can sometimes guess the limit of a quotient of two functions by inspection. For example, as $x \to \infty$, $((\log x)/x) \to 0$ since for large x, $\log x$ is relatively small; hence the quotient has a denominator that becomes larger much faster than the numerator. Note however, that the quotient of the individual limits yields another indeterminant form ∞/∞.

The following rule gives us a systematic method for computing the limit of a quotient in certain cases when the quotient of the individual limits yields an indeterminant form like $0/0$ or ∞/∞.

Theorem 29.1 *L'Hospital's Rule* CASE 1 Let f and g be differentiable in a deleted neighborhood of the point a with $g'(x) \neq 0$ in this neighborhood. Suppose that $\lim_{x \to a} f(x) = \lim_{x \to a} g(x)$ with the common limit being zero, $+\infty$, or $-\infty$. Then $\lim_{x \to a} [f(x)/g(x)] = \lim_{x \to a} [f'(x)/g'(x)]$, provided that the latter limit exists. Also, the limit as x approaches a may be restricted to either the left-hand limit or the right-hand limit, depending on the circumstances.

CASE 2 If in Case 1 we take $a = +\infty$ or $-\infty$, and f and g are differentiable for all x as $x \to +\infty$ or $-\infty$, then under the same conditions as in Case 1 we have that $\lim_{x \to \infty} [f(x)/g(x)] = \lim_{x \to \infty} [f'(x)/g'(x)]$, provided that the latter limit exists. (The similar statement for $x \to -\infty$ is also valid.)

186

Observe that l'Hospital's rule states that under certain conditions, the limit of a quotient of two functions equals the limit of the quotient of the individual derivatives. Geometrically, this means that as $x \to a$, the quotients of the corresponding y coordinates have the same limit as the quotients of the slopes of the corresponding tangent lines to the graphs of f and g.

EXAMPLE 29.1   Find each of the limits (a) $\lim_{x \to 0} (x - \sin x)/2x$ (b) $\lim_{x \to \infty} (x - \sin x)/2x$.

SOLUTION (a)   As $x \to 0$, both the numerator and the denominator approach zero. Differentiating, we obtain the quotient $(1 - \cos x)/2$. As $x \to 0$, $(1 - \cos x) \to 0$; hence, the quotient approaches $0/2 = 0$.

SOLUTION (b)   As $x \to \infty$, both the numerator and the denominator approach ∞. However, the quotient of the derivatives equals $\frac{1}{2} - \frac{1}{2} \cos x$. If x approaches infinity through the numbers $\{k(\pi/2)\}$ for $k = 1, 2, 3, \ldots$, $\cos x$ takes on the values $0, -1, 0, 1$ over and over. Hence, $\frac{1}{2} \cos x$ takes on the values $0, -\frac{1}{2}, 0, \frac{1}{2}$ while $\frac{1}{2} - \frac{1}{2} \cos x$ takes on the values $\frac{1}{2}, 1, \frac{1}{2}, 0$. Therefore, the quotient of the individual derivatives does *not* possess a limit as $x \to \infty$. However, the quotient of the original functions does possess the limit $\frac{1}{2}$ as $x \to \infty$. This follows from the fact that $(x - \sin x)/2x = \frac{1}{2} - \frac{1}{2}[(\sin x)/x]$ and as $x \to \infty$, $|\sin x| \leq 1$; hence, $(\sin x)/x \to 0$ and $(x - \sin x)/2x \to \frac{1}{2}$. In fact, if we let $x \to 0$, we see that $(\sin x)/x \to 1$; hence, $\frac{1}{2} - \frac{1}{2}(\sin x)/x$ approaches $\frac{1}{2} - \frac{1}{2} = 0$. This coincides with the result of part (a) of this example. Finally, part (b) should make it clear that we cannot conclude anything from the fact that the quotient of the derivatives does *not* have a limit, since the original quotient of the two functions still may or may not have a limit. All that l'Hospital's rule allows us to do is to compute the limit of the original quotient whenever the limit of the quotient of the derivatives actually exists.

Note also that l'Hospital's rule may not be applied indiscriminately. We must make certain that the given quotient does in fact lead to an indeterminant form. For example, we may not evaluate $\lim_{x \to 1} [(3x + 1)/(x + 1)]$ by computing $\lim_{x \to 1} [(3x - 1)'/(x + 1)'] = \lim_{x \to 1} (3/1) = 3$, since the original quotient actually has limit 2 as x approaches 1. On the other hand, we cannot conclude that $\lim_{x \to \infty} [(\sin x)/x]$ does not exist because $\lim_{x \to \infty} [(\sin x)'/(x)'] = \lim_{x \to \infty} [(\cos x)/1]$, which does not exist. In this case, $(\sin x)/x \leq (1/x) \to 0$ as $x \to \infty$; therefore, the original quotient does not lead to an indeterminant form.

Often, when graphing a function, we find that l'Hospital's rule is very helpful. The following example relates to Example 28.7.

EXAMPLE 29.2   Find the limit of the function $\sqrt{x} \log x$ as x approaches zero through positive values.

SOLUTION   Rewrite the function in the form $(-1)[(-\log x)/(1/\sqrt{x})]$. Then as $x \to 0$, $x > 0$, we have $-\log x \to +\infty$ and $1/\sqrt{x} \to +\infty$. Differentiating, we obtain

$$\frac{(-1)(-1/x)}{(-\tfrac{1}{2})(x^{-3/2})} = (-2)\sqrt{x}$$

and as $x \to 0$, $(-2)\sqrt{x} \to 0$ through negative values. (See Figure 28.4.) Note that $\sqrt{x} \log x$ is the product of two functions, one of which goes to zero as $x \to 0$ while the other goes to $-\infty$. Hence, we have evaluated the indeterminant form $0(-\infty)$ by transforming the given function into an appropriate quotient. We repeat this technique in the following example.

EXAMPLE 29.3   Compute $\lim\limits_{x \to 0} (\sin x)(\log x)$ where x approaches zero through positive values.

SOLUTION   This is an example of a situation in which we have a product of two functions, one of which goes to zero while the other goes to $-\infty$. To find the limit of this product, we rewrite the function in the form $(-1)[(-\log x)/(1/\sin x)]$. Now, both numerator and denominator approach $+\infty$ as $x \to 0$. Differentiating, we obtain

$$\frac{(-1)(-1/x)}{(-1)(\cos x/\sin^2 x)} = \frac{(-1)(\sin^2 x)}{x \cos x} = (-1)\frac{\sin x}{x} \cdot \sin x \cdot \frac{1}{\cos x}$$

Now as $x \to 0$, $((\sin x)/x) \to 1$, $\sin x \to 0$, and $1/\cos x \to 1$. Therefore, the product has limit zero.

EXAMPLE 29.4   Compute

$$\lim_{x \to \pi/2} \frac{\log(\pi/2 - x)}{\tan(\pi + x)}$$

where x approaches $\pi/2$ through values less than $\pi/2$.

SOLUTION   Both numerator and denominator approach $-\infty$ as $x \to \pi/2$. Differentiating, we obtain

$$\frac{(-1)/[(\pi/2) - x]}{\sec^2(\pi + x)} = -\frac{\cos^2(\pi + x)}{(\pi/2) - x}$$

In this case, however, as $x \to \pi/2$, both numerator and denominator approach zero. Therefore, the quotient of the derivatives yields the indeterminant form $0/0$. To compute this limit, we may again use l'Hospital's rule. Differentiating, we obtain $[2 \cos(\pi + x) \sin(\pi + x)]/ -1$ and as $x \to \pi/2$, the numerator approaches 0 while the denominator remains -1. Hence, the limit is 0 and so the limit of the *original* quotient is 0 also.

The preceding example indicates that we can continue to differentiate whenever the quotient of the first derivatives again yields an indeterminant form. Finally, if the quotient of the second derivatives also yields an indeterminant form, we look at the quotient of the third derivatives, and so on, until one of the succeeding quotients yields a limit. Then this limit, by l'Hospital's rule, is equal to the limit of each of the preceding derivative quotients and hence is equal to the limit of the original quotient. Try this technique on the quotient e^x/x^5 as $x \to \infty$.

EXAMPLE 29.5   Compute $\lim\limits_{x \to \infty} x^{(1/x)}$.

SOLUTION   This problem leads to the indeterminant form ∞^0. In this type of problem, we rewrite the function in the form $e^{(1/x)\log x}$. We find $\lim\limits_{x \to \infty} ((\log x)/x)$ to be 0; hence, the original function has limit $e^0 = 1$ [see also Problem 29.2(h) and (i)].

In the preceding example, we evaluated the indeterminant form ∞^0 by again transforming the given function in such a way that l'Hospital's rule could be applied. There are other indeterminant forms that can be handled in the same way, such as 0^0 and 0^∞ [see Problem 29.1(g) and (h)].

PROBLEMS

29.1   In each of the following, evaluate (if possible) the limit of the given function as $x \to 0$.

(a) $\dfrac{e^{2x} - \cos x}{x}$

(b) $\dfrac{x^2 \cos x}{1 - e^x}$

(c) $xe^{1/x}, \quad x > 0$

(d) xe^{1/x^2}

(e) $\dfrac{\arcsin x}{x}$

(f) $\dfrac{\arctan x}{x}$

(g) x^x, $x > 0$

(h) $x^{1/x}$, $x > 0$

29.2   In each of the following, evaluate (if possible) the limit of the given function as $x \to \infty$

(a) $xe^{1/x}$, $x > 0$

(b) xe^{1/x^2}

(c) $\dfrac{\sin e^{-x}}{\sin(1/x)}$

(d) $x \sin(1/x)$

(e) $x^2 \sin(1/x)$

(f) x^m/e^x, m a positive integer

(g) e^x/x^m, m a positive integer

(h) $\sqrt[x]{x^2}$

(i) $\sqrt[x]{x^n}$, n a positive integer

29.3   In each of the following, evaluate (if possible) the limit of the given function as x approaches the given number.

(a) $\displaystyle\lim_{x \to \pi} \dfrac{\sin x}{x - \pi}$

(b) $\displaystyle\lim_{x \to \pi} \dfrac{\log(\cos 2x)}{(\pi - x)^2}$

(c) $\displaystyle\lim_{x \to \pi/2} (\sin x)^{\tan x}$

(d) $\displaystyle\lim_{x \to \pi/2} \dfrac{1 + \cos 2x}{1 - \sin x}$

(e) $\displaystyle\lim_{x \to \infty} \dfrac{3x^3 - 2x^2 + 7}{x^4 + 1}$

(f) $\displaystyle\lim_{x \to \infty} \dfrac{3x^3 - 2x^2 + 7}{x^2 + 1}$

29.4* Using l'Hospital's rule, prove the following result. If $P(x)$ and $Q(x)$ are polynomials of degree m and n, respectively, show that

$$\lim_{x \to \infty} \frac{P(x)}{Q(x)} = \begin{cases} \text{a nonzero constant, if } m = n \\ \infty, \text{ if } m > n \\ 0, \text{ if } m < n \end{cases}$$

29.5* Using the technique of Example 29.5, prove each of the following.

(a) $\displaystyle\lim_{x \to 0} (1 + \sin x)^{1/x} = e$

(b) $\displaystyle\lim_{x \to 0} (1 + \sin 2x)^{1/x} = e^2$

(c) $\displaystyle\lim_{x \to \infty} (1 + 1/x)^x = e$

(d) $\displaystyle\lim_{x \to \infty} (1 + k/x)^x = e^k$ for any real number k

29.6* Plot the graph of each of the following functions over the interval $0 < x < \infty$.
  (a) $f(x) = x^x$
  (b) $f(x) = x^{1/x}$
29.7* Discuss the continuity and the differentiability of each of the following functions at $x = 0$. Sketch the graphs.
  (a) $f(x) = \log x$, $x \neq 0$ and $f(0) = 0$
  (b) $f(x) = x \log x$, $x \neq 0$ and $f(0) = 0$
  (c) $f(x) = x^2 \log x$, $x \neq 0$ and $f(0) = 0$
  (d) $f(x) = x^3 \log x$, $x \neq 0$ and $f(0) = 0$
29.8* Repeat the preceding problem using the derivatives of the given functions.

Part III

THE INTEGRAL CALCULUS

ANTIDERIVATIVES

In this section, we discuss the antiderivatives of certain algebraic and trans-cendental functions. We also give some simple applications of the antideriva-tive, some of which include the solution to certain problems on the motion of a particle.

Definition 30.1 A function G is called an *antiderivative* of the function f if $G'(x) = f(x)$ for every x in the domain of f. G is also called an *indefinite integral* of the function f.

For example, the function $G(x) = x^3$ is an antiderivative of $f(x) = 3x^2$, since $G'(x) = 3x^2$. Also note that the functions $H(x) = x^3 + 5$ and $K(x) = x^3 - 7$ are antiderivatives of $f(x) = 3x^2$, since $H'(x) = K'(x) = f(x)$. Con-sequently, there are *many* functions that can be called an antiderivative or an indefinite integral of the function $f(x)$. However, each pair of these anti-derivatives must differ by a constant. For suppose that $G(x)$ and $H(x)$ are antiderivatives of $f(x)$. Then the function $H(x) - G(x)$ has a derivative equal to $H'(x) - G'(x) = f(x) - f(x) = 0$. Therefore, $H(x) - G(x)$ must equal a constant C. (See Problem 25.13.) Hence, once we have *one* anti-derivative $G(x)$ of a function $f(x)$, we may express any other antiderivative $H(x)$ of $f(x)$ in the form $H(x) = G(x) + C$ where C is some constant. Then we call the expression $G(x) + C$ *the* antiderivative or *the* indefinite integral of $f(x)$ where $G(x) + C$ represents the set of all antiderivatives of $f(x)$. Hence, *an* indefinite integral is *one* function, while *the* indefinite integral is a *collection* of functions. Therefore, in the preceding example, since we may express every antiderivative of $f(x) = 3x^2$ in the form $x^3 + C$, we call $x^3 + C$ *the* antiderivative or *the* indefinite integral of $3x^2$.

Finally, we see that even though the *derivative* of a given function is *unique*, an antiderivative of a function $f(x)$ may be any one of *infinitely many* func-tions, all of which differ by a constant C from a fixed function $G(x)$.

Theorem 30.1   If k is any constant and r is any *real* number except -1, an antiderivative of kx^r is given by $[k/(r + 1)]x^{r+1}$.

The proof follows from the fact that the derivative of the latter function is kx^r (see the result in Example 28.2).

EXAMPLE 30.1   Using the preceding theorem, we find that the following functions on the left have antiderivatives equal to the functions on the right.

(a) x^2 $x^3/3$
(b) $x^{(1/2)}$ $\frac{2}{3}x^{(3/2)}$
(c) x^{-3} $-\frac{1}{2}x^{-2}$
(d) x^π $x^{\pi+1}/(\pi + 1)$

We now give some simple results concerning antiderivatives.   They follow from the corresponding results on derivatives.

Theorem 30.2   (a)   An antiderivative of a sum of functions is the sum of antiderivatives of the functions; that is, if $F(x)$ and $G(x)$ are antiderivatives of $f(x)$ and $g(x)$, respectively, then an antiderivative of $h(x) = f(x) + g(x)$ is $H(x) = F(x) + G(x)$.

(b)   An antiderivative of a constant times a function is the constant times an antiderivative of the function; that is, if $F(x)$ is an antiderivative of $f(x)$, then for any constant k, $kF(x)$ is an antiderivative of $kf(x)$.

EXAMPLE 30.2   The preceding results together with Theorem 30.1 allow us to compute the antiderivatives of each of the following functions on the left, obtaining the expressions on the right.

(a) $4x^3 - 3x^2 + 2x + 1$ $x^4 - x^3 + x^2 + x + C$

(b) $x^7 - x^5 + x^3 - x$ $\dfrac{x^8}{8} - \dfrac{x^6}{6} + \dfrac{x^4}{4} - \dfrac{x^2}{2} + C$

(c) $3x^\pi + x^2 + x^{(1/2)}$ $[3/(\pi + 1)]x^{\pi+1} + (x^3/3) + \frac{2}{3}x^{(3/2)} + C$

Every *differentiation* formula gives rise to a corresponding *antidifferentiation* formula.   In the following example, we mention a few of these formulas.

EXAMPLE 30.3   The following list consists of certain transcendental functions along with their antiderivatives.   The verification once again involves the differentiation of the functions on the right.

Functions	Antiderivatives
(a) $\sin x$	$-\cos x + C$
(b) $\cos x$	$\sin x + C$
(c) e^x	$e^x + C$

(d) $1/x$ $x > 0$ $\log x + C$
(e) $(1 - x^2)^{-1/2}$ $|x| < 1$ $\arcsin x + C$
(f) $1/(1 + x^2)$ $\arctan x + C$

EXAMPLE 30.4   There are certain derivative formulas that give rise to very useful antidifferentiation formulas.   The following is a list of some of them.

(a) By the chain rule, the derivative of a function $g(x) = (f(x))^{n+1}/(n + 1)$ is given by $g'(x) = (f(x))^n f'(x)$ where n is any real number except -1.   Therefore, the corresponding antiderivative formula indicates that the function $(f(x))^n f'(x)$ has its antiderivative given by

$$(f(x))^{n+1}/(n + 1) + C.$$

For example, the antiderivative of $(\sin^5 x)(\cos x) = (\sin^6 x)/6 + C$, the antiderivative of $(x^2 + 1)^7 (2x) = (x^2 + 1)^8/8 + C$, and the antiderivative of $3x^2(x^3 + 1)^{1/2} = \frac{2}{3}(x^3 + 1)^{3/2} + C$.

(b) By the chain rule, the derivative of $\log f(x)$ is $f'(x)/f(x)$.   Therefore, the antiderivative of $f'(x)/f(x)$ is $\log f(x) + C$.   For example, the antiderivative of $\cos x/\sin x$ is $\log(\sin x) + C$, the antiderivative of $(2x + 1)/(x^2 + x + 2)$ is $\log(x^2 + x + 2) + C$.   Note that the antiderivative of $(2x + 1)/(x^2 + x + 2)^2$ is computed by the method of part (a) and is equal to $-(x^2 + x + 2)^{-1} + C$.

(c) By the chain rule, the derivative of $\sin f(x) = (\cos f(x))f'(x)$. Therefore, the antiderivative of $(\cos f(x))f'(x)$ is simply $\sin f(x) + C$. For example, the antiderivative of $5 \cos 5x$ is $\sin 5x + C$ and the antiderivative of $2x \cos x^2$ is $\sin x^2 + C$.

APPLICATIONS OF THE ANTIDERIVATIVE

EXAMPLE 30.5   Find the function $f(x)$ with the property that its derivative is $f'(x) = 4x - 3$ and its graph passes through the point $(1, 3)$.

SOLUTION   If $f'(x) = 4x - 3$, then $f(x)$ must have the form $2x^2 - 3x + C$. But if $(1, 3)$ is on the graph, then $f(1) = 3$.   Therefore, we substitute $x = 1$ into the expression $2x^2 - 3x + C$, obtaining $2 - 3 + C = 3$. Hence, $C = 4$ and $f(x) = 2x^2 - 3x + 4$.

EXAMPLE 30.6 Construct a polynomial function that has a relative maximum at $x = 2$, a relative minimum at $x = -2$, a nonhorizontal inflection point at $x = 0$, and y intercept $(0, 5)$.

SOLUTION   If a polynomial $P(x)$ has a relative maximum and minimum at $x = 2$ and $x = -2$, respectively, then $P'(x)$ must have roots $x = 2$ and $x = -2$.   Therefore, we may assume that $P'(x) = \pm (x - 2)(x + 2) = \pm(x^2 - 4)$ where the correct sign must now be determined.   If $x = 2$ is a relative maximum, then $P''(x) < 0$ when $x = 2$; hence, $\pm(2x) = \pm(4) < 0$ and therefore, the correct sign is the minus sign and $P'(x) = 4 - x^2$.   If we check $x = -2$ in $P''(x)$, we obtain $P''(-2) = -2(-2) > 0$, and therefore, $P(x)$ has a relative minimum at $x = -2$.   Since $P''(x) = -2x = 0$ only if $x = 0$, we obtain a nonhorizontal inflection point at $x = 0$. Finally, antidifferentiating, we obtain $P(x) = 4x - (x^3/3) + C$.   But the graph of $P(x)$ passes through $(0, 5)$; hence, $P(0) = C = 5$.   Therefore, one polynomial satisfying the given conditions is $P(x) = -x^3/3 + 4x + 5$. As an exercise find another polynomial of the third degree and one of higher degree, both satisfying the given conditions.   Finally, as another exercise, solve the preceding problem when $P(x)$ has a horizontal inflection point at $x = 0$.

EXAMPLE 30.7   Suppose a particle is at rest at the point $x = 1$ on a horizontal coordinate line.   If the particle moves along the line to the right with its acceleration at time t given by $a(t) = 2t + 1$, find the position function $s(t)$ (see Example 26.3).

SOLUTION   The antiderivative of $a(t)$ is $v(t) = t^2 + t + C$.   Since the particle starts at rest, its initial velocity is zero.   Therefore, $v(0) = C = 0$ and the velocity equation is $v(t) = t^2 + t$.   Now the antiderivative of $v(t) = s(t) = t^3/3 + t^2/2 + C$.   But the particle starts from the point $x = 1$; therefore, $s(0) = C = 1$.   Consequently, $s(t) = t^3/3 + t^2/2 + 1$ is the required position function.

EXAMPLE 30.8   Suppose an object is thrown upward from a roof that is 18 ft above the ground with an initial velocity of 12 ft/sec.   Find the position function $h(t)$ that gives the distance of the object from the ground at time t (see Example 26.4).

SOLUTION   Since the acceleration $a(t) = -32$ ft/sec/sec, the velocity $v(t) = -32t + C$.   But $v(0) = 12$; hence, $C = 12$ and $v(t) = -32t + 12$. The function $h(t) = -16t^2 + 12t + C$ and since $h(0) = 18$, we have $h(t) = -16t^2 + 12t + 18$.   This is the required equation.   As an exercise, determine the *time* at which the object strikes the ground and find its *speed* at that instant.

PROBLEMS

30.1   (a)   Using the chain rule, find the derivative of the function $e^{f(x)}$ and then compute the corresponding antiderivative formula.

      (b)   Find the antiderivative of each of the functions $5e^{5x}$; e^{-x}; xe^{x^2}; and $e^{\sin x} \cos x$.

30.2   (a)   Using the chain rule, find the derivative of the function $\cos f(x)$ and then compute the corresponding antiderivative formula.

    (b)   Find the antiderivative of each of the functions $x^2 \sin x^3$; $\sin 5x$; and $[\sin(\log x)]/x$.

30.3 (a)   Using the chain rule, find the derivative of the function arctan $f(x)$; then compute the corresponding antiderivative formula.

    (b)   Find the antiderivative of each of the functions $1/(1+4x^2)$; $2x/(1+x^4)$; and $1/(9+x^2)$.

30.4 In each of the following, compute the antiderivative of the given function.

(a) $x^4 - 3x^2 + 2x + 1$

(b) $x(x^2+1)^{1/2}$

(c) $x^2(x^3+1)^5$

(d) $\cos^4 x \sin x$

(e) $(3x^2+2x)/(x^3+x^2+1)$

(f) $(3x^2+2x)/(x^3+x^2+1)^2$

(g) $(3x^2+2x)/(x^3+x^2+1)^3$

(h) $\tan x$

(i) $\sec^2 x$

(j) $x \cos(2x^2)$

(k) $(1-4x^2)^{-1/2}$

(l) $(4-x^2)^{-1/2}$

(m) $(\log x)/x$

(n) $(\arctan x)/(1+x^2)$

(o) $e^{\sqrt{x}}/(2\sqrt{x})$

(p) $x^2(1-3x^3)^{-1/2}$

(q) $(x^3-5x+1)/\sqrt{x}$

30.5 (a)   Find the function $f(x)$ with the property that its derivative is $f'(x) = 3x^2 + 2x + 1$ and its graph passes through the point $(-1, 2)$.

    (b)   Find the function $f(x)$ with the property that its second derivative is $f''(x) = 2x + 1$ and the slope of the tangent line to its graph at the point $(-1, 0)$ is equal to 2.

30.6 (a)   Construct a polynomial function that has a relative maximum at $x = 1$, a relative minimum at $x = -1$, a *nonhorizontal* inflection point at $x = 0$, and y intercept $(0, 2)$.

    (b)   Repeat part (a) with a *horizontal* inflection point at $x = 0$.

    (c)   Construct a polynomial function that has a *horizontal* inflection point at $x = -2$ and y intercept $(0, 5)$.

    (d)   Construct a polynomial that has a *horizontal* inflection point at $x = 0$, a *nonhorizontal* inflection point at $x = 2$, a relative maximum at $x = 3$, and y intercept $(0, 1)$.

30.7 (a)   Suppose a particle is at rest at the origin on a horizontal coordinate line. If the particle moves along the line to the right with its acceleration at time t given by $a(t) = 3t^2 - 2t - 1$, find the position function $s(t)$.

    (b)   Repeat part (a) when the particle is propelled to the right at the rate of 1 ft/sec.

    (c)   In parts (a) and (b), determine the times and the positions at which the particle reverses its direction.

30.8 (a)   Suppose an object is propelled upward from a point 100 ft above the ground with an initial velocity of 60 ft/sec. Find the position function $h(t)$.

    (b)   Repeat part (a) when the object is propelled downward at the same speed.

(c)   In parts (a) and (b), find the time at which the object strikes the ground and its speed at that instant.

30.9   A particle moves along a horizontal coordinate line with a constant acceleration of K ft/sec/sec.   If the particle starts from rest at the origin and attains a speed of 40 ft/sec after traveling 100 ft to the right, find K and find the time at which the particle reaches the point $x = 100$.

30.10   Prove that if f is a function with the property that for some positive integer n, $f^{(n)}(x) = 0$ for all real x, then f is a polynomial function [see Problems 24.6 and 25.13(a)].

THE DEFINITE INTEGRAL

In giving the definition of the *derivative* of a function, we used the *limit* of a particular kind of *quotient* that was formed from the function in a very special way.   Now, in giving the definition of the *definite integral* of a function, we use the limit of a particular type of *sum* that is formed from the function in a rather special way.   We then establish (in Section 32) the connection between antiderivatives and definite integrals and thus make it very clear why *antiderivatives* are called *indefinite integrals*.

We suggest that before proceeding, you read through Section 18, which deals with limits of sequences.   In what follows, we assume that f is a continuous function defined on a finite interval J: $a \leq x \leq b$.

Definition 31.1   A *partition* of J is a division of J into n smaller subintervals J_i where for each $i = 1, 2, \ldots, n$, J_i: $x_{i-1} \leq x \leq x_i$.   We usually specify a partition P of the interval J by listing the endpoints of the subintervals that form the partition; that is

$$P: a = x_0 < x_1 < x_2 < \cdots < x_{n-1} < x_n = b.$$

Denote the length of J_i by Δ_i where $\Delta_i = x_i - x_{i-1}$ for each $i = 1, 2, \ldots, n$ (see Figure 31.1).

Definition 31.2   A *regular partition* of an interval J is a division of J into n subintervals of *equal* length Δ where $\Delta = (b - a)/n$.   It is clear that there

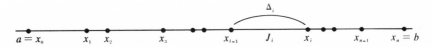

Figure 31.1

201

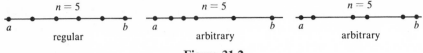

Figure 31.2

exists one and only one *regular* partition corresponding to each natural number n, while there exist infinitely many *arbitrary* partitions corresponding to each natural number n (see Figure 31.2). Consequently, we deal almost exclusively with regular partitions. Hence, for each natural number n, there exists a unique regular partition of J into n subintervals, and we denote this partition by P_n. Observe that P_n is determined by the points $x_0 = a$, $x_1 = a + \Delta$, $x_2 = a + 2\Delta$, ..., $x_{n-1} = a + (n-1)\Delta$, $x_n = a + n\Delta = b$.

Definition 31.3 Let P_n be a regular partition of the interval J into the sub-intervals J_i. Suppose that in each J_i we select an arbitrary point x_i^* where $x_{i-1} \leq x_i^* \leq x_i$. Then the sum

$$\sum_{i=1}^{n} f(x_i^*)\Delta$$

is called a *Riemann sum* of f corresponding to the regular partition P_n.

Since the choice of the points x_i^* is quite arbitrary, there are infinitely many Riemann sums corresponding to each partition P_n. However, if we *specify* how the x_i^* points are to be chosen, we will have a *unique* Riemann sum corresponding to each natural number n for that specification. For example, we could specify that x_i^* be chosen as the midpoint of J_i for each i, or as the left endpoint of J_i, or as the right endpoint of J_i. If we did this, for each n we would have a unique Riemann sum that depends only on n. Therefore each Riemann sum of a function f would itself be a function of the natural number n. Now if we let $n = 1, 2, 3, \ldots$, we obtain a sequence $\{S_n\}$ of Riemann sums where, for each n,

$$S_n = \sum_{i=1}^{n} f(x_i^*)\Delta$$

where $\Delta = (b-a)/n$ for each choice of n.

Definition 31.4 Let f be a continuous function defined on an interval $J: a \leq x \leq b$. Corresponding to each regular partition P_n, let the points x_i^* be chosen in a specific way. Consider the sequence $\{S_n\}$ of the Riemann sums of f. If $\lim_{n \to \infty} S_n = \alpha = $ a finite real number, α is called the *definite integral* of f over the interval from $x = a$ to $x = b$ and we denote α symbolically as $\int_a^b f(x)\, dx$. For the time being, we can take the dx as part of the symbol $\int_a^b dx$ and no significance need be attached to it other than that it tells us the integration is being done with respect to the variable x.

QUESTION   In the preceding definition, note that all that is required of the points x_i^* is that they be specified in some manner. Suppose that after computing α, the points x_i^* are specified in another way *different* from the first way. Will the limit of the new sequence of Riemann sums still be α?

The answer is yes, and the proof depends on the fact that f is *uniformly continuous* on J; that is, given an arbitrarily small $\varepsilon > 0$, there exists $\Delta > 0$ ($\Delta = (b - a)/n$) such that for any two points x^* and z^* in J_i for any i, we have $|f(x^*) - f(z^*)| < \varepsilon$. This means that for very small Δ (or very large n), it makes no difference how you choose the points x_i^*, since the functional values of any two points in J_i closely approximate each other. Hence, the terms $f(x_i^*)\Delta$ and $f(z_i^*)\Delta$ are approximately equal, as are the corresponding Riemann sums. Therefore, if the limit of the sequence of Riemann sums exists and equals α for a *particular* choice of the points x_i^*, the limit exists for *every* choice of the x_i^* and it equals α. Consequently, we may choose the points x_i^* in any convenient way. In practice, definite integrals are rarely computed by finding the limit of a Riemann sum. In Section 32, we will discuss another simpler method for evaluating definite integrals.

REMARK   It turns out that the definite integral of a continuous function f over an interval $a \le x \le b$ *always exists*; that is, the $\lim\limits_{n \to \infty} S_n$ mentioned in Definition 31.4 always exists and is finite. Observe that we have defined the definite integral in a purely analytic manner (as we did with the derivative) and we did not associate this definition with any geometrical situation whatsoever. It turns out, however, that the definite integral can be associated with areas, volumes, arc lengths, and other practical problems. This parallels the treatment of the derivative, when after giving the analytic definition, we proceeded to use it to solve problems involving slopes of tangent lines and velocity and acceleration. We did not then want to present the derivative as the slope of a tangent line, nor do we now want to present the definite integral as the area under a curve. Instead, we prefer to stress in both cases that these concepts can be defined as the limit of a special sequence of real numbers and that they are completely independent of any geometrical situation.

EXAMPLE 31.1   Let $f(x) = c$ where c is any real constant. Prove $\int_a^b f(x)\, dx = c(b - a)$ where a and b are any two real numbers with $a < b$.

SOLUTION   For any choice of a regular partition P_n and for any choice of the points x_i^*, the associated Riemann sum is

$$\sum_{i=1}^{n} f(x_i^*)\Delta = \sum_{i=1}^{n} c\Delta = nc\Delta = nc(b - a)/n = c(b - a)$$

Hence, as $n \to \infty$, the limit trivially exists and is equal to $c(b - a)$.

EXAMPLE 31.2 Let $f(x) = x$. Prove that $\int_a^b x \, dx = \frac{1}{2}(b^2 - a^2)$.

SOLUTION   Let P_n be a regular partition.   If we choose x_i^* to be the *right* endpoint of J_i for each $i = 1, 2, \ldots, n$, the Riemann sum is

$$\sum_{i=1}^n f(x_i^*)\Delta = \sum_{i=1}^n x_i \Delta$$

But for each i, $x_i = a + i\Delta$; therefore, $x_i\Delta = a\Delta + i\Delta^2$ for each i and the Riemann sum is

$$S_n = na\Delta + (1 + 2 + \cdots + n)\Delta^2 = na\Delta + (n/2)(n + 1)\Delta^2$$
$$= na(b - a)/n + [(n^2/2) + (n/2)](b - a)^2/n^2$$
$$= a(b - a) + \tfrac{1}{2}(b - a)^2 + \tfrac{1}{2}(b - a)^2/n.$$

As $n \to \infty$, $S_n \to ab - a^2 + \frac{1}{2}(b^2 - 2ab + a^2) = \frac{1}{2}(b^2 - a^2)$.

As exercises, choose the points x_i^* as the left endpoints of the J_i and then choose them as the midpoints of the J_i; compute the limit of the resulting Riemann sums; and show that the result is $\frac{1}{2}(b^2 - a^2)$ in each case.   Note that in the preceding example, we used the formula $(1 + 2 + 3 + \cdots + n) = (n/2)(n + 1)$.   In the following example we use the formula $(1^2 + 2^2 + 3^2 + \cdots + n^2) = \frac{1}{6}n(n + 1)(2n + 1)$.

EXAMPLE 31.3 Let $f(x) = x^2$. Prove $\int_a^b x^2 \, dx = \frac{1}{3}(b^3 - a^3)$.

SOLUTION   Let P_n be a regular partition.   Choose x_i^* to be the right endpoint of J_i for each $i = 1, 2, \ldots, n$.   Then

$$S_n = \sum_{i=1}^n x_i^2\Delta = \sum_{i=1}^n (a + i\Delta)^2\Delta$$

Now,

$$(a + i\Delta)^2\Delta = a^2\Delta + 2ai\Delta^2 + i^2\Delta^3;$$

hence,

$$S_n = na^2\Delta + 2a\Delta^2(1 + 2 + \cdots + n) + \Delta^3(1^2 + 2^2 + 3^2 + \cdots + n^2)$$
$$= na^2\Delta + 2a\Delta^2(n/2)(n + 1) + \Delta^3\tfrac{1}{6}n(n + 1)(2n + 1).$$

Substituting

$$\Delta = (b - a)/n,$$

we obtain

$$a^2(b - a) + 2a(b - a)^2(n^2/2)(1/n^2) + 2a(b - a)^2(n/2)(1/n^2)$$
$$+ [(b - a)^3/n^3]\tfrac{1}{6}(2n^3 + 3n^2 + n).$$

As $n \to \infty$, $S_n \to a^2(b - a) + 2a(b - a)^2\frac{1}{2} + \frac{1}{3}(b - a)^3 = \frac{1}{3}(b^3 - a^3)$.

PROBLEMS

31.1   In each of the following, compute the definite integral of the given function over the interval indicated using Definition 31.4.
(a) $f(x) = 5$ over $2 \leq x \leq 5$
(b) $f(x) = 2x$ over $a \leq x \leq b$
(c) $f(x) = 2x + 1$ over $0 \leq x \leq 1$
(d) $f(x) = 3x^2$ over $0 \leq x \leq 1$

31.2   Using the results of Section 19 together with Definition 31.4, prove each of the following.

(a) $\displaystyle \int_a^b kf(x)\,dx = k \int_a^b f(x)\,dx$ for any constant k

(b) $\displaystyle \int_a^b (f(x) + g(x))\,dx = \int_a^b f(x)\,dx + \int_a^b g(x)\,dx$

where $g(x)$ is also a continuous function on $J : a \leq x \leq b$.

31.3   Using the results of the two preceding problems, compute each of the following definite integrals.

(a) $\displaystyle \int_a^b 10x\,dx$

(b) $\displaystyle \int_2^5 (2x + 5)\,dx$

(c) $\displaystyle \int_0^1 (3x^2 + 2x + 1)\,dx$

31.4   Repeat Example 31.3, choosing the points x_i^* as follows.
(a) Let x_i^* be the point at which f attains its maximum value in J_i.
(b) Let x_i^* be the point at which f attains its minimum value in J_i.

31.5* Let $f(x) = x^3$.   Prove that $\displaystyle \int_a^b x^3\,dx = \tfrac{1}{4}(b^4 - a^4)$.

[*Hint:*   Use the formula $(1^3 + 2^3 + 3^3 + \ldots + n^3) = [(n/2)(n + 1)]^2$ and the technique exhibited in Examples 31.2 and 31.3.]

THE FUNDAMENTAL THEOREM OF THE CALCULUS

We now consider the connection between antiderivatives and definite integrals by proving the fundamental theorem of the calculus.

Theorem 32.1 Let f be a continuous function defined on an interval $J: a \le x \le b$. Let F be an antiderivative of f on J.   Then

$$\int_a^b f(x)\, dx = F(b) - F(a)$$

PROOF   Let P_n be a regular partition of J (see Definition 31.2).   We are going to choose the points x_i^* in a way that is different from all the ways used earlier.   If F is an antiderivative of f on J, then for every $x \in J$, $F'(x) = f(x)$. Therefore, F is differentiable in J and satisfies the mean value theorem (see Section 25) on each subinterval $J_i: x_{i-1} \le x \le x_i$, for $i = 1, 2, \ldots, n$ where the length of each J_i is $\Delta = (b - a)/n$.   Therefore in each subinterval J_i, there exists a point z_i such that $x_{i-1} < z_i < x_i$ and $[F(x_i) - F(x_{i-1})]/(x_i - x_{i-1}) = F'(z_i)$.   But $F'(z_i) = f(z_i)$; hence, $F(x_i) - F(x_{i-1}) = f(z_i)(x_i - x_{i-1}) = f(z_i)\Delta$. Choose $x_i^* = z_i$ for each $i = 1, 2, \ldots, n$.   Then the Riemann sum is

$$S_n = \sum_{i=1}^n f(x_i^*)\Delta = \sum_{i=1}^n (F(x_i) - F(x_{i-1}))$$
$$= F(x_1) - F(x_0) + F(x_2) - F(x_1) + \cdots + F(x_n) - F(x_{n-1})$$
$$= F(x_n) - F(x_0) = F(b) - F(a). \quad \blacksquare$$

206

As a result of this theorem, we may compute definite integrals of functions, provided we know an antiderivative of the given function. In fact, a continuous function f on an interval $a \le x \le b$ always has an antiderivative on $a \le x \le b$. Therefore, the evaluation of definite integrals boils down to the computation of antiderivatives. The calculation of limits of Riemann sums is hardly ever required.

EXAMPLE 32.1 Prove $\int_a^b x^n \, dx = [1/(n + 1)](b^{n+1} - a^{n+1})$ for every real number $n \ne -1$.

SOLUTION An antiderivative of $f(x) = x^n$ is the function $F(x) = x^{n+1}/(n+1)$. Now $F(b) - F(a) = b^{n+1}/(n + 1) - a^{n+1}/(n+1) = (1/(n+ 1))(b^{n+1} - a^{n+1})$. Compare this result with the ones obtained in Examples 31.2 and 31.3.

EXAMPLE 32.2 Compute $\int_1^b 1/x \, dx$ where $b > 1$.

SOLUTION An antiderivative of $f(x) = 1/x$ is the function $F(x) = \log x$. Consequently, $F(b) - F(1) = \log b - \log 1 = \log b$. *Notice* that the equation $\log b = \int_1^b 1/x \, dx$ is valid for all positive real numbers $b > 1$. Hence, the natural logarithm function can be expressed in the form of a *definite integral* that is a function of its upper limit. In many texts, this is precisely how the natural logarithm function is defined.

EXAMPLE 32.3 Compute $\int_1^2 x/(x^2 + 1)^2 \, dx$.

SOLUTION The integrand can be expressed as $\frac{1}{2}(x^2 + 1)^{-2}(2x)$, which has the form $\frac{1}{2}(f(x))^{-2}f'(x)$ where $f(x) = x^2 + 1$. Therefore, an antiderivative is $-\frac{1}{2}(f(x))^{-1}$ or $-\frac{1}{2}(x^2 + 1)^{-1}$ (see Example 30.4(a)). Evaluating this function at $x = 2$ and at $x = 1$ and subtracting yields $-1/10 - (-\frac{1}{4})$ $= 3/20$.

EXAMPLE 32.4 Compute $\int_1^2 x/(x^2 + 1) \, dx$.

SOLUTION The integrand can be expressed as $\frac{1}{2}[2x/(x^2 + 1)]$, which has the form $\frac{1}{2}f'(x)/f(x)$ where $f(x) = x^2 + 1$. Hence, an antiderivative is $\frac{1}{2} \log f(x)$ or $\frac{1}{2} \log(x^2 + 1)$ (see Example 30.4(b)). Evaluating this function at $x = 2$ and at $x = 1$ and subtracting, we obtain $\frac{1}{2}(\log 5 - \log 2)$.

EXAMPLE 32.5 Compute $\int_0^{\pi/2} \cos x \, dx$.

SOLUTION An antiderivative of $\cos x$ is $\sin x$. The integral is $\sin(\pi/2)$ $- \sin 0 = 1$.

EXAMPLE 32.6 Compute $\int_0^1 (1 + x^2)^{-1}\, dx$.

SOLUTION   An antiderivative of $(1 + x^2)^{-1}$ is arctan x. Therefore, the
integral is arctan 1 − arctan $0 = \pi/4$.

EXAMPLE 32.7 Compute $\int_0^{1/2} (1 - x^2)^{-1/2}\, dx$.

SOLUTION   An antiderivative of $(1 - x^2)^{-1/2}$ is arcsin x. The integral is
therefore arcsin $\frac{1}{2}$ − arcsin $0 = \pi/6$.

Theorem 32.2   Let f be a continuous function on an interval $J: a \le x \le b$ and
let F be an antiderivative of f on J.   Then

(A) $\int_a^a f(x)\, dx = 0$

(B) $\int_a^b kf(x)\, dx = k \int_a^b f(x)\, dx$ for any constant k

(C) $\int_a^b (f(x) + g(x))\, dx = \int_a^b f(x)\, dx + \int_a^b g(x)\, dx$

where g is defined and continuous on J with antiderivative G.

(D) $\int_a^c f(x)\, dx + \int_c^b f(x)\, dx = \int_a^b f(x)\, dx$

where $a < c < b$.

PROOF   (A) follows from the fact that $F(a) - F(a) = 0$.   (B) depends on the
fact that $kF(x)$ is an antiderivative of $kf(x)$ and $kF(b) - kF(a) = k(F(b)
- F(a))$.   (C) depends on the fact that $F(x) + G(x)$ is an antiderivative of $f(x)
+ g(x)$ and $(F(b) + G(b)) - (F(a) + G(a)) = (F(b) - F(a)) + (G(b) - G(a))$.
Finally, (D) follows from the fact that the first integral from a to c is $F(c)
- F(a)$, while the second integral from c to b is $F(b) - F(c)$.   Adding, we ob-
tain $F(b) - F(a)$. ∎
   We are finally in a position to define the definite integral in the case where
the lower limit is greater than the upper limit.

Definition 32.1   Let f be a continuous function on an interval $J: a \le x \le b$
and let F be an antiderivative of f on J.   Then we define $\int_b^a f(x)\, dx$ to be
$-\int_a^b f(x)\, dx$.   Hence, $\int_b^a f(x)\, dx = F(a) - F(b)$ [since $-(F(b) - F(a)) =
F(a) - F(b)$].
   We now state three more properties of the definite integral.

Theorem 32.3 Let f and g be continuous functions defined on an interval $J: a \leq x \leq b$.

(A)   If $f(x) \leq g(x)$ for all $x \in J$, then $\displaystyle\int_a^b f(x)\, dx \leq \int_a^b g(x)\, dx$

(B) $\left| \displaystyle\int_a^b f(x)\, dx \right| \leq \displaystyle\int_a^b |f(x)|\, dx$

(C)   If M is a bound on the function f on the interval J [i.e., $f(x) \leq M$ for all $x \in J$] where M is some positive real number, then $\left| \displaystyle\int_a^b f(x)\, dx \right| \leq M\,|b-a|$. This result means that the absolute value of the integral is less than or equal to a bound on the function times the length of the interval from a to b.

PROOF   (A) follows from a comparison of the Riemann sums corresponding to a regular partition P_n where the points x_i^* are chosen in the same manner for both f and g.

(B) follows from the fact that $f(x) \leq |f(x)|$ for all $x \in J$ and we may replace $g(x)$ in part (A) with the function $|f(x)|$, obtaining $\displaystyle\int_a^b f(x)\, dx \leq \displaystyle\int_a^b |f(x)|\, dx$.   Now if the integral on the left is positive, then

$$\left| \int_a^b f(x)\, dx \right| = \int_a^b f(x)\, dx \leq \int_a^b |f(x)|\, dx$$

and if the integral on the left is negative, its absolute value is equal to

$$-\int_a^b f(x)\, dx = \int_a^b (-f(x))\, dx \leq \int_a^b |f(x)|\, dx$$

[from part (A), since $-f(x) \leq |f(x)|$ for all $x \in J$].

(C) is obtained from parts (A) and (B) in the following way.

$$\left| \int_a^b f(x)\, dx \right| \leq \int_a^b |f(x)|\, dx \leq \int_a^b M\, dx = M \int_a^b dx = M(b-a) = M\,|b-a|$$

provided $a < b$.   (As an exercise, assume $a > b$ and obtain the same result.) ∎

We now complete the *final* link in the chain of relations that involve anti-derivatives and definite integrals.   It turns out that among all the antiderivatives of a continuous function, there is one that can be represented in the form of an integral with a *variable* upper limit; hence, it is called an *indefinite* integral.   However, the antiderivatives of a given function differ from one another by constants; therefore, if *one* antiderivative has the form of an indefinite integral, *every* antiderivative must differ from this integral by at most a constant.   This is the reason *antiderivatives* are called *indefinite integrals* and why the notation $\int f(x)\, dx$ is used for an antiderivative of the function $f(x)$.   *Remember, an indefinite integral is a function, whereas a definite integral is a number.*

Theorem 32.4 Let f be a continuous function defined on an interval $J: a \leq x \leq b$. For each number $z \in J$, the integral $\int_a^z f(x)\, dx$ exists; hence, this indefinite integral is a function G of its *upper limit z*. Then, G is an antiderivative of f on the interval J.

PROOF   Let F be any antiderivative of f on J. Then, for each $z \in J$

$$G(z) = \int_a^z f(x)\, dx = F(z) - F(a).$$

Differentiating both sides of this equation with respect to z yields, for each $z \in J$, $G'(z) = F'(z)$. But $F'(z) = f(z)$ for each $z \in J$ since F is an antiderivative of f. Therefore, $G'(z) = f(z)$ for every $z \in J$ and the function G is an antiderivative of f on J. Observe that we can now write the function F as $F(z) = \int_a^z f(x)\, dx + C$ where $C = F(a)$. But F was *any* antiderivative of f on J; hence, we may now conclude that *every* antiderivative of f on J can be written in the form $\int_a^z f(x)\, dx + C$ and is therefore expressible as an *indefinite* integral *plus* some constant. ■

EXAMPLE 32.8   Express an antiderivative of each of the following functions in the form of an integral.
(A) $f(x) = \cos x$ defined on $J: 0 \leq x \leq 2\pi$.
(B) $f(x) = 1/x^2$ defined on $J: 1 \leq x \leq 2$.

SOLUTION   (A) $F(z) = \int_0^z \cos x\, dx$ is an antiderivative of the function $f(x) = \cos x$ [since $F(z) = \sin z - \sin 0 = \sin z$ for all $z \in J$]. Very often, in this kind of problem, the variable of integration is changed to another variable, say, t, while the upper limit z is changed to x. This results in the expression of the indefinite integral as a function of x, namely, $F(x) = \int_0^x \cos t\, dt$.

(B) $F(x) = \int_1^x 1/t^2\, dt$ is an antiderivative of $f(x) = 1/x^2$ on $1 \leq x \leq 2$ since $F(x) = -1/x - (-1) = 1 - 1/x$ for all $x \in J$ and $F'(x) = 1/x^2$ for each $x \in J$.

PROBLEMS

32.1   Give a complete proof of part (A) of Theorem 32.3, using Riemann sums.
32.2   Evaluate each of the following definite integrals (see Section 30 for the appropriate antiderivatives).

(a) $\int_1^3 (x^2 + 2x - 1)\, dx$

(b) $\int_0^1 x^2/(x^3 + 1)^5\, dx$

(c) $\int_0^1 x^2/(x^3 + 1)\, dx$

(d) $\int_0^{\pi/2} \sin^3 x \cos x\, dx$

(e) $\int_0^1 e^{2x}\, dx$

(f) $\int_0^1 xe^{x^2}\, dx$

(g) $\int_1^{e^2} (\log x)/x\, dx$

(h) $\int_0^{\pi} x \cos(2x^2)\, dx$

(i) $\int_0^{\pi/4} \sec^2 x\, dx$

(j) $\int_0^{\pi/2} (\sin x \cos x)(1 + \sin^2 x)^{1/2}\, dx$

(k) $\int_0^1 x^3(1 + 8x^4)^{1/2}\, dx$

32.3 Evaluate each of the following integrals by using Theorem 32.1.

(a) $\int_{-1}^1 |x|\, dx$

(b) $\int_0^2 |x^2 - 1|\, dx$ (See Figure 13.6.)

32.4 Using part (C) of Theorem 32.3, find a bound on the absolute value of each of the following integrals.

(a) $\int_0^1 e^x(-x^2 + x + 1)\, dx$

(b) $\int_0^1 xe^x \cos x\, dx$

(c) $\int_{\pi/2}^{\pi} x^2 \log x \sin x\, dx$

32.5 Express an antiderivative of each of the following functions in the form of an integral.
   (a) $f(x) = \sin x$ $0 \le x \le 2\pi$
   (b) $f(x) = x + \cos x$ $0 \le x \le 2\pi$
   (c) $f(x) = \sec^2 x$ $0 \le x \le \pi/4$
32.6 Extend the integral definition of $\log b$ given in Example 32.2 to include all real numbers b with $0 < b \le 1$. [*Hint:* Use Theorem 32.2, part (A), and Definition 32.1.]

AREA

We now show that the abstract concept of a definite integral can be applied to the practical problem of finding areas of regions in the plane.

Definition 33.1 Let f be a continuous function defined on an interval $J: a \leq x \leq b$ with $f(x) > 0$ for all $x \in J$. We define the area of the region above the x axis and below the graph of $y = f(x)$ from $x = a$ to $x = b$ as the integral $\int_a^b f(x)\, dx$ (see Figure 33.1).

We now attempt to make this definition a reasonable one. Let P_n be a regular partition of J (see Definition 31.2). Let x_i^* be the midpoint (or any other specified point) of $J_i: x_{i-1} \leq x \leq x_i$ for each $i = 1, 2, \ldots, n$. Then the

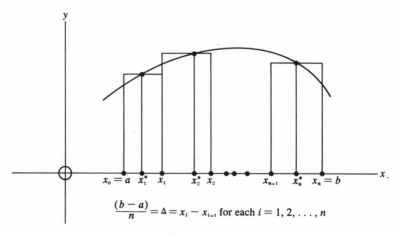

$$\frac{(b-a)}{n} = \Delta = x_i - x_{i-1}, \text{ for each } i = 1, 2, \ldots, n$$

Figure 33.1

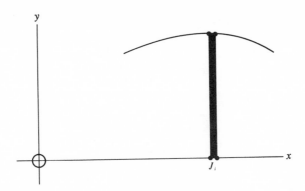

Figure 33.2

Riemann sum $S_n = \sum_{i=1}^{n} f(x_i^*)\Delta$ and as $n \to \infty$, $S_n \to \int_a^b f(x)\, dx$ (see Definition 31.4). We also know that if f has an antiderivative, we may evaluate this integral, using Theorem 32.1. Now for each $i = 1, 2, \ldots, n$, the term $f(x_i^*)\Delta$ in the Riemann sum S_n represents the area of the rectangle with base of length Δ units and height $f(x_i^*)$ units (see Figure 33.1). Consequently, this term of S_n gives an approximation to the area of the region above J_i and below the graph of $y = f(x)$, that is, the area of the region between the vertical lines $x = x_{i-1}$ and $x = x_i$ and below the graph of $y = f(x)$. As $n \to \infty$, $\Delta \to 0$ and the *rectangular* area above each J_i becomes very close to the area above J_i. In fact, if n is taken so large that Δ equals the *width* of a *solid* pencil line constructed with base J_i (see Figure 33.2), the area of the solid line with the square top is not very much different from the area of the solid line with the curved top. Hence S_n, for large n, is just the *sum* of the rectangular areas above each J_i and as $n \to \infty$, S_n becomes a better and better approximation for the actual area under the curve. But as $n \to \infty$, $S_n \to \int_a^b f(x)\, dx$; therefore, we define the *area* as this *integral*.

As exercises, determine the areas associated with Examples 31.1, 31.2, 31.3, and 32.2.

EXAMPLE 33.1   Find the area of the region above the x axis and below the graph of $y = x^n$, where n is any positive integer, between the y axis and the vertical line $x = 1$. What happens to this area as n gets very large? (See Figure 33.3.)

SOLUTION   The area is $\int_0^1 x^n\, dx$, which is equal to $1/(n + 1)$ square units (see Example 32.1). Now as $n \to \infty$, the area goes to zero, since $1/(n + 1) \to 0$.

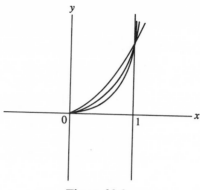

Figure 33.3

EXAMPLE 33.2   Find the area of the region below the graph of $y = 1/x$ and above the x axis, between the vertical lines $x = 1$ and $x = e$.
SOLUTION   The area is $\int_1^e 1/x \, dx$, which equals $\log e = 1$ square unit (see Example 32.2).

EXAMPLE 33.3   Find the area of the region below the graph of $y = \cos x$ and above the x axis, between the y axis and the line $x = \pi/2$.

SOLUTION   The area is $\int_0^{\pi/2} \cos x \, dx = 1$ square unit (see Example 32.5).

EXAMPLE 33.4   Find the area of the region below the graph of $y = 1/(1 + x^2)$ and above the x axis, between the lines $x = -1$ and $x = 1$ (see Figure 33.4).

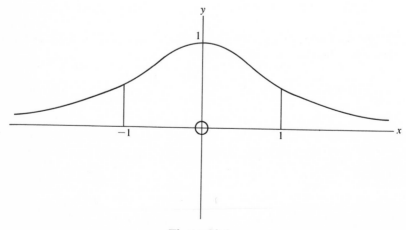

Figure 33.4

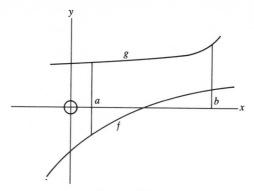

Figure 33.5

SOLUTION   From the diagram, it can be seen that the area to the right of the y axis is the same as the area to the left.   Hence, the area $\int_{-1}^{1}(1+x^2)^{-1}\,dx = 2\int_{0}^{1}(1+x^2)^{-1}\,dx = 2(\arctan 1 - \arctan 0) = 2(\pi/4) = \pi/2$ square units (see Example 32.6).

THE AREA BETWEEN TWO CURVES

We first discuss the general situation of the area of the region enclosed by the nonintersecting graphs of $y = g(x)$ and $y = f(x)$ between the vertical lines $x = a$ and $x = b$ (see Figure 33.5).   Then as a special case, we take $y = g(x) = 0$ and discuss the area of the region above the graph of $y = f(x)$ and below the x axis (see Figure 33.6).   In the situation depicted in Figure 33.5, we simply take a regular partition P_n of the interval $J: a \leq x \leq b$.   We choose the points x_i^* in some specific way (as midpoints, perhaps).   We then observe that the rectangle constructed through each $J_i: x_{i-1} \leq x \leq x_i$ for $i = 1, 2, \ldots, n$, has base of length $\Delta = (b - a)/n$ and height given by $(g(x_i^*) - f(x_i^*))$ (see Figure 33.7).   Consequently, if we take the area of each of the n rectangles and add them, we obtain

$$S_n = \sum_{i=1}^{n}(g(x_i^*) - f(x_i^*))\Delta$$

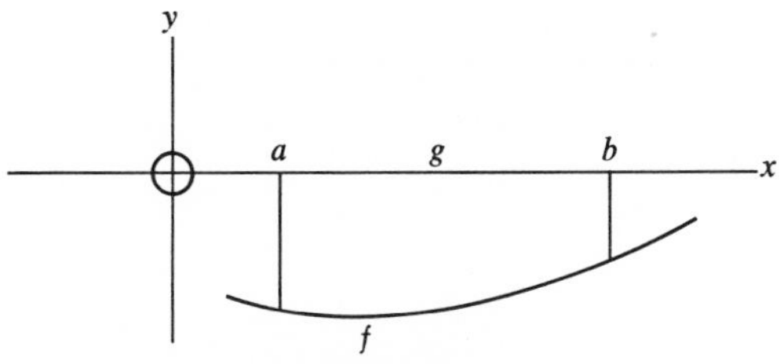

Figure 33.6

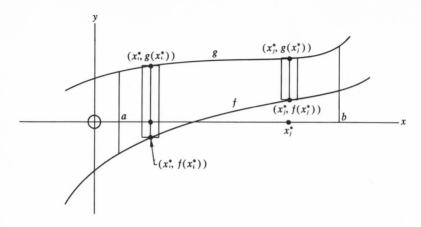

Figure 33.7

which is an *approximation* for the area of the region between the graphs. But this sum is precisely the Riemann sum for the function $g - f$ corresponding to the regular partition P_n. Hence, as $n \to \infty$,

$$S_n \to \int_a^b (g(x) - f(x))\, dx = \int_a^b g(x)\, dx - \int_a^b f(x)\, dx$$

thus, the area between the graphs is defined as either one of these last two expressions.

If we apply these results to the situation depicted in Figure 33.6, we have $g(x) = 0$, and thus

$$S_n = \sum_{i=1}^{n} (-f(x_i^*))\Delta$$

which is the Riemann sum for the function $-f(x)$. Hence, as $n \to \infty$,

$$S_n \to \int_a^b -f(x)\, dx = - \int_a^b f(x)\, dx$$

Now, since $f(x) < 0$ for all $x \in J$, the *integral* of $f(x)$ from a to b will also be negative. However, the area in this case is defined above to be the *negative* of this integral. Hence, the area is *positive* and will always be positive. Consequently, whenever the graph of $y = f(x)$ is entirely below the x axis between $x = a$ and $x = b$, the area it bounds is always given by $- \int_a^b f(x)\, dx$.

EXAMPLE 33.5 (a) Compute $\int_{-1}^{1} x^3\, dx$.

   (b) Compute the area of the region bounded by the x axis and the graph of $y = x^3$ between $x = -1$ and $x = 1$ (see Figure 33.8).

SOLUTION (a) The antiderivative of x^3 is $x^4/4$; hence, the integral is simply $\frac{1}{4} - \frac{1}{4} = 0$.

(b) To find the area to the right of the y axis, we evaluate $\int_0^1 x^3 \, dx = \frac{1}{4}$ square units. To find the area to the left of the y axis, we must deal with a function that has a graph lying wholly below the x axis. Therefore, from the discussion preceding this example, we must evaluate $-\int_{-1}^0 x^3 \, dx$ to find the area. This integral is $-(-\frac{1}{4}) = \frac{1}{4}$ square units; therefore, the total area is $\frac{1}{4} + \frac{1}{4} = \frac{1}{2}$ square units (which is *not* equal to zero).

This example points out that integrals are *not* necessarily areas but can be used, under *certain* restrictions, to find areas. The result of part (a) merely means that the *limit* of a particular *Riemann sum* is zero, it does not mean the associated area is zero. The result of part (b), on the other hand, tells us that the *area* in question is precisely $\frac{1}{2}$, but it does *not* tell us that the corresponding integral is $\frac{1}{2}$.

EXAMPLE 33.6 (a) Compute $\int_0^{3\pi/2} \cos x \, dx$.

(b) Compute the area enclosed by the x axis and the graph of $y = \cos x$ between $x = 0$ and $x = 3\pi/2$ (see Figure 33.9).

SOLUTION (a) As an exercise, show that the integral is equal to -1.

(b) The area above the x axis in Figure 33.9 equals $\int_0^{\pi/2} \cos x \, dx = \sin \pi/2 - \sin 0 = 1$ square unit. The area below the x axis equals $-\int_{\pi/2}^{3\pi/2} \cos x \, dx = -[\sin(3\pi/2) - \sin(\pi/2)] = -(-1-1) = 2$ square units. Therefore, the total area is $1 + 2 = 3$ square units ($\neq -1$) and once again, we see that the area is not equal to the integral in part (a).

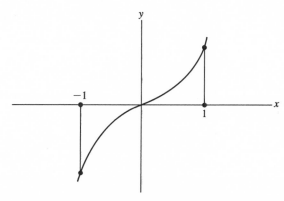

Figure 33.8

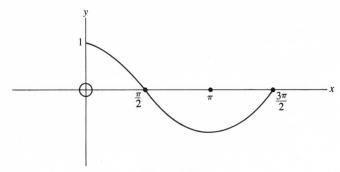

Figure 33.9

EXAMPLE 33.7   Sketch the graphs of $y = x^3$ and $y = -x^2 + 2x$ and compute the areas of each of the regions enclosed by these graphs between the y axis and the line $x = 2$ (see Figure 33.10).

SOLUTION   We will find the areas of the three regions labeled A, B, and C in figure 33.10.

Area A.   This region is bounded above by the graph of $y = -x^2 + 2x$ and below by the graph of $y = x^3$.   Hence, the area is given by

$$\int_0^1 [(-x^2 + 2x) - x^3]\, dx = 5/12 \quad \text{square units}$$

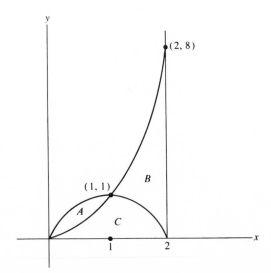

Figure 33.10

Area B. This region is bounded above by the graph of $y = x^3$ and below by the graph of $y = -x^2 + 2x$. Hence, the area is given by

$$\int_1^2 [x^3 - (-x^2 + 2x)]\, dx = \int_1^2 (x^3 + x^2 - 2x)\, dx = 37/12 \quad \text{square units}$$

Area C. This region must be split into two smaller subregions. The first one is bounded above by the graph of $y = x^3$, below by the x axis, and on the right by the line $x = 1$. Hence, the area is simply $\int_0^1 x^3\, dx = \frac{1}{4}$. The second subregion is bounded above by the graph of $y = -x^2 + 2x$, below by the x axis, and on the left by the line $x = 1$. Hence, the area is $\int_1^2 (-x^2 + 2x)\, dx = \frac{2}{3}$. Therefore, the total area of the region C is the sum of the individual areas, namely $\frac{1}{4} + \frac{2}{3} = \frac{11}{12}$ square units.

EXAMPLE 33.8 Find the area of the region below the graph of $y = e^x$, above the graph of $y = (1 + x^2)^{-1}$, between the y axis and the line $x = 1$ (see Figure 33.11).

SOLUTION   Since the graph of $y = e^x$ is above the one of $y = (1 + x^2)^{-1}$ the area is given by the integral

$$\int_0^1 [e^x - (1 + x^2)^{-1}]\, dx = \int_0^1 e^x\, dx - \int_0^1 (1 + x^2)^{-1}\, dx$$

$$= (e - 1) - (\arctan 1 - \arctan 0) = e - 1 - \pi/4 \quad \text{square units}$$

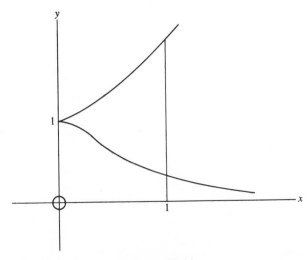

Figure 33.11

As an exercise, find the area of the region bounded by these two graphs between $x = -2$ and $x = -1$.

Finally, we conclude this section with a rather important result which has a very interesting connection with the area under a curve.

Theorem 33.1   *The Mean Value Theorem for Integrals*   Let f be a continuous function defined on an interval $J: a \le x \le b$ with $f(x) \ge 0$ for all x in J. Then there exists at least one point c between a and b with the property that $\int_a^b f(x)\, dx = f(c)(b - a)$.   Observe that this simply means the area under the graph of $y = f(x)$ between $x = a$ and $x = b$ is the same as the area of a rectangle with base J and altitude $f(c)$ where c is some number between a and b (see Figure 33.12).

PROOF   The theorem is certainly true if f is a constant; hence, we may assume f is not constant on J.   Let $A = \int_a^b f(x)\, dx$.   Then A is the area under the curve between $x = a$ and $x = b$.   Let M and m represent values of x that respectively yield the maximum and minimum of the function f on the interval J.   Then $f(m) \le f(x) \le f(M)$ for all x in J.   Also, the rectangle with base J and altitude $f(M)$ must have area *greater* than A while the rectangle with base J and altitude $f(m)$ must have area *less* than A.   Therefore, $f(m)(b - a) < A < f(M)(b - a)$, which implies $f(m) < A/(b - a) < f(M)$. But f is continuous on J; hence, the y coordinates of points on the graph of f must take on every value between $f(M)$ and $f(m)$ as the graph is traced from the point $(M, f(M))$ to the point $(m, f(m))$.   But $A/(b - a)$ is a real number between $f(M)$ and $f(m)$.   Therefore, $A/(b - a)$ must be the y coordinate of at least one point on the graph of $y = f(x)$.   Hence, there exists at least one value c between a and b such that $f(c) = A/(b - a)$.   Therefore, $A = f(c)(b - a)$ and the proof is complete.   ■

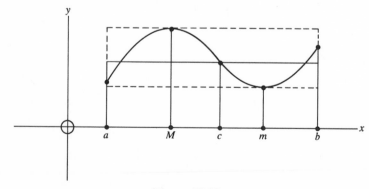

Figure 33.12

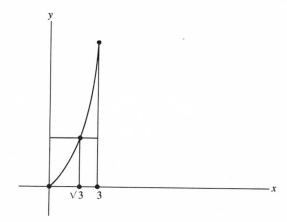

Figure 33.13

EXAMPLE 33.9 Find a point c that satisfies the mean value theorem for integrals for the function $f(x) = x^2$ on the interval $J: 0 \le x \le 3$.

SOLUTION   The area under the curve is $\int_0^3 x^2\, dx = 9$.   Hence $A = 9$, $b - a = 3$, and $A/(b - a) = 3$.   Now, we must find a value c between $x = 0$ and $x = 3$ such that $f(c) = c^2 = 3$.   Therefore, $c = \sqrt{3}$ is the required point (see Figure 33.13).

PROBLEMS

33.1 Find the area of the region above the x axis, below the graph of $y = x^{1/n}$, n any positive integer, between the y axis and the vertical line $x = 1$.   What happens to this area as n becomes large?

33.2 (a) Find the area of the region above the x axis, below the graph of $y = 1/x$ between $x = 1$ and $x = n$, n any positive integer.   What happens to this area as n becomes large?

(b) Repeat part (a) with $y = 1/x^2$.

33.3 (a) Find all the critical values (including inflection points) of the function $y = (1 + x^2)^{-1}$.   Sketch the graph (see Figure 33.4).

(b) Find the area of the region above the x axis, below the graph of $y = (1 + x^2)^{-1}$ between the vertical lines $x = -n$ and $x = n$, when n is any positive integer.   What happens to this area as n becomes large?

33.4 (a) Find the area of the region above the x axis and below the graph of $y = |x|$ between $x = -1$ and $x = 1$.

(b) Find the area of the region above the x axis and below the graph of $y = |x^2 - 1|$ between $x = -2$ and $x = 2$ (see Figure 13.6).

(c) Find the area of the region above the x axis and below the graph of $y = |x^2 - x|$ between $x = 0$ and $x = 2$.

33.5 Using part (C) of Theorem 32.3, find a bound on the area of each of the following regions.

(a) the region above the x axis, below the graph of $y = e^x(-x^2 + x + 1)$, between $x = 0$ and $x = 1$

(b)   the region above the x axis, below the graph of $y = xe^x \cos x$, between $x = 0$ and $x = 1$

33.6 (a)   Compare $\int_{-1}^{2} x^3 \, dx$.

     (b)   Compute the area of the region bounded by the x axis and the graph of $y = x^3$ between $x = -1$ and $x = 2$.

33.7 (a)   Compute $\int_{0}^{2\pi} \sin x \, dx$.

     (b)   Compute the area of the region bounded by the x axis and the graph of $y = \sin x$ between $x = 0$ and $x = 2\pi$.

33.8 Find the area of the region bounded by the x axis and the graph of $y = [x(\log x)^2]^{-1}$ between $x = 2$ and $x = 4$.

33.9 (a)   Sketch the graphs of $y = \sqrt{x}$ and $y = -x^2 + 2$.

     (b)   Compute the areas of each of the three regions above the x axis enclosed by these graphs between the y axis and the line $x = \sqrt{2}$.

33.10 Find the area between the graphs of the equations given in (a) Example 12.1; (b) Example 12.2 from the y axis to the line $x = 4$.

33.11 (a)   Sketch the graphs of $y = \sin x$ and $y = \cos x$ between $x = 0$ and $x = 2\pi$.

     (b)   Find the area of each of the three regions *enclosed* by these graphs.

33.12 (a)   Sketch the graph of $y = x + \sin x$, $0 \le x \le 2\pi$.

     (b)   Find the area of the region enclosed by the graph of part (a) and the line $y = x$.

33.13 (a)   Sketch the graph of $y = x + \cos x$, $0 \le x \le 2\pi$.

     (b)   Find the area enclosed by the graph of part (a) and the line $y = x$, between $x = 0$ and $x = 2\pi$.

33.14 (a)   Evaluate $\int_{-n}^{n} xe^{-x^2} \, dx$, n any positive integer.

     (b)   Sketch the graph of $y = xe^{-x^2}$.

     (c)   Find the area of the region above the x axis, below the graph of $y = xe^{x^2}$, between the lines $x = -n$ and $x = n$, n any positive integer.

     (d)   What happens to this area as n becomes large?

33.15 In each of the following cases find a point c that satisfies the mean value theorem for integrals (see Theorem 33.1).

(a) $f(x) = 3x^2$ $0 \le x \le 2$

(b) $f(x) = 1/x$ $1 \le x \le e$ (See Example 33.2.)

(c) $f(x) = \sin^2 x$ $0 \le x \le \pi$

Section 34

VOLUME

We now consider the application of the definite integral to problems that involve the computation of volumes. We first treat the problem of calculating the volume of a solid that is generated by rotating a plane region about a line. Such a solid is called a *solid of revolution.*

Definition 34.1 Let f be a continuous function defined on an interval $J: a \leq x \leq b$ with $f(x) > 0$ for all $x \in J$. Then the volume of the *solid of revolution* obtained by rotating about the x axis the region determined by the graph of $y = f(x)$ between $x = a$ and $x = b$ is defined as the integral $\pi \int_a^b (f(x))^2 \, dx$. (See Figure 34.1.)

We now attempt to make this definition a plausible one. Let P_n be a regular partition of J (see Definition 31.2). Let x_i^* be the midpoint (or any other specified point) of $J_i: x_{i-1} \leq x \leq x_i$ for each $i = 1, 2, \ldots, n$. Now,

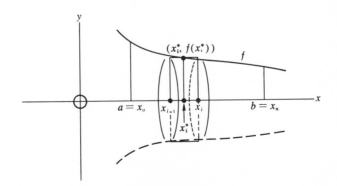

Figure 34.1

223

consider the rectangle above J_i with height $f(x_i^*)$ and base of length $\Delta = (b - a)/n$. This rectangle closely approximates (for large n) the region below the graph of $y = f(x)$ between $x = x_{i-1}$ and $x = x_i$. If we rotate this rectangle about the x axis, we obtain a cylinder of height Δ and radius of the base equal to $f(x_i^*)$. Hence, this cylinder has volume $\pi(f(x_i^*))^2\Delta$. If we do this for each $i = 1, 2, \ldots, n$ and add, we obtain the sum

$$S_n = \sum_{i=1}^{n} \pi(f(x_i^*))^2\Delta$$

This sum is therefore an *approximation* to the volume of the solid of revolution *but* it is also a Riemann sum for the function $\pi(f(x))^2$. As $n \to \infty$, the approximation becomes better and better and

$$S_n \to \int_a^b \pi(f(x))^2 \, dx = \pi \int_a^b (f(x))^2 \, dx$$

Consequently, it is reasonable to define the volume to be this integral.

As exercises, determine the volume of the solid of revolution determined by each of the regions associated with Examples 31.1, 31.2, 31.3, and 32.2.

EXAMPLE 34.1   Find the volume of the solid of revolution generated by rotating about the x axis the region below the graph of $y = 1/x$ between $x = 1$ and $x = 2$.

SOLUTION   The volume is given by

$$\pi \int_1^2 \left(\frac{1}{x}\right)^2 dx = \pi\left(-\frac{1}{x}\bigg|_1^2\right) = \pi\left(-\frac{1}{2} + 1\right) = \frac{\pi}{2} \quad \text{cubic units}$$

EXAMPLE 34.2   Find the volume of the solid of revolution generated by rotating about the x axis the region below the graph of $y = -x^2 + 4$, above the graph of $y = x^2$, between the y axis and the line $x = \sqrt{2}$ (see Figure 34.2).

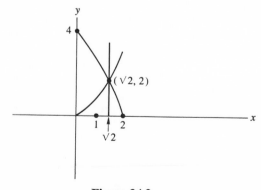

Figure 34.2

SOLUTION   The required volume is the difference between the volumes of the two solids of revolution determined by first rotating about the x axis the entire region under the graph of $y = -x^2 + 4$ between $x = 0$ and $x = \sqrt{2}$, and then rotating about the x axis only the region below $y = x^2$ from $x = 0$ to $x = \sqrt{2}$.   The resulting integrals are

$$\pi \int_0^{\sqrt{2}} (-x^2 + 4)^2 \, dx = \pi \int_0^{\sqrt{2}} (x^4 - 8x^2 + 16) \, dx = \pi\sqrt{2}\left(\frac{172}{15}\right) \text{ cu. units}$$

and

$$\pi \int_0^{\sqrt{2}} x^4 \, dx = \pi\left(\frac{4\sqrt{2}}{5}\right) \quad \text{cubic units}$$

The required volume is $\pi\sqrt{2}(172/15 - \frac{4}{5}) = \pi\sqrt{2}(\frac{32}{3})$ cubic units.

We urge that you take special note of the factor π that precedes the integral for the volume; very often during the calculation of the volume, the factor π outside the integral sign is overlooked.   We also point out that the volume we found in the preceding example *could not* have been found by calculating $\pi \int_0^{\sqrt{2}} [(-x^2 + 4) - x^2]^2 \, dx$; that is, you cannot subtract the functions and then square the difference.   You must square the functions first, integrate, and then subtract; or square the functions, subtract, and then integrate.

EXAMPLE 34.3   Find the volume of the solid of revolution obtained by rotating each of the following regions about the x axis.
  (a)   the region below the graph of $y = e^x$ between $x = 0$ and $x = 1$
  (b)   the region below the graph of $y = \cos x$ between $x = 0$ and $x = \pi/2$

SOLUTION   (a)   The volume is given by the integral

$$\pi \int_0^1 (e^x)^2 \, dx = \pi \int_0^1 e^{2x} \, dx = \frac{\pi}{2} \int_0^1 e^{2x} \, 2dx = \left(\frac{\pi}{2}\right)\left(e^{2x}\Big|_0^1\right)$$

$$= \left(\frac{\pi}{2}\right)(e^2 - 1) \quad \text{cubic units.}$$

  (b)   The volume is given by the integral $\pi \int_0^{\pi/2} (\cos x)^2 \, dx$.   Now, an antiderivative of $(\cos x)^2$ is $\frac{1}{2}(x + \sin x \cos x)$ (check it by differentiation).   Evaluating at $x = \pi/2$ and at $x = 0$ and subtracting yields $\frac{1}{2}(\pi/2)$.   Multiplying by π (from outside the integral sign), we finally obtain $\pi^2/4$ cubic units.

As an exercise, compute the volume of the solid of revolution obtained by rotating about the x axis the region below the graph of $y = e^x$, above the graph of $y = \cos x$, between $x = 0$ and $x = 1$.

ROTATIONS ABOUT THE y AXIS

Suppose we have a situation as depicted in Figure 34.3, where we want to generate a solid of revolution by rotating a given region about the y axis. The method we now explain is valid *only* in cases where $f(x)$ is increasing *or* decreasing over an interval, since we must use the inverse function of f (see Definition 17.5). We repeat what was done in Definition 34.1, only this time we take horizontal instead of vertical rectangles. Hence, we take a regular partition P_n of the interval $J : \alpha \le y \le \beta$ on the y axis. We choose in some specific way a point y_i^* in each subinterval $J_i : y_{i-1} \le y \le y_i$ for each $i = 1, 2, \ldots, n$. Now for each y_i^* there is a corresponding value x_i^* such that $x_i^* = g(y_i^*)$ where g is the *inverse function* of f; that is, $g(y) = x$ if and only if $y = f(x)$. The rectangular area corresponding to each J_i is just $g(y_i^*)\Delta$ where $\Delta = (\beta - \alpha)/n$ (since the horizontal rectangles have height Δ and length of base x_i^*) and the corresponding volume is simply $\pi(g(y_i^*))^2 \Delta$. To approximate the volume of the solid of revolution, we add the volumes associated with the J_i and obtain the sum

$$\sum_{i=1}^n \pi(g(y_i^*))^2 \Delta$$

But this is just the Riemann sum of the function $\pi(g(y))^2$. Hence, this leads us to the integral $\pi \int_\alpha^\beta (g(y))^2 \, dy$ where g is the inverse function of f and the integration is done with respect to y instead of x. Therefore, the volume of the solid of revolution generated by rotating the region about the y axis is *defined* to be this integral.

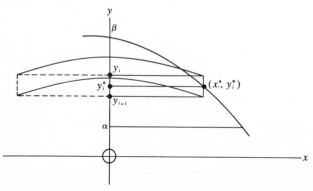

Figure 34.3

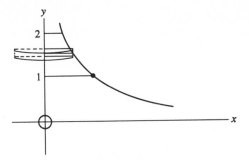

Figure 34.4

EXAMPLE 34.4   Find the volume of the solid of revolution obtained by rotating about the y axis the region between $y = 1$ and $y = 2$ bounded on the left by the y axis and on the right by the graph of $y = 1/x$ (see Figure 34.4).

SOLUTION $f(x) = 1/x$ is a decreasing function and possesses an inverse given by $x = g(y) = 1/y$. Hence, the volume is

$$\pi \int_1^2 \left(\frac{1}{y}\right)^2 dy = \pi\left(-\frac{1}{y}\Big|_1^2\right) = \frac{\pi}{2} \quad \text{cubic units}$$

EXAMPLE 34.5   Find the volume of the solid of revolution generated by rotating about the y axis the region between $y = 2$ and $y = 4$ bounded on the left by the y axis and on the right by the graph of $y = -x^2 + 4$ (see Figure 34.2 again).   To obtain the inverse function of $f(x) = -x^2 + 4$, we solve the equation for x, obtaining $x = \pm(4 - y)^{1/2}$. Since we are dealing only with positive values of x, we take $g(y) = (4 - y)^{1/2}$; thus, $(g(y))^2 = 4 - y$.   Therefore, the volume is $\pi \int_2^4 (4 - y)\, dy = 2\pi$ cubic units.

EXAMPLE 34.6   Find the volume of the solid of revolution obtained by rotating about the y axis the region below the graph of $y = x^2$ between $x = 0$ and $x = \sqrt{2}$ (see Figure 34.2).

SOLUTION   The inverse function of $y = x^2$ is simply $x = \sqrt{y}$ and as x varies from 0 to $\sqrt{2}$, y varies from 0 to 2.   The required volume is the difference between the volumes obtained by first rotating about the y axis the rectangle with vertices $(0, 0)$, $(\sqrt{2}, 0)$, $(\sqrt{2}, 2)$, and $(0, 2)$, and then rotating about the y axis the region above the graph of $y = x^2$ and below the line $y = 2$.   The volume determined by the rotation of the rectangle is the volume of a cylinder with radius of the base $\sqrt{2}$ and height 2.   Hence, the volume is $\pi(\sqrt{2})^2 2 = 4\pi$ cubic units. The second volume is given by $\pi \int_0^2 (\sqrt{y})^2 \, dy = \pi \int_0^2 y \, dy = 2\pi$ cubic units. Therefore, the difference $4\pi - 2\pi = 2\pi$ cubic units is the required volume.

We now give an example in which a solid of revolution is generated by a rotation about a horizontal and a vertical line, neither one of which is a coordinate axis.

EXAMPLE 34.7   Suppose that in the preceding example we rotate the given region (a) about the vertical line $x = \sqrt{2}$ and (b) about the horizontal line $y = 2$. Find the volume of each of the solids generated (see Figure 34.5).

SOLUTION   (a)   Note that in this case the horizontal rectangle depicted in Figure 34.5 is rotated about the line $x = \sqrt{2}$, yielding a cylinder with radius of the base approximately equal to $(\sqrt{2} - \sqrt{y})$ where $(\sqrt{2}, y)$ is a point of the base of the rectangle. Hence, the volume is given by the integral

$$\pi \int_0^2 (\sqrt{2} - \sqrt{y})^2 \, dy = \frac{2\pi}{3}$$

(b)   In this case, we must first compute the volume generated by rotating the rectangle with vertices $(0, 0)$, $(\sqrt{2}, 0)$, $(\sqrt{2}, 2)$, and $(0, 2)$ about the line $y = 2$. This is merely the volume of a cylinder of height $\sqrt{2}$ and radius of the base 2, so it is equal to $\pi(4)\sqrt{2}$. We then compute the volume of the solid generated by rotating about the line $y = 2$

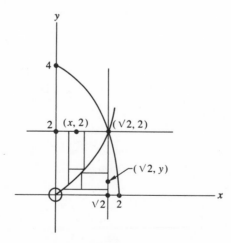

Figure 34.5

the region *above* the graph of $y = x^2$ between the y axis and the line $y = 2$. Finally, we subtract this from the first volume and obtain the required answer. Now, the vertical rectangle in Figure 34.5 is rotated about the line $y = 2$, yielding a cylinder with radius of the base approximately equal to $(2 - x^2)$ where $(x, 2)$ is a point of the base of the rectangle. Hence, the volume is given by the integral

$$\pi \int_0^{\sqrt{2}} (2 - x^2)^2 \, dx = \frac{32\sqrt{2}\pi}{15} \quad \text{cubic units}$$

Subtracting this from $4\sqrt{2}\pi$ yields the required volume $28\sqrt{2}\pi/15$ cubic units.

Finally, as exercises, compute the volume of each of the following solids of revolution.

1. The solid obtained by rotating about the y axis the region above the x axis between $x = \sqrt{2}$ and $x = 2$ and below the graph of $y = -x^2 + 4$. Also, the solids obtained by rotating this same region about the vertical line $x = \sqrt{2}$ and then about the horizontal line $y = 2$.

2. The solid obtained by rotating about the y axis the region enclosed by the two graphs $y = x^2$ and $y = -x^2 + 4$ between $x = 0$ and $x = \sqrt{2}$. Also, the solids obtained by rotating this same region about the vertical line $x = \sqrt{2}$ and the horizontal line $y = 4$.

3. The solid obtained by rotating about the x axis the region enclosed by the graphs of $y = x^2$ and $y = -x^2 + 4$ between $x = -\sqrt{2}$, $x = \sqrt{2}$, $y = 0$, and $y = 4$.

4. The solids obtained by rotating about both the x axis and the y axis the region bounded on the left by the graph of $y = x^2$, on the right by the graph of $y = -x^2 + 4$, and below by the x axis. Also, the solids obtained by rotating this same region about the horizontal line $y = 2$ and the vertical line $x = 2$.

We conclude this section with a second method for computing the volume of a solid of revolution where the rotation is about the y axis.

CYLINDRICAL SHELLS

Let f be a nonnegative continuous function defined on an interval $J: a \leq x \leq b$ with $a \geq 0$. By using Riemann sums, we will show that it is reasonable to define the volume V of the solid of revolution obtained by rotating about the y axis the region below the graph of f between the vertical lines $x = a$ and $x = b$ as $V = 2\pi \int_a^b x f(x) \, dx$.

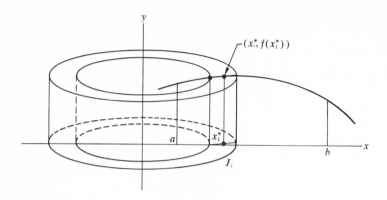

Figure 34.6

REMARK   Observe that the integration is done *with respect to x* and that the condition that f be a one-to-one function is *not* required.   This method is therefore *more* general than the one described earlier in this section for finding volumes of solids obtained by rotations about the y axis.

Let P_n be a regular partition of J and let x_i^* be the midpoint of J_i: $x_{i-1} \leq x \leq x_i$ for each $i = 1, 2, \ldots, n$ (see Figure 34.6).   If a vertical rectangle of width $\Delta = (b - a)/n$ and height $f(x_i^*)$ is rotated about the y axis, it will generate a "cylindrical shell."   If the shell is cut through vertically and then spread out flat, it becomes a thin rectangular sheet with thickness Δ, height $f(x_i^*)$, and width approximately $2\pi x_i^*$ (see Figure 34.7) where x_i^* approximates the outer radius of the cylindrical shell.   Therefore, the volume generated is approximately equal to $2\pi x_i^* f(x_i^*)\Delta$.   If we do this for each $i = 1, 2, \ldots, n$ and add, we obtain a Riemann sum for the function $2\pi x f(x)$.   This leads to the required integral formula.

EXAMPLE 34.8   Repeat Example 34.6 using the method of cylindrical shells.

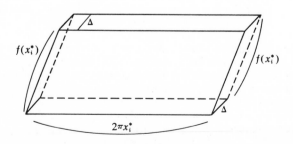

Figure 34.7

SOLUTION Applying the formula for the volume in this case, we obtain

$$V = 2\pi \int_0^{\sqrt{2}} x^3 \, dx = 2\pi \quad \text{cubic units}$$

This is, of course, the same answer we obtained in Example 34.6, but note that we obtained it *directly* in this case. In the earlier example we used the inverse function, we computed *two* volumes, and then subtracted them to obtain the correct answer.

EXAMPLE 34.9 Repeat Example 34.4 using the method of cylindrical shells.

SOLUTION In this case, taking vertical rectangles in Figure 34.4, we obtain a complete rectangle with vertices $(0, 1)$, $(0, 2)$, $(\frac{1}{2}, 2)$, and $(\frac{1}{2}, 1)$ and incomplete rectangles with altitudes given by $(1/x) - 1$ over the interval $\frac{1}{2} \le x \le 1$. Rotating the complete rectangle, we get a volume of $\pi/4$ cubic units. Rotating the incomplete rectangles and using the "cylindrical shells" formula on $\frac{1}{2} \le x \le 1$, we obtain

$$2\pi \int_{1/2}^1 x\left(\frac{1}{x} - 1\right) dx = \frac{\pi}{4} \quad \text{cubic units}$$

Adding the two volumes, we have $\pi/2$ cubic units, which was the answer obtained in Example 34.4 In the earlier example, however, the method was much easier; in the present case we needed a longer and more tricky computation.

As a result of the two preceding examples, we conclude that both methods for computing volumes where the rotations are about the y axis are useful. You must therefore decide for yourself which is easier to use in a given problem.

PROBLEMS

34.1 (a) Compute the volume of the solid of revolution generated by rotating about the x axis the region bounded by the x axis and the graph of $y = x^3$, between $x = -1$ and $x = 2$.
 (b) Repeat part (a) rotating about the y axis.
34.2 Find the volume of the solids generated by rotating each of the following regions first about the x axis, and then about the y axis.
 (a) the region in the first quadrant enclosed by the ellipse $4x^2 + y^2 = 4$
 (b) the region in the first quadrant enclosed by the hyperbola $x^2 - y^2 = 1$ and the vertical line $x = 2$.
34.3 Find the volume of the solid of revolution generated by rotating each of the following regions about the x axis.
 (a) the region above the x axis and below the graph of $y = |x|$, between $x = -1$ and $x = 1$

(b) the region above the x axis and below the graph of $y = |x^2 - 1|$, between $x = -2$ and $x = 2$. (See Figure 13.6.)

34.4 Using part (C) of Theorem 32.3, find a bound on the volume of the solid of revolution generated by rotating each of the following regions about the x axis.

(a) the region above the x axis, below the graph of $y = e^x(-x^2 + x + 1)$, between $x = 0$ and $x = 1$

(b) the region above the x axis, below the graph of $y = xe^x \cos x$, between $x = 0$ and $x = 1$.

34.5 (a) Find the volume of the solid of revolution generated by rotating about the x axis the region below the graph of $y = 1/x$ between $x = 1$ and $x = b, b > 1$.

(b) What happens to this volume as b becomes large?

(c) Repeat parts (a) and (b) using $y = 1/x^2$.

(d) Compare the preceding results with those of Problem 33.2.

34.6 (a) Find the area of the region above the x axis and below the graph of $y = e^{-x}$ between $x = 0$ and $x = b, b > 0$.

(b) What happens to this area as b becomes large?

(c) Find the volume of the solid of revolution generated by rotating about the x axis the region below the graph of $y = e^{-x}$ between $x = 0$ and $x = b, b > 0$.

(d) What happens to this volume as b becomes large?

34.7 (a) Compute the volume of the solid of revolution generated by rotating about the x axis the region bounded by the x axis and the graph of $y = xe^{x^3}$, between the y axis and $x = 1$.

(b) Repeat part (a) with $y = x^2 e^{x^5}$.

34.8 Find the volume of the solid of revolution generated by rotating about the x axis the region above the x axis and below the graph of $y = \sin x$, between the y axis and the line $x = \pi$.

34.9 Find the volume of the solid generated by revolving about the y axis the region between the graph of $y = \log x$, the x axis, the y axis, and the line $y = -1$.

34.10 Find the volume of the solid of revolution obtained by rotating the region below the graph of $y = \cos x$ between $x = 0$ and $x = \pi/2$. (a) about the horizontal line $y = 1$; (b) about the y axis (leave your answer in the form of an integral).

34.11 Find the volume of the solid of revolution obtained by rotating the region below the graph of $y = e^x$ above the line $y = 1$ between $x = 0$ and $x = 1$ (a) about the horizontal line $y = 1$; (b) about the y axis (leave your answer in the form of an integral).

34.12 Consider the regions A, B, and C of Example 33.7 (Figure 33.10). Find the volume of the solids of revolution generated by (a) rotating each one of the three regions about the x axis; (b) both A and C about the y axis; (c) both A and C about the horizontal line $y = 1$; (d) B about the vertical line $x = 1$ and then about $x = 2$.

34.13 *Volumes by Slicing.* Consider a solid erected about the y axis (not necessarily a solid of revolution) with its base perpendicular to the y axis and passing through the origin (see Figure 34.8). Let h be the height of this solid and suppose that for each y between 0 and h the area of a cross-sectional cut [the region determined by the intersection, with the solid, of a plane parallel to the base and passing through the point $(0, y)$] is given by a continuous function $A(y)$.   Using Riemann sums, show that it is reasonable to define the volume of the solid as $\int_0^h A(x)\,dx$.

34.14   Using the result of the preceding problem, show that a pyramid having height h and a rectangular base of area R square units has volume equal to $\frac{1}{3}Rh$.

34.15*   Using the result of Problem 34.13, derive the integral formula for the volume of a *solid of revolution* obtained by a rotation about the y axis. Compare your development with that given in the text immediately preceding Example 34.4.

34.16   Using the method of cylindrical shells, find the volume of the solid of revolution obtained by rotating about the y axis each of the following regions.
  (a)   the region above the x axis, below the graph of $y = -x^2 + 2x$, between $x = 0$ and $x = 2$.   (See Figure 33.10).   Also, rotate this region about the vertical line $x = 3$
  (b)   the region above the x axis, below the graph of $y = xe^{-x^3}$, between $x = 0$ and $x = 1$
  (c)   the region above the x axis, below the graph of $y = \sin x$, between $x = 0$ and $x = \pi$
  (d)   the region below the x axis, above the graph of $y = \cos x$, between $x = \pi/2$ and $x = 3\pi/2$

34.17   Repeat each of the following problems, using the method of cylindrical shells.
  (a)   Problem 34.1(b)
  (b)   Problem 34.11(b)
  (c)   Problem 34.12(b)
  (d)   Problem 34.12(d)

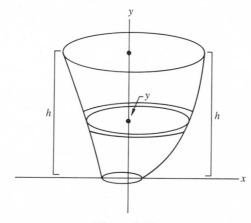

Figure 34.8

Section 35

WORK

We now discuss the application of the definite integral to two basic types of problems involving work. The first involves the work done in stretching a spring; the second involves the work done in filling or emptying water tanks of various shapes and sizes.

Definition 35.1 If a constant force of c pounds (lb) is applied to an object moving it through a distance of d ft, the *work* W done on the object is given by $W = cd$ foot-pounds (ft lb).

For example, if a man lifts a 30-lb rock (at a constant rate) a distance of 2 ft, the amount of work done is simply $(30)(2) = 60$ ft lb, since a force of 30 lb had to be applied to lift the rock.

Suppose that we have a *variable* force acting on an object or particle P; how do we define the work done? We may attempt an answer to this question by first assuming that P is being moved to the right on a horizontal coordinate line (see Figure 35.1). We assume further that when P is at the point x on the line, a force of $f(x)$ pounds is being applied, where f is a continuous function on an interval $J: a \le x \le b$. Now, we can use the calculus to define the work that is done on the particle P as it moves from the point a to the point b on the line. We proceed as follows. Let P_n be a regular partition of J. Choose the points x_i^* as the midpoints of the subintervals J_i where $i = 1, 2, \ldots, n$. For very large n, the *constant* force $f(x_i^*)$ approximates the *variable* force $f(x)$ at each point $x \in J_i$. Therefore, the approximate work done in moving P over J_i from x_{i-1} to x_i is simply $f(x_i^*)\Delta$ where $\Delta = (b - a)/n$. Doing this for each

Figure 35.1

234

$i = 1, 2, \ldots, n$ and adding, we obtain $S_n = \sum_{i=1}^{n} f(x_i^*)\Delta$, which is an *approxima-tion* to the total work W done in moving the particle from a to b. Certainly, for larger and larger n, S_n becomes a better and better approximation to W. But as $n \to \infty$, $S_n \to \int_a^b f(x)\, dx$. This leads us to the following definition.

Definition 35.2 If f is a continuous function defined on an interval $J: a \le x \le b$ with the property that $f(x)$ is the force applied to the particle P at the point $x \in J$, the work W that is done in moving P from a to b is defined as $W = \int_a^b f(x)\, dx$. The function f is called the *force function*.

THE SPRING PROBLEMS

Suppose a spring has natural length L. Then we may attach one end at the point $-L$ on a horizontal coordinate line. Then the free end lies at the origin. Let $f(x)$ be the force that holds the free end of the spring x units to the right of the origin. Hooke's law states that $f(x) = kx$ where k is a constant that depends on the spring (k is called the *spring constant*). There-fore, the work done in moving the free end of the spring from the origin to a point b to the right of the origin is $W = \int_0^b kx\, dx = kb^2/2$. Now, if the free end is already at a point a, where $0 < a < b$, the work done in moving the free end from the point a to the point b is $kb^2/2 - ka^2/2 = (k/2)(b^2 - a^2)$. Also, observe that if the force function is a constant c, $W = \int_a^b c\, dx = c(b-a) =$ force times distance. Hence, the integral definition of work reduces to the elementary definition of work (Definition 35.1).

EXAMPLE 35.1 If it takes a 100-lb force to stretch a spring 2 ft beyond its natural length, how much work is done in stretching it 3 ft beyond its natural length? How much work is done in stretching it from 3 to 5 ft beyond its natural length?

SOLUTION $W = \int_0^3 kx\, dx = k(\frac{9}{2})$, and the spring constant k can be found by using the fact that when $x = 2$, the force function has value 100. Hence, $k = 50$ and $W = 225$ ft lb.

To answer the second question, we compute $(k/2)(b^2 - a^2) = 25(25 - 9)$ $= 400$ ft lb.

THE TANK PROBLEMS

Suppose we have a water tank of some specific shape and we are interested in finding the amount of work needed for pumping water either into or out of the tank. The technique used to solve this kind of problem involves the

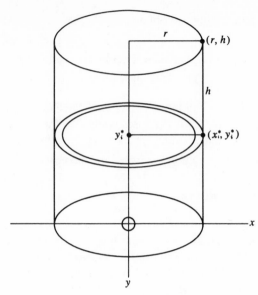

Figure 35.2

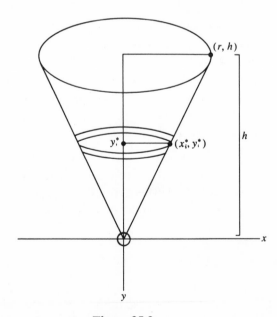

Figure 35.3

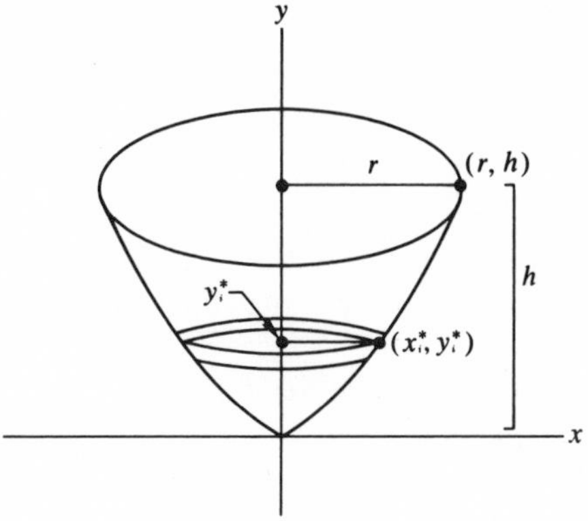

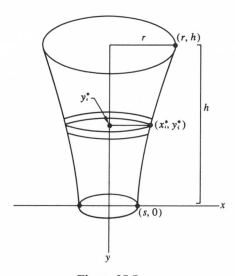

Figure 35.4

approximation of the work required to put in or take out a small disk or pancake of water. We then add all these approximations and obtain the Riemann sum of a certain function. This leads to a corresponding integral definition. We consider here vertical tanks in the shape of a *cylinder*, a *cone*, a *paraboloid of revolution*, and a *hyperboloid of revolution*. Figures 35.2, 35.3, 35.4, and 35.5 represent each of the shapes in the order in which they are named. We have also chosen a convenient way of placing a rectangular coordinate system in each diagram. Now, all of these tanks can be handled in roughly the same way. First, we can assume that the tops are all circular with

Figure 35.5

radius r ft, and that all the tanks are h ft high. Suppose that they are full of water and we would like to pump all the water out at the top. We choose a regular partition P_n of the interval $J: 0 \leq y \leq h$ on the y axis and we choose y_i^* as the midpoint of the subinterval J_i. For each i, we obtain a solid that closely resembles a disk or pancake of height $\Delta = h/n$. The radius of the disk (although it varies in general) can be approximated by the number $x_i^* = g(y_i^*)$ where g is the *inverse function* of f and where $y = f(x)$ is the function corresponding to the graph that generates the solid by a rotation about the y axis (see Section 34).

In the case of the cylinder, the vertical line $x = r$ from $y = 0$ to $y = h$ generates the cylinder; hence, $x = g(y) = r = a$ constant for any y between 0 and h. In the case of the cone, a portion of the slant line with equation $y = mx$ generates, so $x = g(y) = y/m$ for any y between 0 and h. The paraboloid is generated by a function of the type $y = kx^2$ where k is a positive constant; thus, $x = g(y) = (y/k)^{1/2}$ for any y between 0 and h. Finally, the hyperboloid is generated by the graph in the first quadrant of a hyperbola with equation $x^2/a^2 - y^2/b^2 = 1$. Solving for x and taking the positive radical, we obtain the inverse function $x = g(y) = (a/b)(y^2 + b^2)^{1/2}$ for any y between 0 and h.

Suppose that we take $g(y_i^*)$ to be the radius of the disk or pancake. Then, the volume of the pancake for each $i = 1, 2, \ldots, n$ can be approximated by $\pi(g(y_i^*))^2\Delta$ and the weight of the water in this pancake is therefore approximately $(62.5)\pi(g(y_i^*))^2\Delta$ lb (since water weighs 62.5 lb/ft^3). Also, the ith pancake is approximately at the height y_i^*; hence, it must be lifted a total distance of $h - y_i^*$ ft in order to empty it from the top of the tank. Consequently, the work done in moving the ith pancake out of the tank can be approximated by $(62.5)\pi(h - y_i^*)(g(y_i^*))^2\Delta$. Doing this for each $i = 1, 2, \ldots, n$ and adding, we obtain the Riemann sum corresponding to the function $(62.5)\pi(h - y)(g(y))^2$. Hence, we have the following.

Definition 35.3   The total work done in emptying a tank with the preceding specifications is given by $W = (62.5)\pi \int_0^h (h - y)(g(y))^2 \, dy$ where g is the inverse function of the function f whose graph generates the solid of revolution that determines the shape of the water tank.

Before giving some examples, we list a few related problems that can be solved by only slightly changing the preceding integral formula for W.

1.   Suppose that the water tank is full and it is to be emptied by pumping the water through a pipe that extends to a point s ft above the tank. This means that each disk or pancake at a height of y_i^* ft must be lifted $(h + s) - y_i^*$ ft. Consequently, the term $(h - y)$ in the integrand of the integral formula is simply changed to $(h + s - y)$.

2.   Suppose that the water tank is only filled to a level of t ft, where $0 < t < h$, and you want to empty it by pumping the water out at the top.

Then the upper limit in the integral formula is changed to t, since the values of y now lie between $y = 0$ and $y = t$.

3. Suppose that the tank is full and you want to pump out only enough water to lower the water level from h ft to t ft where $0 < t < h$. Then the lower limit in the integral formula is changed to t, since only disks at a height greater than or equal to t are being pumped out.

4. Finally, if you want to pump water *into* an *empty* tank from the bottom, you must consider that a disk that ends up at height y_i^* entered the bottom of the tank and was lifted y_i^* ft. This changes the $(h - y)$ term in the formula for W to simply y and the resulting formula is $W = (62.5)\pi \int_0^h y(g(y))^2\, dy$.

EXAMPLE 35.2 Suppose that a *cylindrical* water tank of height h ft and radius of the base r ft is full of water. Find the work done in pumping all the water out at the top of the tank (see Figure 35.2).

SOLUTION   From the preceding discussion on cylindrical tanks, $g(y) = r$; therefore

$$W = (62.5)\pi \int_0^h (h - y)r^2\, dy = (62.5)\pi r^2 \int_0^h (h - y)\, dy$$

Evaluating the integral, we obtain

$$\left(hy - \frac{y^2}{2}\right)\Big|_0^h = h^2 - \frac{h^2}{2} = \frac{h^2}{2}$$

Therefore, $W = \frac{1}{2}(62.5)\pi r^2 h^2$ is the amount of work required to empty the cylindrical tank.

EXAMPLE 35.3 Repeat Example 35.2 with a conical tank h ft high with radius of its circular top equal to r (see Figure 35.3).

SOLUTION   From the preceding discussion on conical tanks, $g(y) = y/m$ where m is the slope of the slant line that generates the cone. But $m = h/r$; therefore, $g(y) = (r/h)y$. Now, if we place $g(y) = (r/h)y$ in the integral formula for W and evaluate it, we find that $W = \frac{1}{12}(62.5)\pi r^2 h^2$ is the work required to empty the conical tank. We leave the details as an exercise.

EXAMPLE 35.4 Repeat Example 35.2 with a tank in the shape of a paraboloid of revolution where the height of the tank is h ft and the circular top has a radius of r ft (see Figure 35.4).

SOLUTION   From the preceding discussion on paraboloids of revolution, the graph that generates the solid has an equation of the form $y = kx^2$ and passes through the point (r, h). Therefore, $h = kr^2$, $k = h/r^2$, and finally, $x^2 = (g(y))^2 = y/k = (r^2/h)y$. Substituting into the integral formula for W and evaluating, we obtain $W = \frac{1}{6}(62.5\,\pi r^2 h^2$ as the work done in emptying the tank, which has the shape of a paraboloid of revolution.

EXAMPLE 35.5   Repeat Example 35.2 with a tank in the shape of a hyperboloid of revolution where the tank is 10 ft high and has a circular base of radius 2 ft, and a circular top of radius $2\sqrt{2}$ ft (see Figure 35.5 with $r = 2\sqrt{2}$ and $s = 2$).

SOLUTION   From the preceding discussion on hyperboloids of revolution, we have that $g(y) = (a/b)(y^2 + b^2)^{1/2}$ when the hyperbola that generates the solid has equation $x^2/a^2 - y^2/b^2 = 1$.   Consequently, all we need to do is determine the constants a and b.   Now, $(2, 0)$ is evidently a vertex of the hyperbola; hence, $a = 2$, while $(2\sqrt{2}, 10)$ is a point on the graph. Substituting for x, y, and a, we obtain $\frac{8}{4} - 100/b^2 = 1$.   Solving for b, we obtain $b^2 = 100$ and $b = 10$.   Therefore, $g(y) = \frac{1}{5}(y^2 + 100)^{1/2}$. Substituting into the integral formula yields

$$W = (62.5)\left(\frac{\pi}{25}\right) \int_0^h (h - y)(y^2 + 100)\, dy$$

and evaluating, we arrive at $W = (43,750)(\pi/3)$ ft lb.

PROBLEMS

35.1   Find the amount of work required in removing all the water at the top from a vertical *cylindrical* tank that is 6 ft high and has radius of its base equal to 2 ft if (a) the tank is full of water; (b) the tank is half full of water and the water must also pass through a vertical pipe that extends to a point 5 ft above the top of the tank.

35.2   Repeat Problems 35.1(a) and (b) with a vertical *conical* tank 6 ft high having a radius of its top equal to 2 ft.

35.3   Suppose that we have an empty vertical conical tank that is 10 ft high with the radius of its top equal to 2 ft.

(a)   Find the amount of work required to fill the tank with water at the bottom.

(b)   After the tank is filled, how much work is required to pump out at the top enough water to lower the water level from 10 ft to 8 ft?

(c)   After lowering the water level to 8 ft, how much work is now required to empty the tank by pumping the water out at the top?

35.4   A man has a pump that can do a maximum of 31,250 ft lb of work in 1 hour.

(a)   He wants to build a *cylindrical* water tank 10 ft high.   What is the radius of the base of the *largest* tank he can build and still empty at the top in 1 hour?

(b)   He wants to build a *conical* tank with a top of radius 6 ft.   How high can he build the tank and still be able to empty it at the top in 1 hour?

35.5   A water tank has the shape of a paraboloid of revolution.   It is 6 ft high and the radius of its circular top is 2 ft.

(a)   If the tank is full, how much work is required to empty it at the top?

(b) If the tank is empty, how much work is required to fill it at the bottom to a level of 4 ft?

35.6 A water tank has the shape of a hyperboloid of revolution. It is 12 ft high and has a circular top of radius $3\sqrt{2}$ ft and a circular base of radius 3 ft.

(a) If the tank is empty, how much work is required to fill it at the bottom to a level of 8 ft?

(b) After it is filled to a level of 8 ft, how much work is required to empty it at the top?

35.7* Develop a formula for the amount of work required to empty a full tank of water by pumping it out the top if the tank is (a) a hemisphere of radius r ft; (b) a pyramid with a square base s ft by s ft and height h ft; (c) a hyperboloid of revolution of height h, radius of its circular base equal to s ft, and radius of its circular top equal to $s\sqrt{2}$ ft.

35.8 A spring is stretched 3 inches beyond its natural length by a force of 10 lb.

(a) How much work is done in stretching it 7 inches beyond its natural length?

(b) How much work is done in stretching it from 7 inches to 8 inches beyond its natural length?

35.9 A 3000-lb car is lifted from a dock to the deck of a ship 30 ft high by a cable that weighs 5 lb/ft. Find the work done in lifting the car to the deck of the ship.

35.10 By Newton's law of attraction, two objects of weight A and B are attracted to each other by a force of $F(x) = kAB/x^2$ lb, where the objects are x ft apart and k is a constant.

(a) Find the work done in separating the objects from a distance of 2 ft to a distance of 4 ft.

(b) Find the work done in separating the objects from a distance of 1 ft to a distance of d ft, $d > 1$.

(c) Suppose that the objects are 1 ft apart and we start separating them by moving one of them farther and farther away from the other one. How much work must theoretically be done in order to separate the objects *completely*?

35.11* *Force on a Dam.* Suppose we are interested in finding the total force exerted by a liquid of density α against a vertical dam h ft high (see Figure 35.6). Assume that the width of the dam at height y is given by a continuous

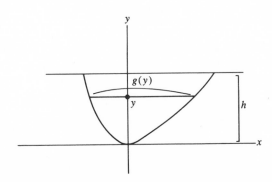

Figure 35.6

function $g(y)$. By using Riemann sums, show that the integral that defines the force on the dam is $\alpha \int_0^h (h-y)g(y)\,dy$.

[*Hint:* You may assume the following. (a) Liquid in a container exerts a force on the bottom equal to its weight.

(b) The force per square foot at the bottom is called the *pressure* at the bottom.

(c) The pressure exerted by a liquid of density α at a point d units below the surface is αd and this pressure is the same in *every* direction.]

35.12 Find the force of water against a triangular dam 100 ft wide (at the water level) and 50 ft deep.

35.13 Find the force of water on a dam shaped like a parabola where the width of the dam at the water level is 100 ft and the depth of the water is 50 ft.

35.14 A cylindrical tank 60 ft long with radius of the base 25 ft is half full of oil. If the tank is lying on its side, find the force exerted by the oil on either end of the tank (assume that the oil weighs 60 lb/ft³).

35.15* Suppose there is a drain pipe below the water surface in the dam of Problem 35.13. Suppose the cover of the drain pipe is circular and has radius 1 ft with its center 6 ft below the surface of the water (see Figure 35.7). Find the force on the cover (you may leave your answer in the form of an integral).

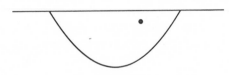

Figure 35.7

Section 36

METHODS OF INTEGRATION

This section should really be named "Methods of Antidifferentiation," since in it we discuss techniques for finding antiderivatives of functions that *do not* have easily recognizable antiderivatives. In Section 30, we covered most of the cases of antiderivative formulas that came from the elementary derivative formulas, and these results have been sufficient up to now (before reading on, review Section 30). There are, however, many functions that require special techniques for finding their antiderivatives. There are basically three such techniques: integration by *parts*, by *substitution*, and finally, by using *partial fractions*. All of these are covered extensively in regular textbooks; in fact, trigonometric substitutions and partial fractions are usually covered quite extensively. Consequently, we discuss in this section an abundance of difficult examples relating to some of these techniques. At the end of the section we include problems that can be solved by using one or more of these various techniques. In any event, all the important antiderivative formulas can be found in the table of integrals in the back of your calculus textbook.

INTEGRATION BY PARTS

Let $u(x)$ and $v(x)$ be two differentiable functions of x defined in some interval J. The product formula for the derivative yields the equation $(uv)' = uv' + u'v$, which leads to $uv' = (uv)' - u'v$. Therefore, the antiderivative of the function uv' equals uv − the antiderivative of the function $u'v$. This means that

$$\int u(x)v'(x)\,dx = u(x)v(x) - \int v(x)u'(x)\,dx + C$$

244   *The Integral Calculus*

This formula frequently allows us to change the form of an integral to one that can be evaluated by known methods.

EXAMPLE 36.1   Find the antiderivative of the function $\sin^2 x$.

SOLUTION   Let $u(x) = \sin x$ and $v'(x) = \sin x$. Then $u'(x) = \cos x$ and $v(x) = -\cos x$ (we choose $C = 0$ to simplify matters). Substituting into the formula, we obtain

$$\int \sin^2 x \, dx = -\sin x \cos x + \int \cos^2 x \, dx$$

But $\cos^2 x = 1 - \sin^2 x$; hence, the integral on the right side of the preceding equation can be rewritten in the form $\int (1) \, dx - \int \sin^2 x \, dx$. This leads to the equation $2 \int \sin^2 x \, dx = x - \sin x \cos x$, which finally reduces to the formula

$$\int \sin^2 x \, dx = \tfrac{1}{2}(x - \sin x \cos x) + C$$

Finally, if in the beginning we had chosen $v(x) = -\cos x + C$, where C is any real constant, we would have obtained the same answer because all of the extra terms would have canceled each other. Verify this.

As an exercise, imitate the preceding solution and show that the antiderivative of the function $\cos^2 x = \tfrac{1}{2}(x + \sin x \cos x) + C$.

EXAMPLE 36.2   Find the antiderivative of the function $2e^x \sin^2 x$.

SOLUTION   Let $u(x) = e^x$ and $v'(x) = 2 \sin^2 x$. Then $u'(x) = e^x$ and $v(x) = x - \sin x \cos x$ (from the preceding example). Substituting into the formula, we obtain

$$\int 2e^x \sin^2 x \, dx = e^x(x - \sin x \cos x) - \int e^x(x - \sin x \cos x) \, dx$$

The latter integral splits into the two integrals $-\int xe^x \, dx + \int e^x \sin x \cos x \, dx$. But xe^x has an antiderivative given by $xe^x - e^x$ [verify this by using the parts method on xe^x with $u(x) = x$ and $v'(x) = e^x$]. Therefore, $-\int xe^x \, dx = -xe^x + e^x$. Now, the other integral involves the function $e^x \sin x \cos x$, which does not look any easier to handle than the original one. However, if we once again use the parts technique, letting $u(x) = e^x$ and $v'(x) = \sin x \cos x$, then $u'(x) = e^x$ and $v(x) = \tfrac{1}{2} \sin^2 x$. Hence,

$$\int e^x \sin x \cos x \, dx = (\tfrac{1}{2})e^x \sin^2 x - \int \tfrac{1}{2}e^x \sin^2 x \, dx$$

Substituting into the first equation, we obtain

$$\int 2e^x \sin^2 x \, dx = e^x(x - \sin x \cos x) - xe^x + e^x$$

$$+ \tfrac{1}{2}e^x \sin^2 x - \tfrac{1}{2}\int e^x \sin^2 x \, dx$$

which leads to the equation

$$\frac{5}{2}\int e^x \sin^2 x \, dx = e^x(1 - \sin x \cos x + \tfrac{1}{2} \sin^2 x)$$

Therefore,

$$\int 2e^x \sin^2 x \, dx = \tfrac{4}{5}e^x(1 - \sin x \cos x + \tfrac{1}{2} \sin^2 x) + C$$

As an exercise, verify this formula by differentiating the right side of the equation and obtaining $2e^x \sin^2 x$.

ANTIDERIVATIVES OF INVERSE FUNCTIONS

In Section 27, we proved a result that enabled us to find the derivative of the inverse function f^* whenever we knew the derivative of f (see Theorem 27.1). We now prove a similar theorem for the antiderivative of an inverse function.

Theorem 36.1   Let f be a continuous one-to-one function defined on the interval $J: a \le x \le b$ with range $J^*: \alpha \le y \le \beta$. Suppose further that f' exists and is nonzero at each point of $a < x < b$. Let f^* be the inverse function of f and let g be an antiderivative of f over J. Then $\int f^*(x) \, dx = xf^*(x) - g(f^*(x)) + C$.
Before proving this theorem, we give some examples.

EXAMPLE 36.3   Using Theorem 36.1, find the antiderivative of each of the functions (a) $\sqrt{x}$; (b) $\log x$; (c) $\arctan x$.

SOLUTION   (a)   The square root function is the inverse of the function $f(x) = x^2$, $x \ge 0$. Hence, we write it as $f^*(x) = x^{1/2}$. An antiderivative $g(x)$ of the function $f(x) = x^2$ is $x^3/3$. Substituting into the formula of Theorem 36.1, we obtain

$$\int x^{1/2} \, dx = x \, x^{1/2} - (x^{1/2})^3/3 = \tfrac{2}{3}x^{3/2} + C$$

(which of course agrees with the answer obtained by our earlier method of computing the antiderivative of $x^{1/2}$ by the power formula).

(b) The log function is the inverse of the exponential function $f(x) = e^x$. Hence, we write it as $f^*(x) = \log x$. An antiderivative $g(x)$ of e^x is e^x. Substituting into the formula of Theorem 36.1, we obtain

$$\int \log x \, dx = x \log x - e^{\log x} = x \log x - x + C$$

As an exercise, find the antiderivative of $\log x$ by using the parts method with $u(x) = \log x$ and $v'(x) = 1$.

(c) The arctangent function is the inverse function of $f(x) = \tan x$, for $-\pi/2 < x < \pi/2$; hence, $f^*(x) = \arctan x$. The antiderivative of $f(x) = \tan x$ is $g(x) = -\log(\cos x)$ [since $\tan x = \sin x/\cos x = (-1)(-\sin x)/\cos x$, which has the form $f'(x)/f(x)$; see Example 30.4(b)]. Substituting into the formula of Theorem 36.1, we obtain

$$\int \arctan x \, dx = x \arctan x - \log(\cos(\arctan x)) + C$$

Now, if $z = \arctan x$, $\cos z = 1/(1 + x^2)^{1/2}$ and $\log(\cos z) = -\frac{1}{2}\log(1+x^2)$. Therefore, $\int \arctan x \, dx = x \arctan x + \frac{1}{2} \log(1 + x^2) + C$.

As an exercise, find the antiderivative of $\arctan x$ by using the parts method with $u(x) = \arctan x$ and $v'(x) = 1$. Finally, as another exercise, use Theorem 36.1 to find the antiderivative of the function $\arcsin x$ and show that it is equal to $x \arcsin x + (1 - x^2)^{1/2} + C$.

We now give the proof of Theorem 36.1.

PROOF Let $y \in J^*$; then $f^*(y) = x$ if and only if $y = f(x)$. From Theorem 27.1, we know that $f^{*'}(y) = 1/f'(x)$ provided $f'(x)$ exists and is not zero. Consider f^* as a function of y defined on J^* and let us find $\int f^*(y) \, dy$ by using the parts method. Let $u(y) = f^*(y)$ and $v'(y) = 1$; then $u'(y) = f^{*'}(y) = 1/f'(x)$ and $v(y) = y$. Therefore, $\int f^*(y) \, dy = yf^*(y) - \int (y/f'(x)) \, dy$. But $y = f(x)$ and $dy = f'(x) \, dx$; hence, the last integral simplifies to $-\int f(x) \, dx = -g(x) = -g(f^*(y))$ where g is an antiderivative of f. Finally, if we exchange y for x and rewrite the preceding equations, we obtain the formula in terms of x. ■

We now complete our discussion of the parts method with one more example.

EXAMPLE 36.4 Find the antiderivative of the function $(\log x)^2$.

SOLUTION Let $u(x) = \log x$ and $v'(x) = \log x$. Then $u'(x) = 1/x$ and $v(x) = x \log x - x$. Therefore,

$$\int (\log x)^2 \, dx = (\log x)(x \log x - x) - \int ((\log x) - 1) \, dx$$

The last integral reduces to

$$- \int \log x \, dx + \int (1) \, dx = -x \log x + x + x$$

Consequently, the antiderivative of $(\log x)^2$ is $2x - 2x \log x + x(\log x)^2 + C$.

As an exercise, verify the preceding result by differentiation.

SUBSTITUTION

At this point, we assume you have read about both rational and trigonometric substitution techniques for finding antiderivatives. We restrict ourselves to a few examples.

EXAMPLE 36.5 Find the antiderivative of each of the functions (a) $e^x \arctan e^x$ (b) $(\log x)/x$.

SOLUTION   (a)   If we let $u = e^x$, then $du = e^x \, dx$ and we obtain

$$\int e^x \arctan e^x \, dx = \int \arctan u \, du = u \arctan u + \tfrac{1}{2} \log(1 + u^2)$$

[see Example 36.3(c)].   Replacing u with e^x, we finally obtain

$$\int e^x \arctan x \, dx = e^x \arctan e^x + \tfrac{1}{2} \log(1 + e^{2x}) + C$$

(b)   If we let $u = \log x$, $du = (1/x) \, dx$ and we obtain

$$\int \frac{\log x}{x} \, dx = \int u \, du = \frac{u^2}{2} = \tfrac{1}{2}(\log x)^2 + C$$

REMARK   In the preceding example, we conformed with the shorthand notation used in most textbooks.   Before going on, however, we exhibit solutions to parts (a) and (b) that emphasize the basis for this technique, namely, the *chain rule*.

(a)   Let $u(x) = e^x$; then $u'(x) = e^x$.   Hence, $e^x \arctan e^x = u'(x) \arctan u(x)$, which has the form $g(u(x))u'(x)$ where the function g is the arctangent function.   Now, if G is an antiderivative of g, $G(u(x))$ has its derivative (by the chain rule) equal to $G'(u(x))u'(x) = g(u(x))u'(x)$.   Therefore, an antiderivative of $g(u(x))u'(x)$ is just $G(u(x))$ where G is an antiderivative of g.   In this problem, g is the arctangent function; therefore, by Example 36.3(c), $G(u(x)) = u(x) \arctan u(x) + \tfrac{1}{2} \log(1 + u^2(x))$.   Substituting e^x for $u(x)$, we obtain the same answer as before.

(b)   Let $u(x) = \log x$;   then $u'(x) = 1/x$.   Therefore, $(\log x)(1/x) = u(x)u'(x)$. But by the chain rule, $(u(x))^2/2$ has its derivative equal to $u(x)u'(x)$. Hence, the antiderivative of $u(x)u'(x)$ is $(u(x))^2/2 + C$. Substituting $\log x$ for $u(x)$, we again obtain $\frac{1}{2}(\log x)^2 + C$.

EXAMPLE 36.6   Find the antiderivative of each of the functions (a) $1/(e^x + e^{-x})$; (b) $\cos(\log x)$; (c) $1/(1 + \sin x)$.

SOLUTION   (a)   Multiplying numerator and denominator by e^x, we obtain $e^x/(1 + e^{2x})$. Letting $u = e^x$, $du = e^x\,dx$, we obtain

$$\int e^x/(1 + e^{2x})\,dx = \int 1/(1 + u^2)\,du = \arctan u = \arctan e^x + C$$

(b)   Let $x = e^t$, $dx = e^t\,dt$; then

$$\int \cos(\log x)\,dx = \int (\cos t)(e^t\,dt)$$

But

$$\int e^t \cos t\,dt = \tfrac{1}{2}e^t(\cos t + \sin t)$$

(verify this by differentiation of the right side of the equation or by using the parts method on $e^t \cos t$ with $u(t) = e^t$ and $v'(t) = \cos t$), and substituting $x = e^t$ and $\log x = t$, we obtain

$$\int \cos(\log x)\,dx = \tfrac{1}{2}x[\cos(\log x) + \sin(\log x)] + C$$

As an exercise, find the antiderivative of $\cos(\log x)$ by using the parts method with $u(x) = \cos(\log x)$ and $v'(x) = 1$.

(c)   Multiplying numerator and denominator by $1 - \sin x$ yields

$$\frac{1 - \sin x}{1 - \sin^2 x} = \frac{1 - \sin x}{\cos^2 x} = \frac{1}{\cos^2 x} - \frac{\sin x}{\cos^2 x} = \sec^2 x + \frac{(-\sin x)}{\cos^2 x}$$

Now, the antiderivative of $\sec^2 x$ is $\tan x$, and letting $u = \cos x$, we obtain $-1/\cos x$ as the antiderivative of $(-\sin x)/(\cos^2 x)$. Hence, $\int 1/(1 + \sin x)\,dx = \tan x - \sec x + C$.

EXAMPLE 36.7   Find the antiderivative of each of the functions (a) $1/(x^2 + 2x + 2)$ (b) $(-x^2 + 4x - 3)^{1/2}$.

SOLUTION   (a)   Completing the square in the denominator, we obtain $1/[1 + (x + 1)^2]$. Let $u = x + 1$; then $du = dx$ and we have $\int 1/(1 + u^2)\,du = \arctan u + c = \arctan(x + 1) + C$.

(b) Completing the square, we obtain $[1 - (x - 2)^2]^{1/2}$. Let $u = x - 2$; then $du = dx$ and we have $\int (1 - u^2)^{1/2} \, du = \frac{1}{2}[\arcsin u + u(1 - u^2)^{1/2}]$ (verify this by differentiating the right side of the equation or by making the substitution $u = \sin t$ on the left and then finding the antiderivative directly)

$$= \tfrac{1}{2}\{\arcsin(x - 2) + (x - 2)[1 - (x - 2)^2]^{1/2}\}$$
$$= \tfrac{1}{2}[\arcsin(x - 2) + (x - 2)(-x^2 + 4x - 3)^{1/2}] + C$$

Observe that in part (a) we took an expression of the form $1/P(x)$ where $P(x)$ was a quadratic polynomial. It turns out that, in general, when this kind of problem comes up, you may be able to solve it by a substitution that changes the given function into one of the form $1/(a^2 + u^2)$. This yields something in terms of the arctangent function. In part (b) we had an expression of the form $(P(x))^{1/2}$ where $P(x)$ was a quadratic polynomial. In this kind of problem, you make a substitution that yields a function of the type $(1 - u^2)^{1/2}$, $(1 + u^2)^{1/2}$, or $(u^2 - 1)^{1/2}$, and these all have known antiderivatives.

We now conclude this section with a problem that can be solved by at least two different methods.

EXAMPLE 36.8   Find the antiderivative of the function $x^2/(1 - x^2)^{1/2}$ over the interval J: $-1 < x < 1$.

SOLUTION 1   *By a Trigonometric Substitution*   Let $x = \sin u$; then $dx = \cos u \, du$, $x^2 = \sin^2 u$, and $(1 - \sin^2 u)^{1/2} = \cos u$. Therefore,

$$\int \frac{x^2}{(1 - x^2)^{1/2}} \, dx = \int \frac{\sin^2 u}{\cos u} \cos u \, du = \int \sin^2 u \, du = \tfrac{1}{2}(u - \sin u \cos u)$$

Now $x = \sin u$ implies $u = \arcsin x$ and $\cos u = (1 - x^2)^{1/2}$. Therefore, we obtain $\frac{1}{2}[\arcsin x - x(1 - x^2)^{1/2}]$.

SOLUTION 2   *By Parts*   Rewrite the function in the form

$$(-\tfrac{1}{2})x(1 - x^2)^{-1/2}(-2x)$$

Let $u(x) = x$ and $v'(x) = (1 - x^2)^{-1/2}(-2x)$; then $u'(x) = 1$ and $v(x) = 2(1 - x^2)^{1/2}$. Now, if we substitute into the parts formula we obtain the same answer as we did in Solution 1.

As an exercise, try to solve this problem by making a rational substitution, namely, let $u^2 = 1 - x^2$; then $x^2 = 1 - u^2$ and $(1 - x^2)^{1/2} = u$. Continue substituting into $\int x^2/(1 - x^2)^{1/2} \, dx$ and see what happens.

PROBLEMS

Find the antiderivative of each of the following functions:
36.1 $(\sin x + \cos x)^2$
36.2 $x \cos x^2$
36.3 $x \sin x^2$
36.4 $1/(1 + \cos x)$
36.5 $(1 + \sin x)^{1/2}$
36.6 xe^{x^2}
36.7 $x^2 e^{x^3}$
36.8 $e^x(\sin x + \cos x)$
36.9 $1/(1 + e^x)$
36.10 $(e^x - e^{-x})/(e^x + e^{-x})$
36.11 $e^{(x + e^x)}$
36.12 2^x
36.13 $a^x, a > 1$
36.14 $xe^x/(1 + x)^2$
36.15 $x \log x$
36.16 $x(\log x)^2$
36.17 $x \log x^2$
36.18 $x^2 \log x^2$
36.19 $(\log x)/x^2$
36.20 $\sin(\log x)$
36.21 $x \arctan x$
36.22 $x \arctan x^2$
36.23 $x \arcsin x$
36.24 $e^x \arcsin e^x$
36.25 $2e^x \cos^2 x$
36.26 $(x - 1)/[(x^2 + 1)(x + 1)]$
36.27 $(3x + 1)/[(x^2 + 1)(x - 1)]$
36.28 $x/(x + 2)^{1/2}$
36.29 $(x^2 - 1)^{1/2}$
36.30 $(-x^2 + 10x - 24)^{1/2}$
36.31 $1/(x^2 + 4x + 5)$
36.32 $1/(-x^2 + 2x)^{1/2}$
36.33 $x^3/(1 + x^2)^{1/2}$
36.34 $x^3(1 - x^2)^{1/2}$
36.35 $x^2/(1 + x^2)^{1/2}$
36.36 $1/(1 + x^2)^2$
36.37 $(x^2 - 4)^{1/2}$
36.38 $(x^2 + 4)^{1/2}$
36.39 $1/(25 - x^2)^{1/2}$
36.40 $x/(1 - x^4)^{1/2}$
36.41* $[(1 + x)/(1 - x)]^{1/2}$
36.42* $[(1 - x)/(1 + x)]^{1/2}$

Evaluate each of the following definite integrals.

36.43 $\int_{\pi/4}^{\pi/2} \cot x \, dx$

36.44 $\int_0^1 10^x \, dx$

36.45 $\int_0^1 1/(1+x^{1/2})\,dx$

36.46 $\int_0^9 x/(1+x^{1/2})\,dx$

36.47 $\int_1^3 x(x^2-1)^{1/2}\,dx$

36.48 $\int_2^{5/2} 1/(-x^2+4x-3)^{1/2}\,dx$

36.49 $\int_0^5 (25-x^2)^{1/2}\,dx$

36.50 $\int_0^9 1/(9+x^2)\,dx$

36.51* $\int_0^2 (1+x^3)^{1/2}\,dx$ (Approximate by Riemann sums.)

36.52* $\int_0^2 (\sin x)/x\,dx$ (Approximate by Riemann sums.)

36.53* By at least *two* different methods, compute the antiderivative of each of the functions
 (a) $x^2/[(1+x^2)^{3/2}]$
 (b) $1/x(1+x^2)^{1/2}$.

36.54* Consider the region between the graphs of the equations given in Example 12.6 from $x=1$ to $x=\sqrt{3}$.
 (a) Find the area of this region.
 (b) Rotate this region about the x axis and find the volume of the solid generated.
 (c) Repeat part (b), taking a rotation about the y axis.

36.55* Consider the region to the right of the y axis enclosed by the graphs of the equations given in Example 12.7.
 (a) Find the area of this region.
 (b) Find the volume of the solid obtained by rotating this region about the y axis.

NUMERICAL
INTEGRATION

In the preceding section, we discussed techniques for finding antiderivatives of arbitrary functions and we eventually derived a multitude of formulas. However, despite this wealth of material, there are many functions that do not have antiderivatives expressible in the form of any of our known algebraic or transcendental functions. For example, $(1 + x^3)^{1/2}$, $(\sin x)/x$, and e^{-x^2} are three such functions. If we do not know an antiderivative of a function, we cannot use the fundamental theorem of the calculus to evaluate a definite integral of the function, *even though we know that the definite integral may exist*. However, we may use the Riemann sums as approximations to the definite integral since if the definite integral of a function exists over some interval J, we know that for large enough n, the Riemann sum corresponding to a regular partition P_n of J will closely approximate it. This method of approximation is one type of *numerical integration*. We now discuss a method that involves the sum of two Riemann sums.

In general, suppose f is a nonnegative continuous function on some interval $J: a \le x \le b$ (see Figure 37.1). Then f determines a region that has an area equal to a limit of Riemann sums and this limit is the integral $\int_a^b f(x)\, dx$. Therefore, if we approximate the area closely enough, we have a good approximation for the integral of $f(x)$ over the interval J. Let P_n be a regular partition of J. Choose x_i^* equal to the left endpoint x_{i-1} of the interval J_i for each $i = 1, 2, \ldots, n$. Then the corresponding Riemann sum is

$$S_n = \sum_{i=1}^{n} f(x_{i-1})\Delta$$

where each term $f(x_{i-1})\Delta$ approximates the *area of the rectangle* constructed over J_i with altitude $f(x_{i-1})$. Now, suppose we choose x_i^* equal to

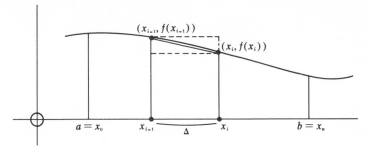

Figure 37.1

the right endpoint x_i of the interval J_i for each $i = 1, 2, \ldots, n$. Then the corresponding Riemann sum is

$$T_n = \sum_{i=1}^{n} f(x_i)\Delta$$

where each term $f(x_i)\Delta$ approximates the *area of the rectangle* constructed over J_i with altitude $f(x_i)$. But for large n, S_n and T_n both approach the area A under the curve of $y = f(x)$ between $x = a$ and $x = b$. Hence, the sequence $\{S_n + T_n\} \to 2A$ and

$$\tfrac{1}{2}(S_n + T_n) = \tfrac{1}{2}\left[f(x_0) + 2\sum_{i=1}^{n-1} f(x_i) + f(x_n)\right]\Delta$$

$$= \left[f(a) + 2\sum_{i=1}^{n-1} f(x_i) + f(b)\right]\frac{(b - a)}{2\,n} \to A$$

where

$$A = \int_a^b f(x)\, dx$$

This method of approximation of the area by the sum $\tfrac{1}{2}(S_n + T_n)$ is called the *trapezoidal rule* since the area of the trapezoid constructed over J_i (see Figure 37.1) is precisely $f(x_i)\Delta$ (the area of the *rectangular* base) *plus* $\tfrac{1}{2}[f(x_{i-1}) - f(x_i)]\Delta$ (the area of the *triangular* top). This sum is just $\tfrac{1}{2}[f(x_{i-1}) + f(x_i)]\Delta$ for each $i = 1, 2, \ldots, n$. Adding these trapezoidal areas gives exactly $\tfrac{1}{2}(S_n + T_n)$. Consequently, we are actually approximating the integral by calculating the areas of *trapezoids* rather than the areas of *rectangles*. There is still another method that calculates the area contained in the region constructed above J_i with a curved *parabolic* top. This method is called *Simpson's rule*, and it is another type of numerical integration that sometimes gives slightly more accurate approximations than the trapezoidal rule.

We now give some examples, beginning with two integrals that can actually be calculated by the fundamental theorem of the calculus. We then compare these answers with the ones obtained by the approximation method.

EXAMPLE 37.1   Use the trapezoidal rule with $n = 4$ to approximate $\int_0^1 1/(1 + x^2)\, dx$ (see Figure 33.6).

SOLUTION

$$\int_0^1 \frac{1}{1 + x^2}\, dx = \arctan x \Big|_0^1 = \arctan 1 - \arctan 0 = \frac{\pi}{4} = 0.7854$$

(correct to four decimal places).   Now, we use the trapezoidal rule.

Let $n = 4$; then $P_4: 0 < \frac{1}{4} < \frac{1}{2} < \frac{3}{4} < 1$ is the corresponding partition and $\frac{1}{2}(S_4 + T_4) = [f(0) + 2f(1/4) + 2f(1/2) + 2f(3/4) + f(1)](1/8) = [1 + 32/17 + 8/5 + 32/25 + 1/2](1/8) = 0.7828$.   Consequently, our approximation is correct to two decimal places.   As an exercise, work out the approximation with $n = 8$ and compare your answer with the preceding two answers.

EXAMPLE 37.2   Use the trapezoidal rule with $n = 4$ to approximate $\int_1^5 1/x\, dx$.

SOLUTION $\int_1^5 1/x\, dx = \log x \Big|_1^5 = \log 5 = 1.6094$ (correct to four decimal places).   Now, using the trapezoidal rule, $P_4: 1 < 2 < 3 < 4 < 5$ is the corresponding partition (see Figure 37.2).   Then $\frac{1}{2}(S_4 + T_4) = [f(1) + 2f(2) + 2f(3) + 2f(4) + f(5)][(5 - 1)/8] = (1 + 1 + \frac{2}{3} + \frac{1}{2} + \frac{1}{5})\frac{1}{2} = 1.6833$.   Consequently, our approximation is correct only to the first decimal place.   As an exercise, work out the approximation with $n = 8$ and compare your answer with the preceding two answers.

EXAMPLE 37.3   Using the trapezoidal rule with $n = 5$, approximate $\int_0^1 e^{-x^2}\, dx$ (see Figure 37.3).

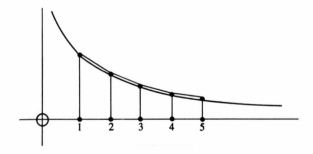

Figure 37.2

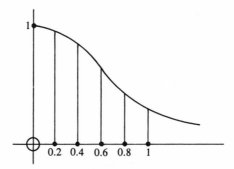

Figure 37.3

SOLUTION . The corresponding partition is $P_5: 0 < 0.2 < 0.4 < 0.6 < 0.8 < 1$. Then $\frac{1}{2}(S_5 + T_5)$
$= [f(0) + 2f(0.2) + 2f(0.4) + 2f(0.6) + 2f(0.8) + f(1)](1/10)$
$= [1 + 2(0.9608) + 2(0.8521) + 2(0.6977) + 2(0.5273) + (0.3679)](1/10)$
$= 0.7444$

As an exercise, compute the area under the graph of $y = e^{-x}$ between $x = 0$ and $x = 1$ by both *direct integration* and the *trapezoidal rule* with $n = 5$. Compare your results with the preceding example.

PROBLEMS

Using the trapezoidal rule with the given n, approximate each of the following definite integrals and check your answer by other means whenever possible.

37.1 $\int_0^2 (1 + x^3)^{1/2} \, dx \qquad n = 4$

37.2 $\int_{\pi/2}^{\pi} (\sin x)/x \, dx \qquad n = 3$

37.3 $\int_0^5 e^{-x^2} \, dx \qquad n = 10$

37.4 $\int_1^9 1/x \, dx \qquad n = 8$

37.5 $\int_0^1 (1 - x^2)^{1/2} \qquad n = 4$

37.6 $\int_0^2 (1 + x^2)^{1/2} \qquad n = 4$

37.7 $\int_0^{\pi} (\sin x)^{1/2} \, dx \qquad n = 4$

37.8 $\int_0^{\pi} (1/(1 + \sin x) \, dx \qquad n = 4$

37.9* If $a > 1$, prove $\lim_{n \to \infty} [(1^a + 2^a + \cdots + n^a)/n^{a+1}] = 1/(a + 1)$. [*Hint:* Look at a convenient Riemann sum of the function $f(x) = x^a$ over the interval $1 \le x \le n$.]

IMPROPER INTEGRALS

Up to now, we have discussed definite integrals of continuous functions f over *finite* intervals $J: a \leq x \leq b$. (See Definition 31.4). In all these cases, f has also been *bounded* on J. If we remove the boundedness condition or the finiteness of the interval J, we obtain what is called an *improper integral*. We have already seen examples where J was infinite and we actually asked for an area or a volume over such an infinite interval (see Problems 32.1, 32.2, 32.3 32.13, 33.5, and 33.6).

In this section we first discuss conditions under which we can integrate a continuous function f over an infinite interval $J: a \leq x < \infty$. Then we relate these results to some of the aforementioned problems.

Definition 38.1 Let f be a continuous function defined on an interval $J: a \leq x < \infty$ and let F be an antiderivative of f on J. Then $\int_a^b f(x)\, dx$ exists for every real number $b > a$ and equals $F(b) - F(a)$ (see Definition 31.4 and Theorem 32.1). Now, if $\lim_{b \to \infty} [F(b) - F(a)]$ exists and is finite, we say that the improper integral $\int_a^\infty f(x)\, dx$ exists and equals this finite limit. There is a corresponding definition for functions f defined on intervals of the type $J: -\infty < x \leq a$, in which case $\int_{-\infty}^a f(x)\, dx$ exists if $\lim_{b \to -\infty} [F(a) - F(b)]$ exists.

EXAMPLE 38.1 Evaluate each of the following improper integrals.

$$\text{(a)} \quad \int_1^\infty 1/x^2 \, dx \qquad \text{(b)} \quad \int_1^\infty 1/x \, dx$$

SOLUTION (a) An antiderivative is $F(x) = -1/x$; hence, the integral of $1/x^2$ from $x = 1$ to $x = b$ is $F(b) - F(1) = -1/b + 1$. Now, $\lim_{b \to \infty} (-1/b + 1) = 1$; therefore, $\int_1^\infty 1/x^2 \, dx = 1$.

Improper Integrals 257

(b) An antiderivative of $1/x$ is $\log x$; therefore, the integral from $x = 1$ to $x = b$ is simply $\log b - \log 1 = \log b$. But $\lim\limits_{b \to \infty} \log b = \infty$; hence, the improper integral does not exist. Relate both of these results to the corresponding areas under the graphs (see Problem 33.2) and to the corresponding volumes of the solids of revolution generated by these two functions (see Problem 34.5).

Theorem 38.1 Let a be any positive real number. Then the improper integral $\int_a^\infty 1/x^\alpha \, dx$ exists for $\alpha > 1$ and does not exist for $\alpha \le 1$.

PROOF If $\alpha > 1$, $1/x^\alpha$ has an antiderivative equal to $[1/(1 - \alpha)](1/x^{\alpha-1})$ where $\alpha - 1 > 0$. Therefore, the integral from a to b is $[1/(1 - \alpha)](1/b^{\alpha-1} - 1/a^{\alpha-1})$, which approaches $[1/(1 - \alpha)](-1/a^{\alpha-1})$ as $b \to \infty$ (since $\alpha - 1 > 0$ implies $1/b^{\alpha-1} \to 0$). Hence, the improper integral exists for $\alpha > 1$. However, for $\alpha = 1$, we know from the preceding example that the improper integral does not exist. Now, for $\alpha < 1$, $1/x^\alpha$ has an antiderivative equal to $[1/(1 - \alpha)](x^{1-\alpha})$ where $1 - \alpha > 0$. Therefore, the integral from a to b is $[1/(1 - \alpha)](b^{1-\alpha} - a^{1-\alpha})$, which approaches ∞ as $b \to \infty$. Hence, the improper integral does not exist. ∎

EXAMPLE 38.2 Evaluate (a) $\int_0^\infty 1/(1 + x^2) \, dx$; (b) $\int_{-\infty}^0 1/1(+ x^2) \, dx$.

SOLUTION (a) An antiderivative is $F(x) = \arctan x$; hence, the integral from 0 to b is given by $\arctan b - \arctan 0 = \arctan b$. Now, $\lim\limits_{b \to \infty} \arctan b = \pi/2$ (since the tangent of an angle approaches infinity as the angle itself approaches $\pi/2$ rad). Therefore, the improper integral has the value $\pi/2$.

(b) In this case, we integrate from b to 0 and obtain $\arctan 0 - \arctan b$. Then $\lim\limits_{b \to -\infty} (-\arctan b) = -\lim\limits_{b \to -\infty} (\arctan b) = -(-\pi/2) = \pi/2$ (since the tangent of an angle approaches $-\infty$ as the angle itself approaches $-\pi/2$ rad). Hence, the improper integral has value $\pi/2$.

Observe that in this example, both the integral from $-\infty$ to 0 and the integral from 0 to ∞ exist. When this occurs, we simply write the sum of the two integrals as

$$\int_{-\infty}^\infty \frac{1}{1 + x^2} \, dx = \frac{\pi}{2} + \frac{\pi}{2} = \pi$$

Finally, relate the preceding results to Problem 33.3(b).

EXAMPLE 38.3 (a) Evaluate $\int_0^\infty xe^{-x^2} \, dx$ and $\int_{-\infty}^0 xe^{-x^2} \, dx$.

(b) Determine the area bounded by the graph of xe^{-x^2} and the x axis from $x = -\infty$ to $x = +\infty$.

SOLUTION   (a)   An antiderivative of xe^{-x^2} is $-\frac{1}{2}e^{-x^2}$.   Hence, the integral from 0 to b is $-\frac{1}{2}(e^{-b^2} - 1)$, which approaches $\frac{1}{2}$ as $b \to \infty$ (since $e^{-b^2} \to 0$).   Therefore, the improper integral from 0 to ∞ has value $\frac{1}{2}$. On the other hand, the second integral from b to 0 is $-\frac{1}{2}(1 - e^{-b^2})$, which approaches $-\frac{1}{2}$ as $b \to -\infty$ (since b^2 is still positive, thus $e^{-b^2} \to 0$). Therefore, the second integral has value $-\frac{1}{2}$.   If we combine these two results, we find that $\int_{-\infty}^{\infty} xe^{-x^2} \, dx = \frac{1}{2} - \frac{1}{2} = 0$.

(b)   The region bounded by the given graphs appears in Figure 38.1. The area to the right of the y axis is given by $\int_0^{\infty} xe^{-x^2} \, dx = \frac{1}{2}$.   The area to the left, however, is below the x axis; therefore, it is equal to $-\int_{-\infty}^0 xe^{-x^2} \, dx = -(-\frac{1}{2}) = \frac{1}{2}$ (see Figure 33.8 and the accompanying discussion).   Therefore, the area is $\frac{1}{2} + \frac{1}{2} = 1$ (of course, from the symmetry, we could also arrive at the fact that the total area is just *twice* the area to the right of the y axis and hence is 1).   Note that the area is *not* given by the value of the integral from $-\infty$ to $+\infty$.   So integrals need *not* be areas, even in the improper case.

Now we discuss the improper integral that arises when we assume that f is *not bounded* near one point in some interval $J: a \le x \le b$.

Definition 38.2   Suppose f becomes unbounded near a and that $\int_z^b f(x) \, dx$ exists for all z, $a < z \le b$. Let F be an antiderivative of f on the interval $a < x \le b$.   Then the integral from z to b is given by $F(b) - F(z)$.   Now, if $\lim_{\substack{z \to a \\ z > a}} [F(b) - F(z)]$ exists and is a finite number, we say that the improper integral $\int_a^b f(x) \, dx$ exists.

In a similar manner, if f becomes unbounded near the point b, we find that the improper integral $\int_a^b f(x) \, dx$ exists if $\lim_{\substack{z \to b \\ z < b}} [F(z) - F(a)]$ exists and is a finite number.

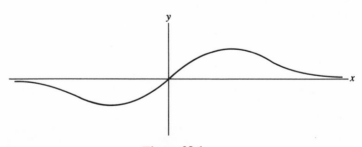

Figure 38.1

Clearly, these are the only two cases that have to be considered, since if f becomes unbounded near some point c where $a < c < b$, we may break up the integral from a to b into the integrals from a to c and from c to b, and then c is an endpoint and is handled according to the previous definitions. Then we say that $\int_a^b f(x)\, dx$ exists if and only if both of the integrals $\int_a^c f(x)\, dx$ and $\int_c^b f(x)\, dx$ exist.

EXAMPLE 38.4   Evaluate $\int_0^1 x^{-1/2}\, dx$.

SOLUTION   This function becomes unbounded near $x = 0$, but for any z, $0 < z < 1$, we have the integral from z to 1 equal to $2\left(x^{1/2}\Big|_z^1\right) = 2(1 - z^{1/2})$. Now as $z \to 0$, $z^{1/2} \to 0$ and therefore the limit is 2. Hence, $\int_0^1 x^{-1/2}\, dx = 2$.

As an exercise, plot the graph of $y = x^{-1/2}$ and relate the preceding result to the area of the appropriate region.

EXAMPLE 38.5   Evaluate $\int_0^1 (1 - x^2)^{-1/2}\, dx$.

SOLUTION   This function becomes unbounded near $x = 1$, but for any z, $0 \le z < 1$, we have

$$\int_0^z (1 - x^2)^{-1/2}\, dx = \arcsin x \Big|_0^z = \arcsin z$$

Now as $z \to 1$, $\arcsin z \to \pi/2$; hence, the value of the improper integral is $\pi/2$.

As an exercise, plot the graph of $y = (1 - x^2)^{-1/2}$ and relate the preceding result to the area of the appropriate region.

EXAMPLE 38.6   Evaluate $\int_0^1 x^{1/2} \log x\, dx$.

SOLUTION   Since $\log x$ becomes unbounded near $x = 0$, this is an improper integral. Now an antiderivative is $F(x) = \frac{2}{3}x^{3/2} \log x - \frac{4}{9} x^{3/2}$ [verify this by differentiating F or by integration by parts with $u(x) = \log x$ and $v'(x) = x^{1/2}$]. Now $F(1) = -\frac{4}{9}$ while $F(z) = \frac{2}{3}z^{3/2} \log z - \frac{4}{9}z^{3/2}$ for any z, $0 < z \le 1$. As $z \to 0$, $F(z) \to 0$ since both $z^{3/2}$ and $z^{3/2} \log z$ go to zero [use l'Hospital's rule on $z^{3/2} \log z$ after rewriting it in the form $\log z/(1/z^{3/2})$]. Therefore, the improper integral has the value $-\frac{4}{9}$.

Plot the graph of $y = x^{1/2} \log x$ and relate the preceding result to the area of the appropriate region (see Figure 28.4).

PROBLEMS

38.1   Prove that the improper integral $\int_0^1 1/x^\alpha \, dx$ exists if and only if $\alpha < 1$.   Compare this result with Theorem 38.1.

38.2   In each of the following, find the area of the region determined by the graph of the given function between the indicated values of x.
(a) $f(x) = x^{-1/2}e^{-\sqrt{x}}$ $x = 0$ to $x = 1$
(b) $f(x) = (\log x)/x^2$ $x = 1$ to $x = \infty$

38.3   In each of the following, find the area of the region and the volume of the corresponding solid of revolution (about the x axis) determined by the graph of the given function between the indicated values of x.
(a) $f(x) = x^{-4/3}$ $x = 1$ to $x = \infty$
(b) $f(x) = x^{-4/3}$ $x = 0$ to $x = 1$
(c) $f(x) = x^{-2/3}$ $x = 1$ to $x = \infty$
(d) $f(x) = x^{-2/3}$ $x = 0$ to $x = 1$
(e) $f(x) = x^{1/2} \log x$ $x = 0$ to $x = 1$
(f) $f(x) = xe^{-x}$ $x = 0$ to $x = \infty$

38.4   Find the value of each of the following integrals, if it exists.

(a) $\displaystyle\int_0^\infty xe^{-x^2/2} \, dx$

(b) $\displaystyle\int_{-\infty}^\infty 1/(e^x + e^{-x}) \, dx$

(c) $\displaystyle\int_0^1 e^{1/x}/x^2 \, dx$

(d) $\displaystyle\int_0^1 (e^{-1/x}/x^2) \, dx$

(e) $\displaystyle\int_0^1 \log x \, dx$

(f) $\displaystyle\int_0^1 x \log x \, dx$

(g) $\displaystyle\int_0^1 (\log x)/x^2 \, dx$

(h) $\displaystyle\int_{-1}^1 x^{-2/3} \, dx$

(i) $\displaystyle\int_0^2 1/(x-2)^2 \, dx$

(j) $\displaystyle\int_0^2 x/(1-x^2)^2 \, dx$

(k) $\displaystyle\int_{-1}^1 (1-x^2)^{-1/2} \, dx$

(l) $\displaystyle\int_0^\infty \cos x \, dx$

(m) $\displaystyle\int_0^{\pi/2} \tan x \, dx$

(n) $\displaystyle\int_0^1 (\sin \sqrt{x})/\sqrt{x} \, dx$

(o) $\displaystyle\int_0^{\pi/2} 1/(1 - \sin x) \, dx$

(p) $\displaystyle\int_0^\infty e^{-x} \cos x \, dx$

38.5 Determine whether the following improper integrals exist [use parts (A) and (B) of Theorem 32.3].

(a) $\int_1^\infty (\sin x)/x^2 \, dx$

(b) $\int_1^\infty e^x/x \, dx$

(c) $\int_0^\infty e^{-x^2} \, dx$

(d) $\int_1^\infty x/(x^3 + 1) \, dx$

38.6* Let $f(x) = 2x \sin(1/x^2) - (2/x) \cos(1/x^2)$ if $x \neq 0$ with $f(0) = 0$, be defined on J: $-1 \leq x \leq 1$. Prove that $f(x)$ has an antiderivative on J, namely, the function $F(x) = x^2 \sin(1/x^2)$ if $x \neq 0$ with $F(0) = 0$. Prove also that $\int_{-1}^1 f(x) \, dx$ does *not* exist.

ARC LENGTH AND SURFACE AREA

We now give two more geometrical applications of the definite integral. First, we discuss a method of finding the *length* of a finite arc of the graph of an arbitrary continuous function; then we explain how to find the *surface area* of a solid of revolution. In many respects, the definite integrals that arise in these two applications are much harder to handle than the integrals that came up when we were dealing with areas and volumes. In fact, the arc length of the graphs of some of the simplest functions is very difficult to compute.

Definition 39.1 Let f be a function that has a continuous derivative on an interval $J: a \leq x \leq b$. Then, the length of the arc of the graph of $y = f(x)$ from the point $(a, f(a))$ to the point $(b, f(b))$ is given by $\int_a^b (1 + (f'(x))^2)^{1/2} \, dx$.

We now attempt to make this definition a reasonable one. Let P_n be a regular partition of the interval J. Draw each of the line segments L_i from $(x_{i-1}, f(x_{i-1}))$ to $(x_i, f(x_i))$ for $i = 1, 2, \ldots, n$ (see Figure 39.1). Now, we may approximate the length of the arc from $(x_{i-1}, f(x_{i-1}))$ to $(x_i, f(x_i))$ by using the length d_i of the line segment L_i, which is, by the distance formula, $[(x_i - x_{i-1})^2 + (f(x_i) - f(x_{i-1}))^2]^{1/2} = [\Delta^2 + (f(x_i) - f(x_{i-1}))^2]^{1/2}$ where $\Delta = (b - a)/n$. Now f is differentiable on each subinterval $J_i: x_{i-1} \leq x \leq x_i$ for $i = 1, 2, \ldots, n$. Therefore, we may apply the mean value theorem, which states that there exists a point $x_i^* \in J_i$ such that

$$f'(x_i^*) = \frac{f(x_i) - f(x_{i-1})}{x_i - x_{i-1}}$$

Hence, $f(x_i) - f(x_{i-1}) = f'(x_i^*)\Delta$. Substituting in the formula for d_i, we

262

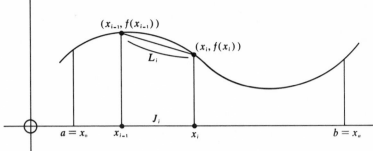

Figure 39.1

obtain $d_i = [\Delta^2 + (f'(x_i^*))^2\Delta^2]^{1/2} = [1 + (f'(x_i^*))^2]^{1/2}\Delta$. Doing this for each $i = 1, 2, \ldots, n$ and adding, we obtain

$$\sum_{i=1}^{n} d_i = \sum_{i=1}^{n} [1 + (f'(x_i^*))^2]^{1/2}\Delta = S_n$$

But S_n is just the Riemann sum of the function $[1 + (f'(x))^2]^{1/2}$ corresponding to the partition P_n. Now, for large n, the segments L_i become shorter and shorter and their lengths d_i become better and better approximations for the lengths of the corresponding arcs. But for large n

$$S_n \to \int_a^b [1 + (f'(x))^2]^{1/2} \, dx$$

since the continuity of $f'(x)$ guarantees the existence of the integral.

EXAMPLE 39.1 Find the length of the arc of the graph of $y = \frac{2}{3}(x - 1)^{3/2}$ from the point $(1, 0)$ to the point $(2, \frac{2}{3})$.

SOLUTION $f'(x) = (x - 1)^{1/2}, (f'(x))^2 = (x - 1)$, and $[1 + (f'(x))^2]^{1/2} = x^{1/2}$. Therefore, the arc length is given by $\int_1^2 x^{1/2} \, dx = \frac{2}{3}(2^{3/2} - 1)$.

EXAMPLE 39.2 Find the length of the arc of the graph of $y = \log x$ from the point where $x = 1$ to the point where $x = 2\sqrt{2}$.

SOLUTION $f'(x) = 1/x, (f'(x))^2 = 1/x^2$, and $[1 + (f'(x))^2]^{1/2} = (1/x) \times (1 + x^2)^{1/2}$. An antiderivative is

$$F(x) = (1 + x^2)^{1/2} + \log\left[\frac{(1 + x^2)^{1/2} - 1}{x}\right]$$

(verify this by differentiation). Evaluating $F(x)$ at $x = 1$ and $x = 2\sqrt{2}$ and subtracting, we obtain $3 - \frac{1}{2}\log 2 - \sqrt{2} - \log(\sqrt{2} - 1)$ as the arclength.

EXAMPLE 39.3   Find the length of the arc of the graph of $y = -\log(\cos x)$ from the point where $x = 0$ to the point where $x = \pi/4$.

SOLUTION $f'(x) = \sin x/\cos x = \tan x$, $(f'(x))^2 = \tan^2 x$, and

$$[1 + (f'(x))^2]^{1/2} = \sec x$$

Therefore, the arc length is $\int_0^{\pi/4} \sec x \, dx = \log(\sec x + \tan x)\Big|_0^{\pi/4} = \log(1 + 2^{1/2})$.

EXAMPLE 39.4   Find the length of the arc in the first quadrant of the circle with equation $x^2 + y^2 = 4$.

SOLUTION   We know that the circumference of the circle is $2\pi(2) = 4\pi$ units. Hence, the arc in the first quadrant from the point $(0, 2)$ to the point $(2, 0)$ is exactly π units in length. We now attempt to verify this by using the integral formula for arc length.

The function that corresponds to the given arc is

$$f(x) = (4 - x^2)^{1/2}$$

hence,

$$f'(x) = \tfrac{1}{2}(4 - x^2)^{-1/2}(-2x) = -x(4 - x^2)^{-1/2}, \quad (f'(x))^2 = x^2(4 - x^2)^{-1}$$

and

$$1 + (f'(x))^2 = 1 + x^2/(4 - x^2) = 4/(4 - x^2)$$

Therefore,

$$[1 + (f'(x))^2]^{1/2} = 2/(4 - x^2)^{1/2}$$

and the arc length is

$$2 \int_0^2 1/(4 - x^2)^{1/2} \, dx$$

But this is an *improper* integral (see Definition 38.2) and equals

$$2 \lim_{\substack{z \to 2 \\ z < 2}} \int_0^z 1/(4 - x^2)^{1/2} \, dx = 2 \lim_{\substack{z \to 2 \\ z < 2}} \left(\arcsin(x/2)\Big|_0^z \right)$$

$$= 2 \lim_{\substack{z \to 2 \\ z < 2}} (\arcsin(z/2) - \arcsin 0) = 2(\arcsin 1) = 2(\pi/2) = \pi$$

THE SURFACE AREA OF A SOLID OF REVOLUTION

Suppose that f is a nonnegative function that has a continuous derivative on an interval $J: a \le x \le b$. Suppose that we rotate about the x axis the region above the x axis, below the graph of $y = f(x)$, between $x = a$ and $x = b$ (see Figure 39.2) and thus obtain a solid of revolution. What is the *lateral surface area* of this solid?

Definition 39.2 The lateral surface area in the preceding situation is given by $2\pi \int_a^b f(x)(1 + (f'(x))^2)^{1/2} \, dx$ (the integral exists since f' is continuous).

We now attempt to make this definition plausible. Let P_n be a regular partition of J (see Figure 39.2). If for each i we draw the line segments L_i joining the points $(x_{i-1}, f(x_{i-1}))$ and $(x_i, f(x_i))$, and then rotate the resulting trapezoid above each J_i about the x axis, we get a solid that is a part of a cone (frustum). This part is obtained by slicing through the cone two planes parallel to the base of the cone and Δ units apart where $\Delta = (b - a)/n$. This solid then has bases of radius $f(x_{i-1})$ and $f(x_i)$, respectively, and slant height d_i (the length of L_i). It is known that the lateral area A_i of such a solid is given by the circumference of the midsection times the slant height d_i. But from our discussion of the *arc length* formula, we found that we could use the mean value theorem on J_i to find a point x_i^* such that $d_i = (1 + (f'(x_i^*))^2)^{1/2}\Delta$ for each $i = 1, 2, \ldots, n$. Now, for large enough n, x_i^* is close enough to the midpoint of J_i so that the radius of the midsection of the frustum about J_i can be very closely approximated by $f(x_i^*)$. Therefore, the lateral surface area A_i is approximately equal to $2\pi f(x_i^*)d_i$. Adding these areas for $i = 1, 2, \ldots, n$, we obtain the Riemann sum, corresponding to the partition P_n, of the function $2\pi f(x)[1 + (f'(x))^2]^{1/2}$. Therefore, the definite integral of this function over the interval from a to b is defined to be the lateral surface area.

Observe that if we required the total surface area (the lateral area plus the area of the top and bottom), we would add the value of the preceding integral to the areas of the circular top and bottom, namely, $\pi(f(a))^2$ and $\pi(f(b))^2$.

Finally, if $y = f(x)$ satisfies all the preceding conditions along with the additional property that it is a one-to-one function, we may rotate the corresponding region about the y axis (see Figure 39.3) and the lateral surface area

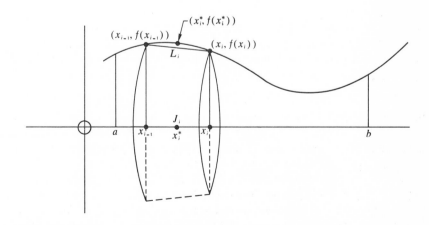

Figure 39.2

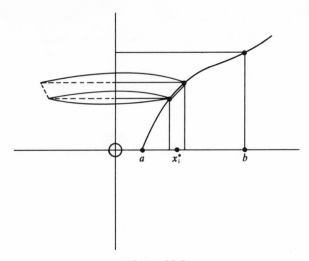

Figure 39.3

of the solid of revolution obtained is given by the integral

$$2\pi \int_a^b x[1 + (f'(x))^2]^{1/2} \, dx$$

In this case, after taking a regular partition P_n of $J: a \leq x \leq b$, we find that the trapezoids are *horizontal* instead of *vertical* and the frustum with slant height d_i has the radius of its midsectional cut approximately equal to x_i^* instead of $f(x_i^*)$.

REMARK   When computing the volume of a solid of revolution (see the discussion following Definition 34.1), we rotated rectangles and used the volumes of the cylinders that were generated.   However, in both of the preceding discussions on surface area, we rotated trapezoids and used the volumes of the frustums of cones that were generated.   In fact, if we had used cylinders in the discussion following Definition 39.2, the lateral area A_i would have been approximated by $2\pi f(x_i^*)\Delta$ (circumference of the base times the height).   This would have led to the formula $2\pi \int_a^b f(x) \, dx$, which is just 2π times the area under the curve.   This formula is clearly different from the one given in Definition 39.2 and is therefore incorrect.   The question is, why is it incorrect?   The reason hinges on the fact that in generating solids of revolution, for large n, the small region between the slant line top and curved top of the strip above J_i (see Figures 39.1 and 39.2) accounts for a negligible error in volume.   On the other hand, in dealing with the surface area generated by this small region, it is precisely the curved top, and not the slant line top, that accounts for *all* of the surface area.   Hence, a small error in the length of the generating arc is critical and causes a significant error in the corresponding

Riemann sum and eventually in the integral formula.   Consequently, the use of the slant height d_i (instead of just Δ) is absolutely necessary if the resulting approximation is to accurately reflect the true surface area.

EXAMPLE 39.5   Consider the two regions A and B (see Figure 39.4) determined by the graph of $y = x^2$ between the coordinate axes and the lines $x = 2$ and $y = 4$.   Find the lateral surface area of the solid of revolution obtained by rotating (a) region A about the x axis; (b) region B about the y axis.

SOLUTION   (a) $f(x) = x^2$, $f'(x) = 2x$, $(f'(x))^2 = 4x^2$, and

$$f(x)[1 + (f'(x))^2]^{1/2} = x^2(1 + 4x^2)^{1/2}$$

An antiderivative is

$$F(x) = (x/16)(1 + 4x^2)^{3/2} - (x/32)(1 + 4x^2)^{1/2} - \tfrac{1}{64}\log(2x + (1 + 4x^2)^{1/2})$$

The surface area is therefore

$$2\pi[F(2) - F(0)] = 2\pi[\tfrac{1}{8}(17)^{3/2} - \tfrac{1}{16}(17)^{1/2} - \tfrac{1}{64}\log(4 + 17^{1/2})]$$

(b)   Since the rotation is about the y axis, we use $2\pi \int_0^2 x(1 + 4x^2)^{1/2}\, dx$ to compute the required surface area.   An antiderivative is $F(x) = \tfrac{2}{3}\tfrac{1}{8}(1 + 4x^2)^{3/2} = \tfrac{1}{12}(1 + 4x^2)^{3/2}$ and $F(2) - F(0) = \tfrac{1}{12}(17^{3/2} - 1)$. Therefore, the surface area is $(\pi/6)(17^{3/2} - 1)$.

EXAMPLE 39.6   Find the lateral surface area of the solid of revolution obtained by rotating about the x axis the region below the graph of $y = \cos x$ between $x = 0$ and $x = \pi/2$.

SOLUTION $f(x) = \cos x$, $f'(x) = -\sin x$, $(f'(x))^2 = \sin^2 x$, and

$$f(x)[1 + (f'(x))^2]^{1/2} = \cos x(1 + \sin^2 x)^{1/2}$$

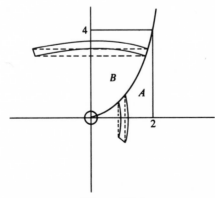

Figure 39.4

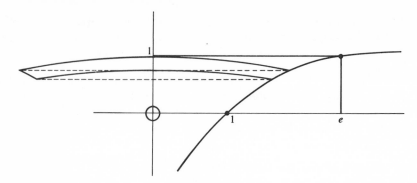

Figure 39.5

An antiderivative is

$$F(x) = \tfrac{1}{2}[\sin x(1 + \sin^2 x)^{1/2} + \log(\sin x + (1 + \sin^2 x)^{1/2})]$$

(by making the substitution $u = \sin x$). Therefore, the surface area is
$2\pi[F(\pi/2) - F(0)] = 2\pi\tfrac{1}{2}[\sqrt{2} + \log(1 + \sqrt{2})] = \pi[\sqrt{2} + \log(1 + \sqrt{2})].$

EXAMPLE 39.7   Find the lateral surface area of the solid of revolution obtained by rotating about the y axis the region in the first quadrant bounded by the coordinate axes, the graph of $y = \log x$, and the line $y = 1$ (see Figure 39.5).

SOLUTION   Since the rotation is about the y axis, we use

$$2\pi \int_1^e x[1 + (f'(x))^2]^{1/2}\, dx$$

Now, $f(x) = \log x$, $f'(x) = 1/x$, $(f'(x))^2 = 1/x^2$, and $x[1+(f'(x))^2]^{1/2} = x[(1 + x^2)/x^2]^{1/2} = (1 + x^2)^{1/2}$. An antiderivative is

$$F(x) = \tfrac{1}{2}[x(1 + x^2)^{1/2} + \log(x + (1 + x^2)^{1/2})]$$

and the required surface area is $2\pi[F(e) - F(1)]$.

PROBLEMS

39.1   In each of the following, find the length of the arc of the given graph between the indicated points.
(a) $y = x^{3/2}$ $x = 0$ to $x = 1$ and $x = 0$ to $x = 4$
(b) $y = x^2$ $x = 0$ to $x = 1$
(c) $y = (\tfrac{2}{3})(x^2 + 2)^{3/2}$ $x = 0$ to $x = 1$ (Leave your answer in the form of an integral.)
(d) $y = (16 - x^2)^{1/2}$ $x = 0$ to $x = 4$
(e) $y = (1 - 4x^2)^{1/2}$ $x = 0$ to $x = \tfrac{1}{2}$ (Leave your answer in the form of an integral.)
(f) $y = \log x$ $x = 1$ to $x = 5$

(g) $y = \log(\sin x)$ $x = \pi/3$ to $x = \pi/2$
(h) $y = -\log(\cos x)$ $x = 0$ to $x = \pi/6$
(i) $y = \log(1 - x^2)$ $x = 0$ to $x = \frac{3}{4}$
(j)* $y = (\frac{1}{2})(e^x + e^{-x})$ $x = -1$ to $x = 1$

39.2 In each of the following, estimate the arc length by using the trapezoidal rule (see Section 37), with a convenient choice of n, to approximate the integral for the arc length.

(a) $y = x^{1/2}$ $x = 0$ to $x = 4$
(b) $y = x^3$ $x = 0$ to $x = 2$
(c) $y = 1/x$ $x = 1$ to $x = 5$
(d) $y = e^{-x}$ $x = 0$ to $x = 2$
(e) $y = (1 + x^2)^{1/2}$ $x = 0$ to $x = 1$

39.3 Find the lateral surface area of the solid of revolution obtained by rotating about the x axis each of the regions bounded by the graphs of the following functions.

(a) $y = 8x$ between $x = 0$ to $x = 1$
(b) $y = \frac{1}{3}x^3$ between $x = 0$, $y = 0$, and $x = 1$
(c) $y = \sin x$ between $x = 0$, $x = \pi$, and $y = 0$
(d) $y = \cos x$ between $x = \pi/2$, $x = 3\pi/2$, and $y = 0$
(e) $y = \log x$ between $x = 1$, $x = 2\sqrt{2}$, and $y = 0$ (Leave your answer in the form of integral.)
(f) $y = e^{-x}$ between $x = 0$, $x = \log 2$, and $y = 0$
(g) $y = \frac{1}{2}(e^x + e^{-x})$ between $x = 0$, $x = 1$, and $y = 0$ (Leave your answer in the form of an integral.)

39.4 Find the lateral surface area of the solid of revolution obtained by rotating each of the following regions about the y axis.

(a) the region below the x axis, to the right of the y axis, above the line $y = -1$ and to the left of the graph of $y = \log x$
(b) the region to the right of the y axis, below the line $y = \pi/2$, and to the left of the graph of $y = \arcsin x$ (Leave your answer in the form of an integral.)
(c) the region in the first quadrant bounded by the graph of $x^2 + y^2 = 1$ (The solid is a hemisphere with a surface area equal to 2π square units.)

39.5 (a) Using the integral formula, find the surface area of the solid of revolution obtained by rotating the upper half of the unit circle about the x axis.   (The solid is, of course, a sphere with surface area 4π square units.)
       (b) Repeat part (a) with the upper half of the ellipse with equation $4x^2 + y^2 = 1$.

39.6 (a) Find the *total* surface area of the solid of revolution obtained by rotating about the x axis the region below the graph of $y = 4\sqrt{x}$, between $x = 0$ and $x = 4$.
       (b)* Find the total surface area of the solid of revolution obtained by rotating about the y axis the region above the graph of $y = 4\sqrt{x}$ and below the line $y = 8$.

39.7 Using the trapezoidal rule (see Section 37), with a convenient choice of n, approximate the lateral surface area of the solid of revolution obtained by rotating about the x axis each of the following regions.

(a) the region above the x axis, below the graph of e^{-x^2}, between the y axis and the line $x = 1$
(b) the region above the x axis, below the graph of $y = x^4$, between the y axis and the line $x = 1$

PARAMETRIC EQUATIONS, TANGENTS, AND ARC LENGTH

In this section, we discuss the representation of plane curves by parametric equations. Because we assume you are somewhat familiar with this topic, our introductory discussion is brief. A method for finding the tangent line at a point on such a curve is developed and a technique for determining the arc length of such curves is explained. At the end of this section, we also include some problems on area and volume associated with curves that are represented parametrically.

It is not always possible to represent an arbitrary curve by an equation $y = f(x)$, since many curves have the property that a vertical line intersects the curve in more than one point. This means that for some number x, we cannot have a unique image y (see Definitions 15.1 and 15.2). For example, the circle with equation $x^2 + y^2 = 1$, the parabola with equation $y^2 = x$, and the hyperbola with equation $x^2 - y^2 = 1$ cannot be represented by an equation in the form $y = f(x)$. In these cases, we usually overcome this difficulty by decomposing the curve into smaller curves, each of which is representable by an equation $y = f(x)$ [see Remark 3(a) and (b) following Example 16.9]. Now, suppose that instead of considering y as a function of x, we consider *both* x and y as functions of an independent variable t (called a parameter) in such a way that the point (x, y) traverses the curve as t varies over some interval $\alpha \le t \le \beta$. Then we may represent the curve parametrically by the two equations $x = x(t)$, $y = y(t)$ where $\alpha \le t \le \beta$. It turns out that such a representation is not unique; we will have more to say about this after the following example.

EXAMPLE 40.1 Express parametrically each of the curves (a) $x = 2$; (b) $y^2 = x$; (c) $x^2 + y^2 = 1$; (d) $x^2 - y^2 = 1$.

SOLUTION (a) Every point on the line $x = 2$ has x coordinate 2, while the y coordinate may be any real number. Therefore, we may represent this vertical line parametrically by the equations $x = x(t) = 2$ and $y = y(t) = t$ where t varies over the interval $-\infty < t < \infty$.

(b) The parabola $y^2 = x$ may be represented parametrically by the equations $x = x(t) = t^2$ and $y = y(t) = t$ where t varies over the interval $-\infty < t < \infty$. Note that for negative values of t, we obtain points on the graph of $y^2 = x$ that are *below* the x axis (see Figure 40.1). Hence, the parametric representation gives the *entire* parabola $y^2 = x$ rather than the two-part representation given by the equations $y = \sqrt{x}$ and $y = -\sqrt{x}$ [see Example 13.2(a)].

(c) The circle $x^2 + y^2 = 1$ may be represented parametrically by the equations $x = \cos t$, $y = \sin t$, where $0 \le t < 2\pi$ and where t is the *angle* measured in radians from the positive x axis to the radial line drawn to the point (x, y). Observe that for t between π and 2π, we obtain the semicircle *below* the x axis [see Example 13.2(b)].

(d) The hyperbola $x^2 - y^2 = 1$ may be represented parametrically by the equations $x = \sec t$, $y = \tan t$, where t varies over each of the two intervals $J_1: -\pi/2 < t < \pi/2$ and $J_2: \pi/2 < t < 3\pi/2$. Note that the parametric representation yields the complete graph of the hyperbola and not merely the portion of the graph above the x axis [see Example 13.2(c)].

Also, observe that every equation in the form $y = f(x)$ can be put into a parametric form by simply letting $x = t$ and $y = f(t)$. For example,

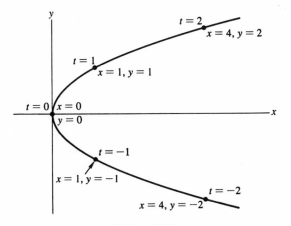

Figure 40.1

$y = x^2 + x + 1$ has a parametric form $x = t$, $y = t^2 + t + 1$.   On the other hand, in some parametric representations we may eliminate the parameter t and obtain an equation in the form $y = f(x)$.   We do this by solving one of the parametric equations for t and then substituting into the other equation. For example, if the parametric representation is $x = t^3$, $y = t^2$, we solve the first equation for t, obtaining $t = x^{1/3}$.   Substituting $t = x^{1/3}$ into $y = t^2$, we obtain $y = x^{2/3}$.   More problems of this type appear at the end of this section.

Very often, t corresponds to the time at which a particle P, moving in the plane, is at the point (x, y).   Hence, the parametric equations tell us *where* the particle is at a specific time t, whereas the equation $y = f(x)$ tells us only the path on which the particle P moves (see Problem 26.11).   For example, con-sider the line $y = x$ in the first quadrant.   Suppose we parametrize this line by letting $x = t$ and $y = t$ with $0 \le t < \infty$ where t represents time measured in seconds.   Then when $t = 1$, the particle P is at the point $(1, 1)$; when $t = 2$, it is at the point $(2, 2)$; after 3 sec, P is at the point $(3, 3)$; and so on.   However, if we parametrize this line by letting $x = t^2$ and $y = t^2$ with $0 \le t < \infty$, then P still moves on the line $y = x$, but in this case when $t = 2$, the particle P is at the point $(4, 4)$, and after 3 sec, P is at the point $(9, 9)$.   Consequently, the particle is now moving along $y = x$ at a faster rate than it was moving under the first parametric representation.   In general, we can parametrize the line $y = x$ by choosing any real-valued function $f(t)$ defined on $0 \le t < \infty$ and letting $x = f(t)$ and $y = f(t)$.   Then after t_0 sec, P is at the point $(f(t_0),$ $f(t_0))$.   Hence, there are *infinitely many* distinct parametrizations of this same line, and each one describes a different physical situation when it is interpreted as giving the location of a particle P at time t.

Finally, arbitrary curves can also be represented parametrically in infinitely many ways.   For instance, the parabola $y^2 = x$ of Example 40.1(b) can be represented by the pair of equations $x = (t + 1)^2$, $y = t + 1$, $-\infty < t < \infty$, *or* by the pair $x = t^6$, $y = t^3$, $-\infty < t < \infty$.   Observe, however, that it cannot be represented by the pair $x = t^4$, $y = t^2$, $-\infty < t < \infty$, since no real value of t yields a point below the x axis.

As exercises, find at least two distinct parametrizations of each of the curves given in Example 40.1.

TANGENT LINES

We now discuss a method for determining the slope of a tangent line to a point P on a curve that is represented parametrically by the functions $x(t)$ and $y(t)$, where both $x(t)$ and $y(t)$ have continuous derivatives on an interval $\alpha \le t \le \beta$.   Recall that in Definition 14.1 a tangent line was defined as the limiting line of secant lines drawn from P to nearby points on the curve (see

Figure 14.3). Using this idea together with the definition of the derivative, we defined the slope of a tangent line to the graph of $y = f(x)$ as the value of the derivative of f at the point in question (see Problem I following Example 22.5). Now, in the case of a parametric representation of a curve, we do not have an expression $y = f(x)$ to differentiate directly. However, we may still use the essential idea that a tangent line T at a point $P = (x(t_0), y(t_0))$ is the limiting line of secant lines S_i drawn from nearby points $P_i = (x(t_i), y(t_i))$ (see Figure 40.2). If we denote the slope of S_i by m_i, then

$$m_i = \frac{y(t_i) - y(t_0)}{x(t_i) - x(t_0)}$$

which may be rewritten as

$$\frac{[y(t_i) - y(t_0)]/(t_i - t_0)}{[x(t_i) - x(t_0)]/(t_i - t_0)}$$

Then as $t_i \to t_0$, $P_i \to P$, $S_i \to T$, and $m_i \to m$ where m is the slope of the tangent line T. Hence m may be defined as

$$\lim_{t \to t_0} \frac{[y(t) - y(t_0)]/(t - t_0)}{[x(t) - x(t_0)]/(t - t_0)} = \frac{\lim_{t \to t_0}[(y(t) - y(t_0)]/(t - t_0)}{\lim_{t \to t_0}[(x(t) - x(t_0)]/(t - t_0)} = \frac{y'(t_0)}{x'(t_0)}$$

This leads us to the following definition.

Definition 40.1 Let C be a curve represented parametrically by the functions $x(t)$ and $y(t)$ defined on an interval $J: \alpha \le t \le \beta$, where both $x(t)$ and $y(t)$ have continuous derivatives on J. Then we define the slope of the tangent line to C at the point $P = (x(t_0), y(t_0))$ by the quotient $y'(t_0)/x'(t_0)$. Note that if we are dealing with the graph of $y = f(x)$, we may use the parametric form $x = t$

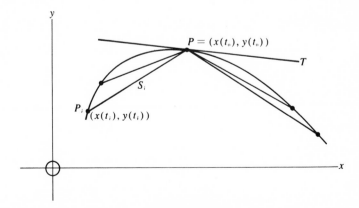

Figure 40.2

and $y = f(t)$. Then $x'(t) = 1$, $y'(t) = f'(t)$, and the quotient $y'(t)/x'(t)$ reduces to $f'(t)$.

REMARK   Observe that we may make further use of the derivatives $x'(t)$ and $y'(t)$ in plotting the curve C.   Since x is a function of t, $x'(t) > 0$ over the interval J: $\alpha \le t \le \beta$ implies that the values of x are increasing as t varies from α to β.   Hence, the curve must be traced to the right as t varies from α to β. [A similar analysis when $x'(t) < 0$ tells us that the curve is drawn to the left as t varies from α to β.]   On the other hand, if $y'(t) > 0$ on J, the values of y are increasing as t varies from α to β, so the curve must be traced in an upward direction.   [A similar analysis when $y'(t) < 0$ tells us that the curve is drawn in a downward direction as t varies from α to β.]   Combining the preceding results, we obtain the following summary.

on J	the curve is traced
$x'(t) > 0$, $y'(t) > 0$	up and to the right
$x'(t) > 0$, $y'(t) < 0$	down and to the right
$x'(t) < 0$, $y'(t) > 0$	up and to the left
$x'(t) < 0$, $y'(t) < 0$	down and to the left

For an application of these results see Example 40.3.

EXAMPLE 40.2 (a)   Find the slope of the tangent line to the curve $x = \log t$, $y = 1 + t^2$, at the points where $t = 1$ and $t = e^{1/2}$.

(b)   Find the slope of the tangent line to the curve $x = a \cos t$, $y = b \sin t$, at the points where $t = 5\pi/4$ and $t = \pi/6$.

SOLUTION   (a) $x'(t) = 1/t$, $y'(t) = 2t$, and $y'(t)/x'(t) = (2t)/(1/t) = 2t^2$. Hence, the required slopes are 2 and $2e$, respectively.

(b) $x'(t) = -a \sin t$, $y'(t) = b \cos t$, and $y'(t)/x'(t) = (-b/a)(\cos t/\sin t)$ Hence, the required slopes are $(-b/a)[\cos(5\pi/4)/\sin(5\pi/4)] = -b/a$ and $(-b/a)[\cos(\pi/6)/\sin(\pi/6)] = (-b/a)\sqrt{3}$.   Note that this curve for $0 \le t \le 2\pi$ is just the ellipse with equation $x^2/a^2 + y^2/b^2 = 1$.   Hence, $(-b/a)(\cos t/\sin t)$ yields the slope of the tangent line to this ellipse at every point except the vertices $(\pm a, 0)$.   As an exercise, find the preceding slopes by using the technique of implicit differentiation (see Section 24) on the equation $x^2/a^2 + y^2/b^2 = 1$.

EXAMPLE 40.3 (a)   Sketch the curve defined by $x(t) = t^2 - 1$, $y(t) = t^3 - t$, $-\infty < t < \infty$.

(b)   Find the slope of the tangent line at the points where $t = -1$, 0, and 1.

SOLUTION   (a)   For large negative t, x is large and positive, while y must be large and negative.   Hence, the graph is in the fourth quadrant for large negative t.   Now, $x'(t) = 2t$ and $y'(t) = 3t^2 - 1$.   Therefore, when $t < 0$, $x'(t) < 0$, so the curve is drawn to the left as t varies over the interval $-\infty < t < 0$.   On the other hand, when $-\infty < t < -\sqrt{3}/3$, $y'(t) > 0$, so the curve is drawn in an upward direction.   Consequently, for $-\infty < t < -\sqrt{3}/3$, the curve is drawn upward and to the left. When $-\sqrt{3}/3 < t < 0$, $x'(t) < 0$ and $y'(t) < 0$; hence, the curve is traced downward and to the left and there must be a relative maximum at $t = -\sqrt{3}/3$.   When $0 < t < \sqrt{3}/3$, $x'(t) > 0$ and $y'(t) < 0$; hence, the curve is traced downward and to the right.   Finally, for $\sqrt{3}/3 < t < \infty$, $x'(t) > 0$ and $y'(t) > 0$; hence, the curve is now traced upward and to the right and there must be a relative minimum at $t = \sqrt{3}/3$.   A sketch appears in Figure 40.3.

SOLUTION   (b)   To find the required slopes, we use $y'(t)/x'(t)$, which is $(3t^2 - 1)/(2t)$.   When $t = -1$, the curve passes through the origin and is traced upward and to the left, while the tangent line T_1 has slope -1. When $t = 0$, the curve passes through the point $(-1, 0)$, and $y'(t)/x'(t)$ has a nonzero numerator and a zero denominator, which indicates a vertical tangent line at $(-1, 0)$.   In fact, $(-1, 0)$ is a vertical inflection point.   Finally, when $t = 1$, the curve again passes through the origin, but this time the curve is traced upward and to the right, and the tangent line T_2 has slope 1.   Consequently, this curve possesses *two* distinct tangent lines at the origin, where the tangent lines indicate the direction in which the curve is traced on each occasion that it passes through the origin.   Thus, we have a phenomenon that cannot occur on graphs of functions $y = f(x)$, since in that case a tangent line at a point (if it exists)

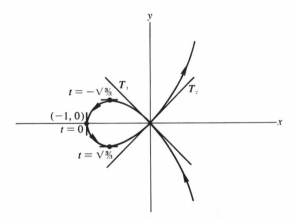

Figure 40.3

must be *unique*. As an exercise, draw in Figure 40.3 the sequence of secant lines that have limiting line T_1 and the sequence of secant lines that have limiting line T_2.

Finally, there is the question of concavity. We know it in the vicinity of the relative maximum and minimum points, but how can we be certain that there are no inflection points in the first or fourth quadrants? To settle this question, we must obtain a formula that gives the second derivative of y with respect to x.

In general, if $y = f(x)$ with both x and y as functions of t, we have $y(t) = f(x(t))$. Differentiating with respect to t and using the chain rule, we obtain $y'(t) = f'(x(t))x'(t)$. Hence, $f'(x(t)) = y'(t)/x'(t)$, which is the formula developed prior to Definition 40.1. Now, to find f'', we differentiate both sides again with respect to t, obtaining $f''(x(t))x'(t) = [y'(t)/x'(t)]'$. This implies $f''(x(t)) = [y'(t)/x'(t)]'[1/x'(t)]$. Hence, for a given value $t = t_0$, $f''(x(t_0))$ is the second derivative of y with respect to x at the point $(x(t_0), y(t_0))$. Therefore, it is this formula that enables us to decide questions of concavity. For instance, in the preceding example, $f''(x(t)) = [(3t^2 - 1)/(2t)]'[1/(t^2 - 1)'] = (3t^2 + 1)/(4t^3)$, which is *not* zero for any real number t. Therefore, there exist no inflection points in the first and the fourth quadrants. Note that $f''(x(t))$ does not exist at $t = 0$ and this relates to the vertical inflection point at $(-1, 0)$.

ARC LENGTH

If a curve C is represented parametrically by the equations $x = x(t)$ and $y = y(t)$ over the interval $J: \alpha \leq t \leq \beta$, where both $x(t)$ and $y(t)$ have continuous derivatives, the arc length may be defined by imitating the method used in Definition 39.1. Choosing a regular partition of J into n subintervals J_i, we induce a partition (not necessarily regular) of the interval $x(\alpha) \leq x \leq x(\beta)$ on the x axis. Then we reproduce Figure 39.1, labeling the diagram to emphasize the present situation (see Figure 40.4). Observe that the length d_i of L_i is given by $[(x(t_i) - x(t_{i-1}))^2 + (y(t_i) - y(t_{i-1}))^2]^{1/2}$. Applying the mean value theorem to the function $x(t)$ on the interval $J_i: t_{i-1} \leq t \leq t_i$, we obtain a point t_i^* in J_i with the property that $x(t_i) - x(t_{i-1}) = x'(t_i^*)(t_i - t_{i-1}) = x'(t_i^*)\Delta$ where $\Delta = (\beta - \alpha)/n$ (since we have a regular partition of J). In a similar way, we obtain a point t_i^{**} (not necessarily equal to t_i^*) with the property that $y(t_i) - y(t_{i-1}) = y'(t_i^{**})\Delta$. Hence d_i may be approximated by $[(x'(t_i^*))^2\Delta^2 + (y'(t_i^{**}))^2\Delta^2]^{1/2} = [(x'(t_i^*))^2 + (y'(t_i^{**}))^2]^{1/2}\Delta$. Now, for large enough n, t_i and t_{i-1} are very close to each other. Hence, if we choose t_i' to be the midpoint of the interval from t_i^* to t_i^{**}, $x'(t_i')$ closely approximates $x'(t_i^*)$ and $y'(t_i')$ closely approximates $y'(t_i^{**})$. Therefore, d_i may be estimated

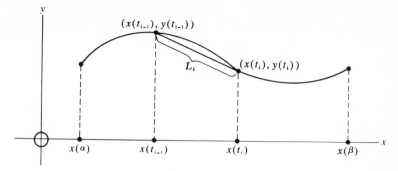

Figure 40.4

by the expression $[(x'(t_i'))^2 + (y'(t_i'))^2]^{1/2}\Delta$ for each $i = 1, 2, \ldots, n$. Consequently, the total arc length, which may be approximated by $\sum_{i=1}^{n} d_i$, can be estimated by

$$\sum_{i=1}^{n} \{[(x'(t_i'))^2 + (y'(t_i'))^2]^{1/2}\Delta\}$$

But this is just a Riemann sum of the function of t given by

$$[(x'(t))^2 + (y'(t))^2]^{1/2}$$

Therefore, the following definition is a reasonable one.

Definition 40.2 If a curve C is represented by the parametric equations $x = x(t)$ and $y = y(t)$ over the interval $\alpha \leq x \leq \beta$ where $x(t)$ and $y(t)$ both have continuous derivatives, the arc length is defined as

$$\int_{\alpha}^{\beta} [(x'(t))^2 + (y'(t))^2]^{1/2} \, dt$$

(Note that the continuity of the derivatives implies the continuity of the integrand, which implies the existence of the integral.) Observe also that whenever we deal with the graph of an equation $y = f(x)$, where $\alpha \leq x \leq \beta$, we may use the parametric equations $x = t$ and $y = f(t)$ over $\alpha \leq t \leq \beta$. Then, by the preceding integral formula, the arc length is $\int_{\alpha}^{\beta} [1 + (f'(t))^2]^{1/2} \, dt$, which agrees with the formula given in Definition 39.1 after replacing t by x, α by a, and β by b.

EXAMPLE 40.4 Find the length of the curve: $x = 2t - 1$, $y = t^2 + 2$, $0 \leq t \leq 1$ (see Figure 40.5).

SOLUTION $x'(t) = 2$, $y'(t) = 2t$, $(x'(t))^2 = 4$, and $(y'(t))^2 = 4t^2$. Therefore, the arc length is $\int_0^1 (4 + 4t^2)^{1/2} \, dt$. Now, $(4 + 4t^2)^{1/2} = 2(1 + t^2)^{1/2}$

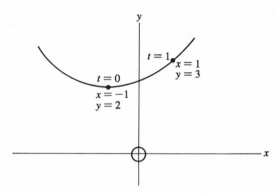

Figure 40.5

and an antiderivative is $2\frac{1}{2}[t(1 + t^2)^{1/2} + \log(t + [1 + t^2]^{1/2})]$. Evaluating at $t = 1$ and $t = 0$ and subtracting, we obtain $\sqrt{2} + \log(1 + \sqrt{2})$, which is the required arc length.

EXAMPLE 40.5 Find the length of the curve: $x = t^2 + 2t + 3$, $y = t^2 - 2t + 1$, $0 \le t \le 1$.

SOLUTION $x'(t) = 2t + 2$, $y'(t) = 2t - 2$, $(x'(t))^2 = 4t^2 + 8t + 4$, $(y'(t))^2 = 4t^2 - 8t + 4$, and $(x'(t))^2 + (y'(t))^2 = 8t^2 + 8 = 8(1 + t^2)$. Hence, the arc length is given by $2\sqrt{2} \int_0^1 (1 + t^2)^{1/2} \, dt$. Using the same antiderivative as in the preceding example, we obtain $2 + \sqrt{2} \log(1 + \sqrt{2})$, which is the required arc length.   As an exercise, plot the curve using the technique described in the Remark that follows Definition 40.1.

EXAMPLE 40.6   Find the length of the curve: $x = e^t \cos t$, $y = e^t \sin t$, $0 \le t \le \pi$.

SOLUTION $x'(t) = -e^t \sin t + e^t \cos t$, $y'(t) = e^t \cos t + e^t \sin t$. After squaring both $x'(t)$ and $y'(t)$ and adding, we obtain (after some simplification) $2e^{2t}$. Hence, the arc length is

$$\int_0^\pi (2e^{2t})^{1/2} \, dt = (\sqrt{2})e^t \Big|_0^\pi = \sqrt{2}(e^\pi - 1)$$

As an exercise, plot the curve using the technique described in the Remark that follows Definition 40.1.

PROBLEMS

40.1   In each of the following, plot the curves and, whenever possible, eliminate the parameter and obtain y as a function of x.

(a) $x = t + 1$, $y = 2t - 2$, $-\infty < t < \infty$

(b) $x = t + 1, y = t^2, -\infty < t < \infty$

(c) $x = \sqrt{t}, y = t^2 + 1/t, 0 < t \leq 4$

(d) $x = t^2 - 1, y = t^3 - t, -1 \leq t \leq 0$ (See Example 40.3.)

(e) $x = t^2 - 1, y = t^3 - t, 0 \leq t \leq 1$ (See Example 40.3.)

(f) $x = e^t \cos t, y = e^t \sin t, 0 \leq t \leq \pi$ (See Example 40.6.)

(g)* $x = \cosh t, y = \sinh t, -\infty < t < \infty$ (*Hint:* See Problem 28.6.)

(h)* $x = (3t)/(t^3 + 1), y = (3t^2)/(t^3 + 1), -1 < t < \infty$ (*Hint:* see Figure 16.7 and the accompanying discussion.)

40.2 Find the equation of the tangent line to each of the following curves at each of the given points. Sketch the curves and the tangent lines.

(a) $x = t + 1, y = t^2, -\infty < t < \infty$, at the points $(1, 0), (2, 1)$, and $(0, 1)$

(b) $x = t^2 - 1, y = t^3 - t, -\infty < t < \infty$, at the points $(0, 0), (-1, 0), (3, 6)$, and $(-\frac{2}{3}, 2\sqrt{3}/9)$ (See Example 40.3.)

(c) $x = 4 \cos t, y = 3 \sin t, 0 \leq t \leq 2\pi$, at the points $(4, 0), (0, 3), (2, 3\sqrt{3}/2)$, and $(2, -3\sqrt{3}/2)$ [See Example 40.2(b).]

(d) $x = e^t \cos t, y = e^t \sin t, 0 \leq t \leq \pi$, at the points $(1, 0), (-e^\pi, 0)$, and $(0, e^{\pi/2})$ (See Example 40.6.)

40.3 Find the points at which each of the following curves has a horizontal tangent.

(a) $x = t^2 - 1, y = t^3 - t, -\infty < t < \infty$ (See Example 40.3.)

(b) $x = t - 3, y = t^2 \perp 1, -\infty < t < \infty$

(c) $x = e^t \cos t, y = e^t \sin t, 0 \leq t \leq 2\pi$ (See Example 40.6.)

40.4 In each of the following, find the length of the given arc.

(a) $x = t + 1, y = t^2, 0 \leq t \leq 2$

(b) $x = t^2, y = t^3, 1 \leq t \leq 2$

(c) $x = t^3, y = t^2, 1 \leq t \leq 2$

(d) $x = \cos 3t, y = \sin 3t, 0 \leq t \leq \pi/3$

(e) $x = \cos^2 t, y = \sin^2 t, 0 \leq t \leq \pi/3$

(f)* $x = \log t, y = (1 + t)/t, 1 \leq t \leq 2$

40.5* (a) Let $x = x(t), y = y(t), \alpha \leq t \leq \beta$, represent a curve similar to the one shown in Figure 40.6. Assume also that both $x(t)$ and $y(t)$ have continuous derivatives on $\alpha \leq t \leq \beta$. Show that the area between this curve and the x axis may be defined as $\int_\alpha^\beta y(t)x'(t)\, dt$. [*Hint:* Choose a regular partition of the interval $\alpha \leq t \leq \beta$. This induces a partition (not necessarily regular) of the interval $x(\alpha) \leq x \leq x(\beta)$ on the x axis. Then estimate the area by imitating the method described in Definition 33.1.]

(b) Use the integral formula of part (a) to compute the area above the x axis and below the curve $x = t^2 - 1, y = t^3 - t, 1 \leq t \leq 2$. Then

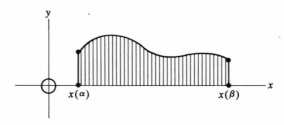

Figure 40.6

eliminate the parameter and find this area by integrating the appropriate function of x (see Example, 40.3).

40.6 In each of the following, use the integral formula of Problem 40.5(a) to compute the area above the x axis and below the given curve.

(a) $x = t + 1, y = 2t - 2, 1 \leq t \leq 3$
(b) $x = t + 1, y = t^2, 0 \leq t \leq 2$
(c) $x = t^2 - 1, y = t^3 - t, -1 \leq t \leq 0$ (See Example 40.3.)
(d) $x = \cos t, y = \sin t, 0 \leq t \leq \pi$
(e) $x = \cos^2 t, y = \sin^2 t, 0 \leq t \leq \pi/2$
(f) $x = a \cos t, y = b \sin t, 0 \leq t \leq \pi/2, \qquad a > 0, b > 0$
  [*Note:* Multiply your answer by 4 and you have the area enclosed by an ellipse with vertices $(a, 0)$, $(0, b)$, $(-a, 0)$, and $(0, -b)$; see Example 40.2(b).]

40.7* Let $x = x(t), y = y(t), \alpha \leq t \leq \beta$ represent a curve similar to the one shown in Figure 40.6 Assume also that both $x(t)$ and $y(t)$ have continuous derivatives on $\alpha \leq x \leq \beta$.   Show that the volume of the solid of revolution obtained by rotating the shaded region of Figure 40.6 about the x axis may be defined as $\pi \int_\alpha^\beta y^2(t)x'(t)\, dt$.   [*Hint:* Use the hint given in Problem 40.5(a), then estimate the volume by imitating the method described in Definition 34.1.]

40.8 Use the formula given in Problem 40.7 to compute the volume of the solid of revolution obtained by rotating about the x axis each of the regions determined by the following curves.

(a) $x = 2t, y = t^2 + 1, 0 \leq t \leq 1$
(b) $x = t^2 - 1, y = t^3 - t, 1 \leq t \leq 2$ (See Example 40.3.)
(c) $x = t^2 - 1, y = t^3 - t, -1 \leq t \leq 0$ (See Example 40.3.)
(d) $x = 3 \cos t, y = 2 \sin t, 0 \leq t \leq \pi/2$ [See Example 40.2(b).]

40.9* Let $x = x(t), y = y(t), \alpha \leq t \leq \beta$, represent a curve similar to the one shown in Figure 40.6. Assume also that both $x(t)$ and $y(t)$ have continuous derivatives on $\alpha \leq t \leq \beta$. Show that the *surface area* of the solid generated by rotating the shaded region of Figure 40.6 about the x axis may be defined as $2\pi \int_\alpha^\beta y(t)[(x'(t))^2 + (y'(t))^2]^{1/2}\, dt$.   [*Hint:* Use the hint given in Problem 40.5(a), then estimate the surface area by imitating the method described in Definition 39.2.]

40.10 Use the formula given in Problem 40.9 to compute the surface area of the solid generated by rotating about the x axis each of the regions determined by the following curves.

(a) $x = 2t, y = t^2 + 1, 0 \leq t \leq 1$
(b) $x = t^2 - 1, y = t^3 - t, 1 \leq t \leq 2$ (See Example 40.3; leave your answer in the form of an integral.)
(c) $x = 2t^3/3, y = t^2 - 1, 2 \leq t \leq 3$
(d) $x = \cos t, y = \sin t, 0 \leq t \leq \pi$
(e) $x = \cos^2 t, y = \sin^2 t, 0 \leq t \leq \pi/2$
(f)* $x = e^t \cos t, y = e^t \sin t, 0 \leq t \leq \pi/2$ (See Example 40.6.)

POLAR COORDINATES, TANGENTS, ARC LENGTH, AND AREA

In Section 1, when we discussed the rectangular coordinate system, we mentioned that we were able to locate points in the plane by attaching number coordinates to these points in a special way. The method depended on knowing the distance of a point from the horizontal and the vertical axes. We now discuss another way of locating points in the plane. This method depends on knowing the distance of the point from a fixed point called the *pole* and on knowing the direction of the half line or ray from the pole to the point. Suppose that we choose a fixed point in the plane, call it the *pole*, and designate it by O. Draw the horizontal line that extends from O to the right and call this line the *polar axis*. Note that if we superimpose our rectangular coordinate system, placing the pole at the origin, the polar axis is the positive x axis (see Figure 41.1). Now, suppose that we are given a point P in the plane different from O. Let r be the distance from O to P. Then P lies on the circumference of a circle C with center O and radius r. Draw the ray L through the points O and P. Then P is the intersection point of the ray L and the circle C. Consequently, instead of locating P by taking the intersection of two lines (one vertical and one horizontal) as we did in the rectangular coordinate system, we can also locate P by taking the intersection of a ray and a circle.

If the circle C has radius r and if the angle from the polar axis to L is θ, we can assign to the point P the *polar coordinates* (r, θ)(see Figure 41.2). For example, if P has rectangular coordinates $(1, 1)$, then $r = \sqrt{2}$ and $\theta = 45°$. Therefore, P has polar coordinates $(\sqrt{2}, 45°)$ or, if we use radian measure for

281

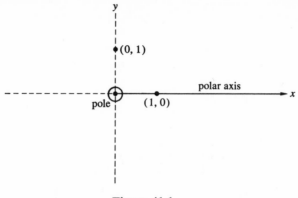

Figure 41.1

the angle θ, $(\sqrt{2}, \pi/4)$. The latter is the more useful in practice, since a radian (unlike a degree) is a pure real number (see the Appendix); thus, the polar coordinates are actually real number coordinates. Figure 41.3 contains some examples of points in the plane with their corresponding rectangular and polar coordinates. Note that this particular association, as it stands, is one-to-one since each point $P \neq O$ lies on only one circle C with center at O and the angle θ is unique, taking values from $0°$ up to and not including $360°$ or from 0 rad up to and not including 2π rad. (Note that for $P = O$, $r = 0$ and θ may be any angle.) However, there is a convention in use that complicates matters somewhat since it removes the uniqueness property of the

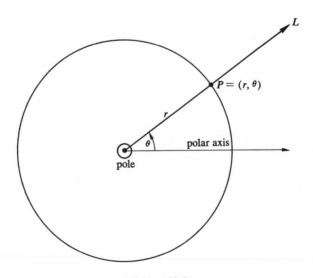

Figure 41.2

polar coordinates.   Since this convention is widely used, we will make a few brief remarks concerning it.   The convention involves the use of negative numbers for the first polar coordinate.   For example, we might locate the point $(-1, -1)$ by taking the ray L from O through the point $(1, 1)$ and then measuring $\sqrt{2}$ units backward on the extension of L through $(-1, -1)$ (see Figure 41.4).   When we do this, we attach a minus sign to the number $\sqrt{2}$ and we write $(-1, -1)$ as $(-\sqrt{2}, \pi/4)$ in polar coordinates.   Hence, $(\sqrt{2}, \pi/4)$ represents the point $(1, 1)$ while $(-\sqrt{2}, \pi/4)$ represents the point $(-1, -1)$. Since $(-1, -1)$ can also be represented by the polar coordinates $(\sqrt{2}, 5\pi/4)$ we see that the polar coordinates of a point are *not* unique.   Of course, if we neglect to use this convention or if we modify it slightly to include the condition that θ be chosen such that $0 \le \theta < \pi$, we preserve the one-to-one association between points in the plane and their polar coordinates.   Consequently, we will usually adhere to the conditions $r \ge 0, 0 \le \theta < 2\pi$, and on occasion,

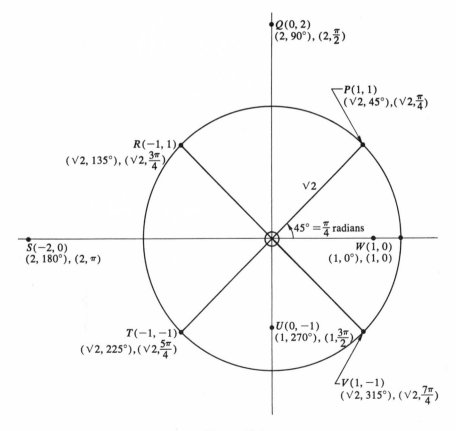

Figure 41.3

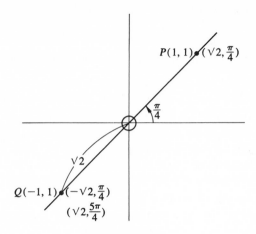

Figure 41.4

when allowing for negative r, θ will be restricted to $0 \le \theta < \pi$.   We will rarely let θ be unrestricted since we would then obtain *infinitely* many representations of any point.   For example, the point $(\sqrt{2}, \pi/4)$ may be written as $(\sqrt{2}, \pi/4 + 2k\pi)$ where k is *any* integer.

We come now to the relationship between the polar coordinates of a point and its rectangular coordinates.   Once again, we superimpose our rectangular system on the polar system.   Let P be a point in the plane with rectangular coordinates (x, y) and polar coordinates (r, θ) (see Figure 41.5).   Observe that $\cos \theta = x/r$ and $\sin \theta = y/r$.   Therefore, $x = r \cos \theta$ and $y = r \sin \theta$ where $r = (x^2 + y^2)^{1/2}$.   These three equations fully relate the rectangular

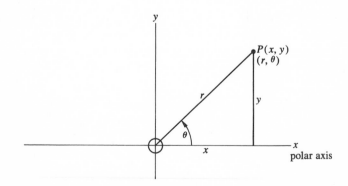

Figure 41.5

coordinates of a point with its polar coordinates. For example, if we are given the point $(2, \pi/3)$ in polar coordinates, we find the rectangular coordinates from the equations

$$x = r \cos \theta = 2 \cos \frac{\pi}{3} = 2 \cdot \tfrac{1}{2} = 1$$

$$y = r \sin \theta = 2 \sin \frac{\pi}{3} = 2 \cdot \frac{\sqrt{3}}{2} = \sqrt{3}$$

Therefore, the rectangular coordinates are $(1, \sqrt{3})$.

On the other hand, if we are given the point with rectangular coordinates $(2\sqrt{3}, 2)$, we may use the equation $r = (x^2 + y^2)^{1/2} = (12 + 4)^{1/2} = 4$ together with $\cos \theta = x/r = 2\sqrt{3}/4 = \sqrt{3}/2$ and $\sin \theta = y/r = \tfrac{1}{2}$. Therefore, θ must be $\pi/6$ rad and the polar coordinates are $(4, \pi/6)$.

In the rectangular coordinate system, we had equations in x and y that gave rise to graphs in the plane. In the polar coordinate system, we have equations in r and θ that also give rise to graphs in the plane. We now study a few equations and their graphs.

EXAMPLE 41.1 Determine the graphs of polar equations of the form $r = k$ where k is a positive constant.

SOLUTION The set of points (r, θ) with first coordinate k all lie on the *circle* about the origin having radius k. The corresponding equation in rectangular coordinates is $x^2 + y^2 = k^2$.

EXAMPLE 41.2 Determine the graphs of polar equations of the form $\theta = k$ where k is a positive constant.

SOLUTION The set of points (r, θ) with second coordinate k all lie on the ray from O that makes an angle of k rad with the polar axis. If we allow for negative as well as positive values of r, we obtain the entire line through O and not just the ray or half line.

EXAMPLE 41.3 Graph the polar equation $r = \sin \theta$, $0 \le \theta \le \pi$.

SOLUTION We make a table of values for θ and r and then plot the points in Figure 41.6.

$\theta = 0$	$\pi/6$	$\pi/3$	$\pi/2$	$2\pi/3$	$5\pi/6$	π
$r = \sin \theta = 0$	$\tfrac{1}{2}$	$\sqrt{3}/2$	1	$\sqrt{3}/2$	$\tfrac{1}{2}$	0

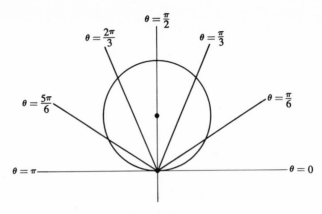

Figure 41.6

This is a circle of radius $\frac{1}{2}$ with center at $(0, \frac{1}{2})$. We may verify this analytically by multiplying $r = \sin \theta$ by r, obtaining $r^2 = r \sin \theta$. But $r^2 = x^2 + y^2$, $r \sin \theta = y$, and therefore, $x^2 + y^2 = y$ is the corresponding rectangular equation. Completing the square yields $x^2 + y^2 - y + \frac{1}{4} = \frac{1}{4}$, which has the standard form $x^2 + (y - \frac{1}{2})^2 = (\frac{1}{2})^2$ and this is the equation of the circle with center $(0, \frac{1}{2})$ and radius $\frac{1}{2}$.

EXAMPLE 41.4 Find the graph of the equation $r = 2\theta$, $0 \le \theta \le 2\pi$.

SOLUTION Once again, we make a table of values; then we plot the points in Figure 41.7, obtaining a spiral.

$\theta = 0$	$\pi/4$	$\pi/2$	$3\pi/4$	π	$5\pi/4$	$3\pi/2$	$7\pi/4$	2π
$r = 2\theta = 0$	$\pi/2$	π	$3\pi/2$	2π	$5\pi/2$	3π	$7\pi/2$	4π

As an exercise, graph $r = 2\theta$ where $-2\pi \le \theta \le 0$.

EXAMPLE 41.5 Find the graph of the equation $r = \sin 2\theta$, $0 \le \theta \le 2\pi$.

SOLUTION Making a table of values, we obtain the following.

$\theta = 0$	$\pi/4$	$\pi/2$	$3\pi/4$	π	$5\pi/4$	$3\pi/2$	$7\pi/4$	2π
$2\theta = 0$	$\pi/2$	π	$3\pi/2$	2π	$5\pi/2$	3π	$7\pi/2$	4π
$\sin 2\theta = 0$	1	0	-1	0	1	0	-1	0

Plotting the points, we obtain the graph shown in Figure 41.8.

REMARK In Example 41.3, the graph of $r = \sin \theta$ had only one loop, whereas in the preceding example, the graph of $r = \sin 2\theta$ has four loops. Now, if we graph $r = \sin 3\theta$, we obtain a graph with only three loops (see

Figure 41.9). This leads us eventually to the general statement that the equation $r = \sin n\theta$ yields a graph with $2n$ loops if n is even and n loops if n is odd. Verify this for $n = 4$ and $n = 5$.

EXAMPLE 41.6 Find the graph of the equation $r = 1 + \cos\theta$, $0 \le \theta \le 2\pi$.

SOLUTION We use the following table of values.

$\theta = 0$	$\pi/4$	$\pi/2$	$3\pi/4$	
$\cos\theta = 1$	$\sqrt{2}/2$	0	$-\sqrt{2}/2$	
$1 + \cos\theta = 2$	$1 + \sqrt{2}/2$	1	$1 - \sqrt{2}/2$	
$\theta = \pi$	$5\pi/4$	$3\pi/2$	$7\pi/4$	2π
$\cos\theta = -1$	$-\sqrt{2}/2$	0	$\sqrt{2}/2$	1
$1 + \cos\theta = 0$	$1 - \sqrt{2}/2$	1	$1 + \sqrt{2}/2$	2

Plotting these points, we obtain a graph that is called a *cardiod* (see Figure 41.10).

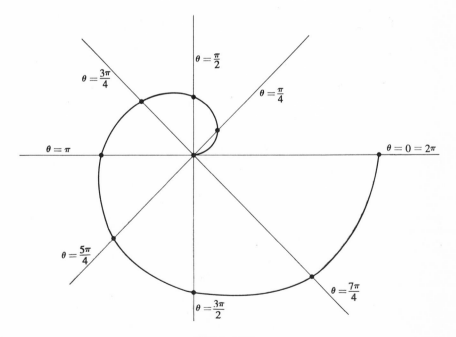

Figure 41.7

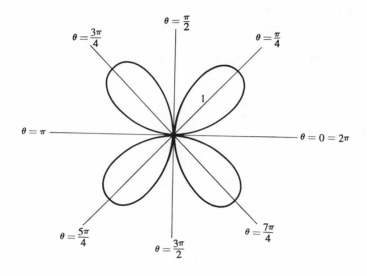

Figure 41.8

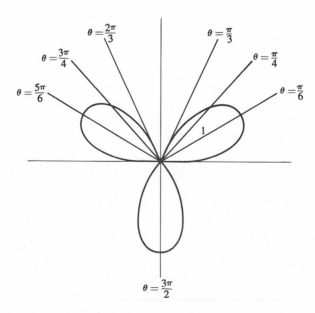

Figure 41.9

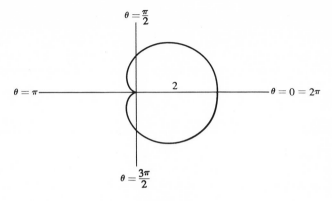

Figure 41.10

TANGENT LINES

In the preceding examples, we considered the graphs of some elementary polar equations having the form $r = f(\theta)$. We now use the calculus to determine the slope of the tangent line at a point on a polar graph. Observe that the slope of the tangent line at a point (r, θ) on an arbitrary polar graph with equation $r = f(\theta)$ is *not* $f'(\theta)$. This can be seen from Example 41.1, where $r = f(\theta) = k$ implies $f'(\theta) = 0$ at *every* point (k, θ). But the graph is a circle and must have nonhorizontal tangent lines at all points except $(k, \pi/2)$ and $(k, 3\pi/2)$. Also, in Example 41.4, $r = f(\theta) = 2\theta$ implies $f'(\theta) = 2$ at *every* point $(2\theta, \theta)$. But this graph is a spiral and clearly does not possess a tangent line having slope 2 at every point. Hence, finding the slope of a tangent line at a point (r, θ) on a polar graph is not simply a matter of computing $f'(\theta)$. However, we may compute the slope by a method based on Definition 40.1.

In the discussion associated with Figure 41.5, we saw that polar coordinates were related to rectangular coordinates by the equations $x = r \cos \theta$, $y = r \sin \theta$, where $r = (x^2 + y^2)^{1/2}$. Now, if we have a polar graph with equation $r = f(\theta)$, an arbitrary point (r, θ) on the graph has corresponding rectangular coordinates given by $x = f(\theta) \cos \theta$, $y = f(\theta) \sin \theta$. But this is just the parametric representation (with parameter θ) of the polar graph in rectangular coordinates. Therefore, the polar equation $r = f(\theta)$ can be converted into a parametric representation by the equations $x = f(\theta) \cos \theta$, $y = f(\theta) \sin \theta$, where the point $(x(\theta), y(\theta))$ traverses the curve as the parameter θ varies over an interval of the type $\alpha \le \theta \le \beta$. For example, the polar equation $r = \sin \theta$, with $0 \le \theta \le \pi$, can be transformed into $x = \sin \theta \cos \theta$, $y = \sin^2 \theta$, $0 \le \theta \le \pi$, while the polar equation $r = 2\theta$, with $0 \le \theta \le 2\pi$, can be transformed into $x = 2\theta \cos \theta$, $y = 2\theta \sin \theta$, $0 \le \theta \le 2\pi$. Consequently, *we may now use all the machinery of Section 40 on these polar graphs.* In the special case of the slope of a tangent line to a point on a polar graph, we may now use Definition 40.1 to obtain a formula for this slope.

Theorem 41.1 Let $r = f(\theta)$ be the equation of a polar graph where $\alpha \leq \theta \leq \beta$. Then the slope of the tangent line to the graph at an arbitrary point (r, θ) is given by

$$\frac{f'(\theta) \sin \theta + f(\theta) \cos \theta}{f'(\theta) \cos \theta - f(\theta) \sin \theta}$$

PROOF Expressing $r = f(\theta)$ in parametric form, we obtain $x = x(\theta) = f(\theta) \cos \theta$, $y = y(\theta) = f(\theta) \sin \theta$. From Definition 40.1, we have that the slope of the tangent line at the point $(x(\theta), y(\theta))$ is $y'(\theta)/x'(\theta)$. Therefore, using the product rule, we differentiate $y(\theta) = f(\theta) \sin \theta$ and $x(\theta) = f(\theta) \cos \theta$. We then take the quotient of these derivatives, obtaining the required result. ■

Observe that for $r = f(\theta) = k$, the formula reduces to $(k \cos \theta)/(-k \sin \theta) = -(x/y)$, which agrees with our earlier result on the slope of a tangent to a circle [see Problem 24.5(a)].

EXAMPLE 41.7 (a) Find the slope of the tangent line to the graph of $r = 2\theta$, $0 \leq \theta \leq 2\pi$, at the points $(0, 0)$, $(\pi, \pi/2)$, and $(2\pi, \pi)$. (See Figure 41.7.)

(b) Find the slope of the tangent line to the graph of $r = 1 + \cos \theta$, $0 \leq \theta \leq 2\pi$, at the points $(2, 0)$, $(1, \pi/2)$, and $(0, \pi)$. (See Figure 41.10.)

SOLUTION (a) If $r = f(\theta) = 2\theta$, then $f'(\theta) = 2$. Substituting into the formula of Theorem 41.1, we obtain

$$\frac{y'(\theta)}{x'(\theta)} = \frac{2 \sin \theta + 2\theta \cos \theta}{2 \cos \theta - 2\theta \sin \theta} = \frac{\sin \theta + \theta \cos \theta}{\cos \theta - \theta \sin \theta}$$

Evaluating this at $\theta = 0$, we obtain 0, indicating a horizontal tangent at the origin. At $\theta = \pi/2$, the formula yields the value $-2/\pi$, which is the slope of the tangent line at $(\pi, \pi/2)$. Finally, at $\theta = \pi$, the formula yields the value π, which is the slope of the tangent line at $(2\pi, \pi)$. Draw each of these three tangent lines in Figure 41.7. Where is the tangent line vertical?

(b) If $r = f(\theta) = 1 + \cos \theta$, then $f'(\theta) = -\sin \theta$. Substituting, we obtain

$$\frac{y'(\theta)}{x'(\theta)} = \frac{-\sin^2 \theta + (1 + \cos \theta) \cos \theta}{-\sin \theta \cos \theta - (1 + \cos \theta) \sin \theta} = \frac{-\sin^2 \theta + \cos \theta + \cos^2 \theta}{-2 \sin \theta \cos \theta - \sin \theta}$$

Evaluating at $\theta = 0$, we find that the slope is undefined, since $x'(\theta) = 0$. In this case, we have a vertical tangent at $(2, 0)$. Evaluating $y'(\theta)/x'(\theta)$ at $\theta = \pi/2$, we obtain the value 1; hence, the slope of the tangent line at the point $(1, \pi/2)$ is 1. Draw these two tangent lines in Figure 41.10. Finally, evaluating at $\theta = \pi$, we obtain $y'(\theta) = 0$ and $x'(\theta) = 0$. Hence, the slope of the tangent line is undefined, but in this case the tangent is not vertical, it simply does not exist. To see why, draw a sequence of secant lines to the origin in the second quadrant. Then draw a sequence of secant lines to the origin in the third quadrant. Are the limiting lines of these two sequences the same?

EXAMPLE 41.8 Find the slope of the tangent line to the graph of $r = \sin 2\theta$, $0 \leq \theta \leq 2\pi$, at the points where $\theta = 0$, $\pi/2$, π, $3\pi/2$, and 2π.

SOLUTION   For each of the given values of θ, the curve passes through the origin (see Figure 41.8); hence, we may have five distinct tangent lines at the origin (compare this with Example 40.3). Since $r = f(\theta) = \sin 2\theta$, $f'(\theta) = 2 \cos 2\theta$ and

$$\frac{y'(\theta)}{x'(\theta)} = \frac{2 \cos 2\theta \sin \theta + \sin 2\theta \cos \theta}{2 \cos 2\theta \cos \theta - \sin 2\theta \sin \theta}$$

When $\theta = 0$, the curve passes through the origin for the *first* time. The slope of the tangent line is $y'(0)/x'(0) = 0$; hence, we have a horizontal tangent, namely, the x axis. When $\theta = \pi/2$, the curve passes through the origin for the *second* time. However, $x'(\pi/2)$ is zero and the slope of the tangent is undefined. In this case, we have a vertical tangent, namely, the y axis. As exercises, verify each of the following statements.

1.   When $\theta = \pi$, the curve passes through the origin for the *third* time and the tangent to the curve is the x axis.

2.   When $\theta = 3\pi/2$, the curve passes through the origin for the *fourth* time and the tangent to the curve is the y axis.

3.   When $\theta = 2\pi$, the curve passes through the origin for the *fifth* time and the tangent to the curve is the x axis.

Hence, the graph of $r = \sin 2\theta$, $0 \leq \theta \leq 2\pi$, passes through the origin five times and has two distinct tangent lines at the origin.

ARC LENGTH

Just as in the preceding discussion on tangent lines, the equation $r = f(\theta)$ may be represented parametrically by the equations $x = f(\theta) \cos \theta$ and $y = f(\theta) \sin \theta$. We may then apply Definition 40.2 and obtain the following result.

Theorem 41.2   Let $r = f(\theta)$ be the equation of a polar graph where $\alpha \leq \theta \leq \beta$. Then the arc length is given by $\int_{\alpha}^{\beta} [(f(\theta))^2 + (f'(\theta))^2]^{1/2} \, d\theta$.

PROOF   From Definition 40.2 (replacing t with θ), we know that the arc length is given by $\int_\alpha^\beta [(x'(\theta))^2 + (y'(\theta))^2]^{1/2}\, d\theta$.   But $x(\theta) = f(\theta) \cos \theta$; hence, $x'(\theta) = -f(\theta) \sin \theta + f'(\theta) \cos \theta$ and $(x'(\theta))^2 = (f(\theta))^2 \sin^2 \theta - 2f(\theta) f'(\theta) \sin \theta \cos \theta + (f'(\theta))^2 \cos^2 \theta$.   On the other hand, $y(\theta) = f(\theta) \sin \theta$; hence, $y'(\theta) = f(\theta) \cos \theta + f'(\theta) \sin \theta$; hence, $(y'(\theta))^2 = (f(\theta))^2 \cos^2 \theta + 2f(\theta) f'(\theta) \sin \theta \cos \theta + (f'(\theta))^2 \sin^2 \theta$.   Therefore, $(x'(\theta))^2 + (y'(\theta))^2 = f((\theta))^2 (\sin^2 \theta + \cos^2 \theta) + (f'(\theta))^2 (\cos^2 \theta + \sin^2 \theta) = (f(\theta))^2 + (f'(\theta))^2$ and the proof is completed. ∎

EXAMPLE 41.9 (a)   Find the length of the arc determined by $r = \sin \theta$, $0 \le \theta \le \pi$.   (See Figure 41.6.)
(b)   Find the length of the arc determined by $r = 2\theta$, $0 \le \theta \le 2\pi$. (See Figure 41.7.)

SOLUTION   (a) $r = f(\theta) = \sin \theta$; hence,

$$f'(\theta) = \cos \theta \quad \text{and} \quad (f(\theta))^2 + (f'(\theta))^2 = \sin^2 \theta + \cos^2 \theta = 1$$

Therefore, the arc length is $\int_0^\pi 1\, d\theta = \pi$ (which was to be expected, since the length of the arc is the circumference of a circle of radius $\frac{1}{2}$).
(b) $r = f(\theta) = 2\theta$; hence, $f'(\theta) = 2$ and $(f(\theta))^2 + (f'(\theta))^2 = 4\theta^2 + 4 = 4(1 + \theta^2)$.   Therefore, the arc length is $\int_0^{2\pi} 2(1 + \theta^2)^{1/2}\, d\theta$. An antiderivative of $2(1 + \theta^2)^{1/2}$ is $2\frac{1}{2}[\theta(1 + \theta^2)^{1/2} + \log(\theta + [1 + \theta^2]^{1/2})]$. Evaluating at $\theta = 2\pi$ and at $\theta = 0$ and subtracting, we obtain

$$[2\pi(1 + 4\pi^2)^{1/2} + \log(2\pi + [1 + 4\pi^2]^{1/2})]$$

as the required arc length.

EXAMPLE 41.10   Find the length of the arc determined by $r = 1 + \cos \theta$, $0 \le \theta \le 2\pi$ (see Figure 41.10).

SOLUTION   By the symmetry of this graph, we need only compute the arc length from $\theta = 0$ to $\theta = \pi$ and then double it.   Now, $f(\theta) = 1 + \cos \theta$, $f'(\theta) = -\sin \theta$, and $(f(\theta))^2 + (f'(\theta))^2 = 1 + 2 \cos \theta + \cos^2 \theta + \sin^2 \theta = 2 + 2 \cos \theta = 2(1 + \cos \theta)$.   Therefore, the arc length is

$$2 \int_0^\pi \sqrt{2(1 + \cos \theta)^{1/2}}\, d\theta$$

But $(1 + \cos \theta)^{1/2} = \sqrt{2} \cos(\theta/2)$ for all θ between 0 and π [see the Appendix, Corollary 3 (a)], and an antiderivative of $\cos(\theta/2) = 2 \sin(\theta/2)$. Hence, the arc length is

$$2 \int_0^\pi (\sqrt{2})(\sqrt{2}) \cos\left(\frac{\theta}{2}\right) d\theta = 4 \int_0^\pi \cos\left(\frac{\theta}{2}\right) d\theta = 8 \left(\sin\left(\frac{\theta}{2}\right) \Big|_0^\pi \right) = 8$$

AREA

Suppose that we are given a polar graph like the one shown in Figure 41.11, which is represented by an equation $r = f(\theta)$ where f is continuous over the interval $J: \alpha \le \theta \le \beta$. We would like to find the area enclosed by the graph of $r = f(\theta)$ between the ray $\theta = \alpha$ and the ray $\theta = \beta$. We proceed just as we did in our discussion of area in rectangular coordinates (see Definition 33.1), except that we now approximate the required area by adding the areas of *circular sectors* instead of *rectangles*.

Let $P: \alpha = \theta_0 < \theta_1 < \theta_2 < \cdots < \theta_n = \beta$ be a regular partition of the interval J into n subintervals of length $\Delta = (\beta - \alpha)/n$. Then for each $i = 1$, $2, \ldots, n$ there is a corresponding region bounded by the graph of $r = f(\theta)$ and the rays $\theta = \theta_{i-1}$ and $\theta = \theta_i$ (see the shaded area in Figure 41.11). Now, if for each i we choose a number θ_i^* such that $\theta_{i-1} < \theta_i^* < \theta_i$, we may approximate the shaded area by computing the area of the circular sector of radius $f(\theta_i^*)$ bounded by the rays $\theta = \theta_{i-1}$ and $\theta = \theta_i$. But such a circular sector has area equal to $\frac{1}{2}$ *times* the radius squared *times* the angle (measured in radians) from θ_{i-1} to θ_i (since, in a circle of radius r, a central angle of 2π rad corresponds to a circular sector having area πr^2; hence, for a central angle of θ rad, we have $\pi r^2/2\pi = A/\theta$ where A is the area of the circular sector corresponding to θ). Hence, the shaded area may be approximated by $\frac{1}{2}f^2(\theta_i^*)(\theta_i - \theta_{i-1}) = \frac{1}{2}f^2(\theta_i^*)\,\Delta$, and the total area may then be approximated by $\sum_{i=1}^{n} \frac{1}{2}f^2(\theta_i^*)\,\Delta$. But this is just a Riemann sum for the function $\frac{1}{2}f^2(\theta)$ corresponding to the

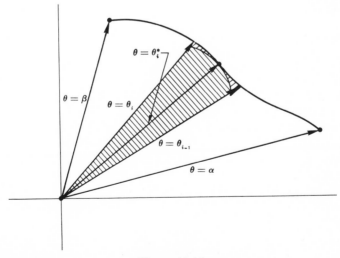

Figure 41.11

regular partition P.   Now, when n is large, the shaded area between $\theta = \theta_{i-1}$ and $\theta = \theta_i$ is nearly equal to the area of the circular sector of radius $f(\theta_i^*)$ between $\theta = \theta_{i-1}$ and $\theta = \theta_i$.   Hence, it is reasonable to make the following definition.

Definition 41.1   The area enclosed by the graph of $r = f(\theta)$ between the rays $\theta = \alpha$ and $\theta = \beta$ is defined as $\frac{1}{2}\int_{\alpha}^{\beta} f^2(\theta)\, d\theta$.

EXAMPLE 41.11   In each of the following, find the area enclosed by the graph of the given function between the indicated rays.

(a) $r = \sin\theta$, $\theta = 0$ to $\theta = \pi/2$ (See Figure 41.6.)
(b) $r = 2\theta$, $\theta = 0$ to $\theta = \pi$ (See Figure 41.7.)
(c) $r = \sin 2\theta$, $\theta = 0$ to $\theta = \pi/2$ (See Figure 41.8.)
(d) $r = 1 + \cos\theta$, $\theta = 0$ to $\theta = \pi$ (See Figure 41.10.)

SOLUTION   (a) $r = f(\theta) = \sin\theta$, $f^2(\theta) = \sin^2\theta$, and an antiderivative of $\sin^2\theta$ is $\frac{1}{2}(\theta - \sin\theta\cos\theta)$.   Hence, the required area is

$$\frac{1}{2}\int_0^{\pi/2}\sin^2\theta\, d\theta = \frac{1}{2}\left[\frac{1}{2}(\theta - \sin\theta\cos\theta)\Big|_0^{\pi/2}\right] = \frac{1}{4}\left(\frac{\pi}{2}\right) = \frac{\pi}{8}$$

(which was to be expected, since we have actually found the area enclosed by a semicircle of radius $\frac{1}{2}$).

(b) $r = f(\theta) = 2\theta$, $f^2(\theta) = 4\theta^2$; thus, the required area is

$$\frac{1}{2}\int_0^{\pi}(4\theta^2)\, d\theta = 2\left(\frac{\theta^3}{3}\Big|_0^{\pi}\right) = \frac{2\pi^3}{3}$$

(c) $r = f(\theta) = \sin 2\theta$, $f^2(\theta) = \sin^2 2\theta$, and an antiderivative of $\sin^2 2\theta$ is $\frac{1}{4}(2\theta - \sin 2\theta\cos 2\theta)$.   Hence, the required area is

$$\frac{1}{2}\int_0^{\pi/2}\sin^2 2\theta\, d\theta = \frac{1}{2}\cdot\frac{1}{4}\left[(2\theta - \sin 2\theta\cos 2\theta)\Big|_0^{\pi/2}\right] = \frac{1}{8}(\pi) = \frac{\pi}{8}$$

Note that this is the area enclosed by *one* loop of the graph of $r = \sin 2\theta$ $0 \le \theta \le 2\pi$.   Hence, the *total* area enclosed by the four loops is $\pi/2$.

(d) $r = f(\theta) = 1 + \cos\theta$, $f^2(\theta) = 1 + 2\cos\theta + \cos^2\theta$;   thus, the required area is

$$\frac{1}{2}\int_0^{\pi}(1 + 2\cos\theta + \cos^2\theta)\, d\theta$$

$$= \frac{1}{2}\int_0^{\pi}1\, d\theta + \frac{1}{2}\int_0^{\pi}2\cos\theta\, d\theta + \frac{1}{2}\int_0^{\pi}\cos^2\theta\, d\theta$$

The first integral is simply $\pi/2$, the second integral is $\sin \theta \left.\right|_0^{\pi} = 0$. Since an antiderivative of $\cos^2 \theta$ is $\frac{1}{2}(\theta + \sin \theta \cos \theta)$, the third integral is $\frac{1}{4}\left[(\theta + \sin \theta \cos \theta) \left.\right|_0^{\pi}\right] = \pi/4$. Hence, the required area is $\pi/2 + \pi/4 = 3\pi/4$. Note that this is the area enclosed in the upper portion of the cardiod in Figure 41.10; therefore, the total area enclosed by this cardiod is $3\pi/2$.

PROBLEMS

41.1 Plot the following points in the polar coordinate system.
(a) $(3, \pi/3)$ (b) $(6, -\pi/6)$ (c) $(2, \pi)$
(d) $(3, 4\pi/3)$ (e) $(4, -3\pi/2)$ (f) $(1, 5\pi/6)$
(g) $(-2, \pi/4)$ (h) $(2, 5\pi/4)$ (i) $(2, 3\pi/4)$

41.2 Find the corresponding rectangular coordinates for each of the points given in the preceding problem.

41.3 Find polar coordinates for each of the following points given in the usual rectangular coordinate system.
(a) $(0, 3)$ (b) $(-1, 0)$ (c) $(-4, 3)$
(d) $(5, 12)$ (e) $(0, 0)$ (f) $(\sqrt{2}, \sqrt{2})$

41.4 Graph the following equations.
(a) $r = 4$ $r = 7$
(b) $\theta = \pi/4$ $\theta = \pi/2$
(c) $r = \theta$ $r = 3\theta$
(d) $r\theta = \pi$
(e) $r = \cos \theta$ $r = \cos 2\theta$ $r = \cos 3\theta$
(f) $r = \sin 3\theta$ $r = \sin 4\theta$
(g) $r = \sin \theta + \cos \theta$ $r = \sin \theta - \cos \theta$
(h) $r = 1 + \sin \theta$ $r = 3 + \sin \theta$ $r = 1 + 3 \sin \theta$
(i) $r = 2 + 4 \cos \theta$
(j) $r^2 = \sin 2\theta$ $r^2 = \cos 2\theta$

41.5 In each of the following, find the slope of the tangent line to the given graphs at the indicated points.
(a) $r = \sin \theta, 0 \le \theta \le \pi$, at the points $(0, 0)$, $(\frac{1}{2}, \pi/6)$, and $(\sqrt{2}/2, \pi/4)$ (See Figure 41.6.)
(b) $r = 3\theta, 0 \le \theta \le 2\pi$, at the points $(0, 0)$, $(\pi, \pi/3)$, and $(3\pi, \pi)$
(c) $r = 1 + \cos \theta, 0 \le \theta \le 2\pi$, at the points $(\frac{3}{2}, \pi/3)$ and $(1, 3\pi/2)$ [See Figure 41.10 and Example 41.7(b).]
(d) $r = \sin 3\theta, 0 \le \theta \le 2\pi$, at the points $(1, \pi/6)$, $(1, 3\pi/2)$, $(0, \pi/3)$, $(0, 2\pi/3)$, and $(0, \pi)$ (See Figure 41.9.)
(e) $r = e^\theta, 0 \le \theta \le \pi$, at the points $(1, 0)$, $(2, \log 2)$, and (e^π, π)

41.6* Let (r, θ) be an arbitrary point on the graph of $r = f(\theta)$. Let L_1 be the ray through (r, θ), let L_2 be the tangent line to the graph at (r, θ), and let β be the angle measured from L_1 to L_2 (see Figure 41.12). Prove that $\tan \beta = f(\theta)/f'(\theta)$. [*Hint*: Use Theorem 4.1 with $m_1 = \tan \theta$ and $m_2 = \tan \alpha$. Replace $\tan \theta$ with $\sin \theta/\cos \theta$, and $\tan \alpha$ with $y'(\theta)/x'(\theta)$, as given in Theorem 41.1, and then simplify.]

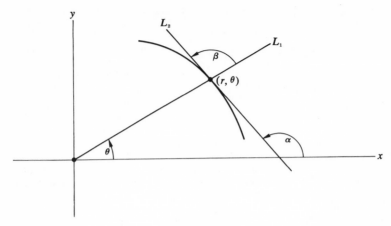

Figure 41.12

41.7   In each of the following, using the result of the preceding problem, find
tan β at each of the points on the given curve.
(a) $r = \sin\theta$, $0 \le \theta \le \pi$, at $(\frac{1}{2}, \pi/6)$ and $(\sqrt{2}/2, \pi/4)$ (See Figure 41.6.)
(b) $r = 3\theta$, $0 \le \theta \le 2\pi$, at $(\pi, \pi/3)$, $(3\pi, \pi)$, and at any point (r, θ)
(c) $r = 1 + \cos\theta$, $0 \le \theta \le 2\pi$, at $(\frac{3}{2}, \pi/3)$ and $(1, 3\pi/2)$ [See Figure 41.10 and
Example 41.7(b).]
(d) $r = \sin 3\theta$, $0 \le \theta \le 2\pi$, at $(1, \pi/6)$ and $(\sqrt{2}/2, \pi/4)$ (See Figure 41.9.)
(e) $r = e^{\theta}$, $0 \le \theta \le \pi$, at every point on the graph
41.8   In each of the following, sketch the given curve and find its length.
(a) $r = \cos\theta$, $0 \le \theta \le \pi/2$
(b) $r = 3\theta$, $0 \le \theta \le 2\pi$
(c) $r = e^{\theta}$, $0 \le \theta \le \log 2$
(d) $r = e^{-\theta}$, $0 \le \theta \le \log 2$
(e) $r = \sin 2\theta$, $0 \le \theta \le \pi/4$ (Leave your answer in the form of an integral.)
41.9   In each of the following, sketch the given curve and find the area enclosed
by the curve between the indicated rays.
(a) $r = \cos\theta$, $\theta = 0$, $\theta = \pi/6$
(b) $r = 2\theta$, $\theta = 0$, $\theta = \pi/4$; $\theta = \pi/2$, $\theta = 3\pi/2$ (See Figure 41.7.)
(c) $r = \sin 3\theta$, one loop   (See Figure 41.9.)
(d) $r = \cos 3\theta$, one loop
(e) $r = \cos 2\theta$, one loop
(f) $r = 1 + \cos\theta$, $\theta = \pi/2$, $\theta = \pi$ (See Figure 41.10.)
(g) $r = 1 + \sin\theta$, $\theta = 0$, $\theta = \pi/2$
(h) $r = e^{\theta}$, $\theta = 0$, $\theta = \log 2$
(i) $r = \sec\theta$, $\theta = 0$, $\theta = \pi/3$
(j) $r = \sin^2\theta$, $\theta = 0$, $\theta = \pi/2$
41.10*   Let $r = f(\theta)$, $\alpha \le \theta \le \beta$, represent a curve bounding a region like the one
shaded in Figure 41.13.   Prove that the *volume* of the solid revolution ob-
tained by rotating this region about the x axis may be defined as

$$\pi \int_{\beta}^{\alpha} f^{2}(\theta)\sin^{2}\theta[f'(\theta)\cos\theta - f(\theta)\sin\theta]\,d\theta$$

[*Hint*: Represent $r = f(\theta)$ parametrically and then apply the formula given in Problem 40.7. Note also that $x(\theta)$ is a *decreasing* function here, and this is the reason we integrate from β to α.]

41.11 Using the formula given in the preceding problem, find the volume of the solid of revolution obtained by rotating about the x axis the regions bounded below by the x axis and above by each of the following curves.

(a) $r = \sin \theta,\ \pi/4 \le \theta \le 3\pi/4$ (See Figure 41.6.)
(b) $r = \cos \theta,\ 0 \le \theta \le \pi/2$
(c) $r = 2\theta,\ \pi/2 \le \theta \le \pi$ (See Figure 41.7; leave your answer in the form of an integral.)
(d) $r = 1 + \cos \theta,\ 0 \le \theta \le \pi/2$ (See Figure 41.10.)

41.12* Let $r = f(\theta),\ \alpha \le \theta \le \beta$, represent a curve bounding a region like the one shaded in Figure 41.13. Prove that the *surface area* of the solid of revolution obtained by rotating this region about the x axis may be defined as

$$2\pi \int_\alpha^\beta f(\theta) \sin \theta [(f(\theta))^2 + (f'(\theta))^2]^{1/2}\, d\theta$$

[*Hint*: Represent $r = f(\theta)$ parametrically, apply the formula given in Problem 40.9, and then imitate the proof of Theorem 41.2.]

41.13 Using the formula given in the preceding problem, find the surface area of the solid of revolution obtained by rotating about the x axis the regions bounded below by the x axis and above by each of the following curves.

(a) $r = \sin \theta,\ \pi/4 \le \theta \le 3\pi/4$ (See Figure 41.6.)
(b) $r = \cos \theta,\ 0 \le \theta \le \pi/2$
(c) $r = 2\theta,\ \pi/2 \le \theta \le \pi$ (See Figure 41.7; leave your answer in the form of an integral.)
(d) $r = 1 + \cos \theta,\ 0 \le \theta \le \pi/2$ (See Figure 41.10.)
(e) $r = e^\theta,\ \pi/2 \le \theta \le \pi$

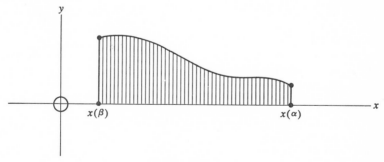

Figure 41.13

Appendix

TRIGONOMETRY

Recall that an angle α with vertex at the origin and initial side on the positive x axis (see Figure A.1) may be measured in degrees or radians. An angle has 1 *degree* if it subtends an arc equal in length to 1/360 of the circumference of the circle and an angle has 1 *radian* if it subtends an arc equal in length to the radius of the circle (see Figure A.2). Hence a straight angle has 180° and π rad (see Figure A.3). This leads to the conversion formulas

$$1° = \frac{\pi}{180} \text{ rad}$$

$$1 \text{ rad} = \left(\frac{180}{\pi}\right)°$$

Observe also that an angle of 2π rad subtends an arc that is the entire circle and therefore has length $2\pi r$. Hence, if an angle of α rad subtends an arc

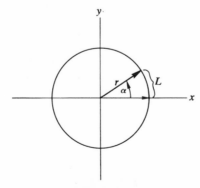

Figure A.1

299

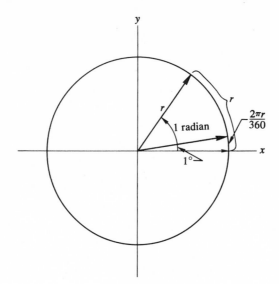

Figure A.2

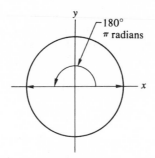

Figure A.3

of length L, we must have $\alpha/L = 2\pi/2\pi r = 1/r$. Therefore, $\alpha = L/r$. Consequently, when measured in radians, α is the quotient of two linear quantities L and r and is therefore a pure or dimensionless real number (see Figure A.4). It is for this reason that radian measure is used almost exclusively when dealing with the trigonometric functions.

THE TRIGONOMETRIC FUNCTIONS

We now define six real-valued functions of a real variable, each one having a domain that is a set of angles measured in radians. Let α be any angle with its vertex at the origin, initial side along the positive x axis, and terminal

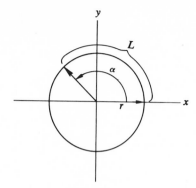

Figure A.4

side intersecting a circle about the origin of radius r at the point (a, b) (see Figure A.5). Now there are three real numbers a, b, and r that represent the lengths of the sides of the right triangle containing α. We first define functions of the angle α in such a way that the values of the functions are ratios of the lengths of two distinct sides of this right triangle. There are precisely six distinct ratios: b/r, a/r, b/a, r/b, r/a, and a/b. Note that if we choose a *smaller* circle of radius r' or a *larger* one of radius r'', and if the terminal side of α intersects these circles at the points (a', b') and (a'', b'') respectively, then the three right triangles in Figure A.6 are *similar*. Hence, $b/r = b'/r' = b''/r''$ and so on for the other five ratios. Therefore, we need only consider the ratios in any fixed circle like the one in Figure A.5.

Now, using the preceding ratios, we define the following functions.

sine: $\sin \alpha = b/r$
cosine: $\cos \alpha = a/r$

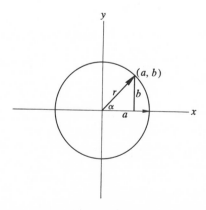

Figure A.5

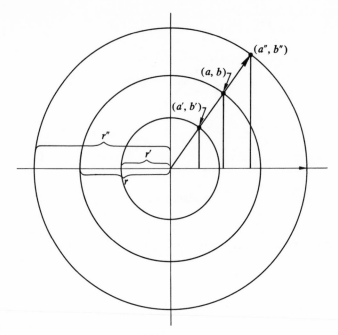

Figure A.6

tangent: tan $\alpha = b/a = \sin \alpha/\cos \alpha$
cosecant: csc $\alpha = r/b = 1/\sin \alpha$
secant: sec $\alpha = r/a = 1/\cos \alpha$
cotangent: cot $\alpha = a/b = \cos \alpha/\sin \alpha = 1/\tan \alpha$

To extend these definitions to angles α that do not have terminal sides in the first quadrant, we simply let (a, b) represent the coordinates of the point of intersection of the terminal side of α with the circle of radius r about the origin.   However, a and b may now be negative numbers and the ratios are just fractions composed of either the two coordinates of the point (a, b) or the radius r with one of the coordinates of the point (a, b).   Hence, we can no longer consider these fractions as ratios of the lengths of the sides in a right triangle.

Note that sin $\alpha > 0$ for any α with terminal side in the first or second quadrant since r is always positive and since b is positive whenever the terminal side is above the x axis.   On the other hand, sin $\alpha < 0$ for any α with terminal side in the third or fourth quadrant.   The signs of the other functions in each of the four quadrants may be determined from the preceding definitions and verified by using Table 1 of Section 27.

The actual values of the six functions at $\alpha = \pi/6$, $\alpha = \pi/4$, and $\alpha = \pi/3$ may be determined from Table 2 of Section 27.

Observe also that $\sin 0 = \sin \pi = 0$ (since $b = 0$ in each case), while $\sin(\pi/2) = 1$ (since $b = r$ in this case) and $\sin(3\pi/2) = -1$ (since $b = -r$ in this case). The values of the other five functions at $\alpha = 0$, $\alpha = \pi/2$, $\alpha = \pi$, and $\alpha = 3\pi/2$ may be determined from Table 3 of Section 27. With regard to the tangent at $\pi/2$ and $3\pi/2$, it is clear that when α is $\pi/2$ or $3\pi/2$, $a = 0$ and the ratio b/a is *undefined*. However, as $\alpha \to \pi/2$, $b \to r$, $a \to 0$, and therefore, $b/a = \tan \alpha \to \infty$. Hence, for values of α close to $\pi/2$ rad we obtain large positive values for $\tan \alpha$. On the other hand, as $\alpha \to 3\pi/2$, b is negative and $b \to -r$, while $a \to 0$ and $b/a = \tan \alpha \to -\infty$. Hence, for values of α close to $3\pi/2$ rad, we obtain large negative values for $\tan \alpha$.

THE DOMAINS AND RANGES

The *domain* of the sine and cosine functions is the set of all angles with vertices at the origin and initial sides along the positive x axis. Since we are measuring angles in radian measure, and since we may continue to rotate the terminal side of α around and around in either the clockwise or counterclockwise direction, the set of all angles corresponds to the set of all real numbers. In fact from Figure A.5, a rotation of the terminal side of α 2π rad in either direction, yields an angle with the same sine and cosine as α since (a, b) remains the point of intersection of the terminal side of the new angle with the circle. Hence, $\sin(\alpha + 2\pi) = \sin(\alpha - 2\pi) = \sin \alpha$ and $\cos(\alpha + 2\pi) = \cos(\alpha - 2\pi) = \cos \alpha$. In a similar way, we obtain the relations $\sin(\alpha + 2k\pi) = \sin \alpha$ and $\cos(\alpha + 2k\pi) = \cos \alpha$ where k is any integer. The same relations hold for the other four functions at each angle α in their respective domains. Consequently the trigonometric functions are clearly *not* one-to-one. In fact, they are infinitely many-to-one, since, for example, every angle of the type $k\pi$, where k is any integer, is mapped to the real number 0 by the sine function.

The *range* of the sine and cosine functions is the set of all real numbers y such that $-1 \leq y \leq 1$ since $\sin \alpha = b/r$ and $|\sin \alpha| = |b|/r \leq r/r = 1$, while $\cos \alpha = a/r$ and $|\cos \alpha| = |a|/r \leq r/r = 1$.

Note also that all the angles of the form $\pi/2 + k\pi$, where k is any integer, must be eliminated from the domain of the tangent and secant functions, while all angles of the form $k\pi$, where k is any integer, must be eliminated from the domain of the cotangent and cosecant functions.

Finally, observe that the range of the tangent and the cotangent functions is the set of *all* real numbers (recall earlier we showed that $\tan \alpha \to \infty$ as $\alpha \to \pi/2$, and $\tan \alpha \to -\infty$ as $\alpha \to 3\pi/2$), while the range of the secant and the cosecant functions is the set of all real numbers y such that $|y| \geq 1$. This follows from the fact that $|1/\cos \alpha| \geq 1$ (since $|\cos \alpha| \leq 1$) and $|1/\sin \alpha| \geq 1$ (since $|\sin \alpha| \leq 1$). The six graphs appear in Figure A.7. In each case the function is graphed over its *principal* domain.

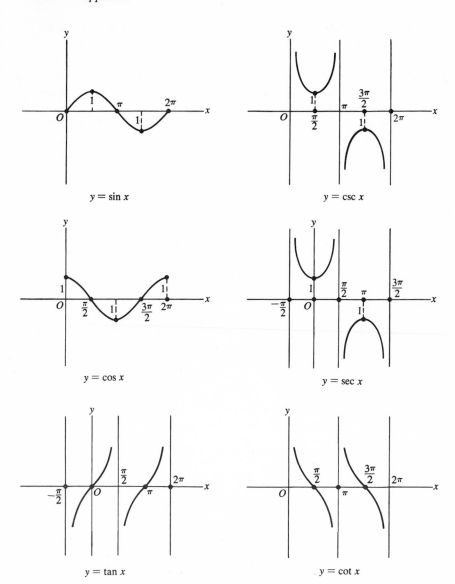

Figure A.7

PROPERTIES

Each of the following properties may be verified by using Figure A.8.
1. (a) $\sin(-\alpha) = -\sin\alpha$
    (b) $\cos(-\alpha) = \cos\alpha$

2. (a) $\sin \alpha = \cos(\pi/2 - \alpha)$
   (b) $\cos \alpha = \sin(\pi/2 - \alpha)$
3. (a) $\sin \alpha = \sin(\pi - \alpha)$
   (b) $\cos \alpha = -\cos(\pi - \alpha)$
4. $\sin^2 \alpha + \cos^2 \alpha = 1$. This follows from the fact that $\sin^2 \alpha + \cos^2 \alpha = b^2/r^2 + a^2/r^2 = (b^2 + a^2)/r^2 = r^2/r^2 = 1$.
5. $1 + \tan^2 \alpha = \sec^2 \alpha$. This follows from the fact that $1 + \tan^2 \alpha = 1 + b^2/a^2 = (a^2 + b^2)/a^2 = r^2/a^2 = \sec^2 \alpha$.

We now conclude by proving the addition formulas for sine and cosine along with numerous important corollaries.

Theorem A.1 For any two angles α and β,

$$\cos(\alpha + \beta) = \cos \alpha \cos \beta - \sin \alpha \sin \beta$$

PROOF From the discussion associated with Figure A.6, it is clear that we may use the unit circle throughout this proof in order to simplify our computations.

Let α and β be two angles with the following properties (see Figure A.9).
(A) The initial side of α passes through the point $P = (1, 0)$.
(B) The terminal side of α is the initial side of β.
(C) The terminal side of β intersects the unit circle at the point $Q = (a, b)$.

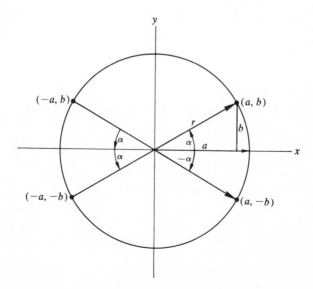

Figure A.8

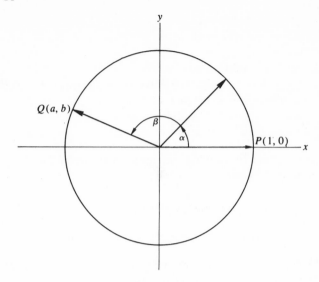

Figure A.9

Then, by definition, $\sin(\alpha + \beta) = b/1 = b$ (since $\alpha + \beta$ has terminal side passing through Q). In a similar way, we obtain $\cos(\alpha + \beta) = a$, hence $Q = (\cos(\alpha + \beta), \sin(\alpha + \beta))$.

Now, suppose we rotate the coordinate axes through an angle α in the counterclockwise direction, obtaining a new x', y' coordinate system (see Figure A.10). Then the initial side of β is on the positive x' axis, while the

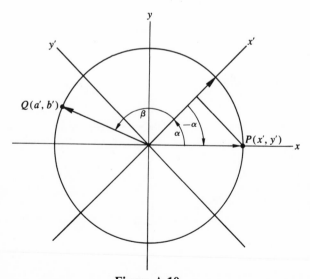

Figure A.10

terminal side of β passes through the point Q. But if Q has coordinates (a', b') in the x', y' system, then by definition, $\cos \beta = a'$ and $\sin \beta = b'$. Hence, $Q = (\cos \beta, \sin \beta)$ in this new coordinate system.

Now, in the x', y' system, the terminal side of the angle $-\alpha$ passes through the point P. If P has coordinates (x', y'), then by definition, $\cos(-\alpha) = x'$ and $\sin(-\alpha) = y'$. But $\cos(-\alpha) = \cos \alpha$ and $\sin(-\alpha) = -\sin \alpha$; therefore, $P = (\cos \alpha, -\sin \alpha)$ in the x', y' coordinate system.

Summarizing our results thus far, we have in the x, y coordinate system

$$P = (1, 0) \qquad Q = (\cos(\alpha + \beta), \sin(\alpha + \beta))$$

in the x', y' coordinate system

$$P = (\cos \alpha, -\sin \alpha) \qquad Q = (\cos \beta, \sin \beta)$$

But the distance D between P and Q does not change under a rotation of axes. Hence, we may compute the distance between P and Q in each coordinate system and we must obtain the same answer D.

In the x, y system

$$D^2 = [\cos(\alpha + \beta) - 1]^2 + \sin^2(\alpha + \beta)$$
$$= \cos^2(\alpha + \beta) - 2\cos(\alpha + \beta) + 1 + \sin^2(\alpha + \beta) = -2\cos(\alpha + \beta) + 2$$

In the x', y' system

$$D^2 = (\cos \beta - \cos \alpha)^2 + (\sin \beta - (-\sin \alpha))^2$$
$$= \cos^2 \beta - 2\cos \alpha \cos \beta + \cos^2 \alpha + \sin^2 \beta + 2\sin \alpha \sin \beta + \sin^2 \alpha$$
$$= 2 - 2(\cos \alpha \cos \beta - \sin \alpha \sin \beta)$$

Therefore, $-2\cos(\alpha + \beta) + 2 = 2 - 2(\cos \alpha \cos \beta - \sin \alpha \sin \beta)$ and canceling, we obtain the required addition formula.

Corollary A.1   (a) $\cos 2\alpha = 2\cos^2 \alpha - 1$
                              (b) $\cos 2\alpha = 1 - 2\sin^2 \alpha$

Corollary A.2   (a) $\cos^2 \alpha = (\tfrac{1}{2})(1 + \cos 2\alpha)$
                              (b) $\sin^2 \alpha = (\tfrac{1}{2})(1 - \cos 2\alpha)$

Corollary A.3   (a) $(1 + \cos \alpha)^{1/2} = \sqrt{2}\cos(\alpha/2)$ for $0 \le \alpha \le \pi$ [to prove this, simply state Corollary A.2(a) for an angle α', then let $\alpha' = \alpha/2$].

(b) $(1 - \cos \alpha)^{1/2} = \sqrt{2}\sin(\alpha/2)$ for $0 \le \alpha \le 2\pi$.

Theorem A.2   For any two angles α and β,

$$\sin(\alpha + \beta) = \sin \alpha \cos \beta + \cos \alpha \sin \beta$$

PROOF $\sin(\alpha + \beta) = \cos(\pi/2 - [\alpha + \beta])$ [write α' in Property 2(a) and then let $\alpha' = \alpha + \beta$].   Now,

$$\cos(\pi/2 - [\alpha + \beta]) = \cos([\pi/2 - \alpha] + [-\beta])$$
$$= \cos(\pi/2 - \alpha) \cos(-\beta) - \sin(\pi/2 - \alpha) \sin(-\beta)$$

(by Theorem A.1).   But $\cos(\pi/2 - \alpha) = \sin \alpha$ [by Property 2(a)]; $\cos(-\beta)$ $= \cos \beta$, $\sin(\pi/2 - \alpha) = \cos \alpha$ [by Property 2(b)]; and $\sin(-\beta) = -\sin \beta$. Therefore, the last equation may be rewritten as $\sin(\alpha + \beta) = \cos(\pi/2 - [\alpha + \beta]) = \sin \alpha \cos \beta - \cos \alpha(-\sin \beta) = \sin \alpha \cos \beta + \cos \alpha \sin \beta$ and we have the required addition formula.

Corollary A.4 $\sin 2\alpha = 2 \sin \alpha \cos \alpha$.

PROBLEMS

A.1   Prove that $\sec \alpha + \tan \alpha = 1/(\sec \alpha - \tan \alpha)$ where $0 \leq \alpha \leq 2\pi$, $\alpha \neq \pi/2$, $\alpha \neq 3\pi/2$.

A.2   Prove that $(\sin \alpha - \cos \alpha)^3 = 3 \sin \alpha - 2 \sin^3 \alpha + 2 \cos^3 \alpha - 3 \cos \alpha$.

A.3   Prove that $\sin^2 \alpha = \tan^2 \alpha/(1 + \tan^2 \alpha)$ where $0 \leq \alpha \leq 2\pi$.   Is this a valid equation for $\alpha = \pi/2$ or $\alpha = 3\pi/2$?   Explain your answer.

A.4   (a)   Let α be any angle with $0 < \alpha < \pi/2$.   Using Theorems A.1 and A.2, prove that if $\beta = \pi/2 + \alpha$, $\sin^2 \beta + \cos^2 \beta = 1$.
    (b)   Using the result of part (a), prove that $\sin^2 \gamma + \cos^2 \gamma = 1$ where $\gamma = \pi + \alpha$ and $\gamma = 3\pi/2 + \alpha$.

A.5   (a)   Prove that for any two angles α and β such that α, β, and $\alpha + \beta$ are all in the domain of the tangent function, we have

$$\tan(\alpha + \beta) = (\tan \alpha + \tan \beta)/(1 - \tan \alpha \tan \beta)$$

[*Hint*: Write $\tan(\alpha + \beta)$ as $\sin(\alpha + \beta)/\cos(\alpha + \beta)$, use Theorems A.1 and A.2, divide the resulting numerator and denominator by $(\cos \alpha \cos \beta)$, and then simplify.]

A.6   Write formulas for (a) $\sin(\alpha - \beta)$; (b) $\cos(\alpha - \beta)$; (c) $\tan(\alpha - \beta)$.

A.7   Prove that for any angle α with $0 < \alpha < \pi$, $\tan(\alpha/2)$ may be expressed in each of the following ways.

(a) $\left(\dfrac{1 - \cos \alpha}{1 + \cos \alpha}\right)^{1/2}$

(b) $\dfrac{1 - \cos \alpha}{\sin \alpha}$

(c) $\dfrac{\sin \alpha}{1 + \cos \alpha}$

A.8   Look up a proof for the law of sines which states that in any triangle with angles α, β, γ and opposite sides of length a, b, c, respectively (see Figure A.11), we have $\dfrac{a}{\sin \alpha} = \dfrac{b}{\sin \beta} = \dfrac{c}{\sin \gamma}$.

A.9   Look up a proof for the law of cosines, which states that in any triangle with angles α, β, γ and opposite sides of length a, b, c, respectively (see Figure A.11), we have

$c^2 = a^2 + b^2 - 2ab \cos \gamma$
$b^2 = a^2 + c^2 - 2ac \cos \beta$
$a^2 = b^2 + c^2 - 2bc \cos \alpha$

A.10   Prove that $\cos 40° + \cos 80° + \cos 160° = 0$.

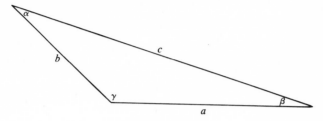

Figure A.11

BIBLIOGRAPHY

ANALYTIC GEOMETRY

G. Fuller, *Analytic Geometry* (Addison-Wesley, Reading, Mass.).
C. Love and E. Rainville, *Analytic Geometry* (MacMillan, New York).
N. McCoy and R. Johnson, *Analytic Geometry* (Holt, Rinehart, and Winston, New York).
E. Purcell, *Analytic Geometry* (Appleton-Century-Crofts, New York).
P. Rees, *Analytic Geometry* (Prentice-Hall, Englewood Cliffs, N.J.).

CALCULUS

L. Bers, *Calculus* (Holt, Rinehart, and Winston, New York).
J. Chover, *Calculus for the Social Sciences* (W. A. Benjamin, Inc., New York).
R. Courant, *Differential and Integral Calculus*, Vol. 1 (Interscience, New York).
R. Courant and F. John, *Introduction to Calculus and Analysis*, Vol. 1 (Interscience, New York).
R. Fisher and A. Ziebur, *Calculus and Analytic Geometry* (Prentice-Hall, Englewood Cliffs, N.J.).
N. Haaser, J. LaSalle, and J. Sullivan, *Introduction to Analysis* (Ginn, Boston).
E. Hille and S. Salas, *First Year Calculus* (Blaisdell, Waltham, Mass.).
R. Johnson and F. Kiokemeister, *Calculus with Analytic Geometry* (Allyn and Bacon, Boston).
S. Lang, *A First Course in Calculus* (Addison-Wesley, Reading, Mass.).
L. Leithold, *The Calculus with Analytic Geometry* (Harper & Row, New York).
J. Olmsted, *Intermediate Analysis* (Appleton-Century-Crofts, New York).

CALCULUS

M. Protter and C. Morrey, *College Calculus with Analytic Geometry* (Addison-Wesley, Reading, Mass.).

G. Sherwood and A. Taylor, *Calculus* (Prentice-Hall, Englewood Cliffs, N.J.).

S. Stein, *Calculus in the First Three Dimensions* (McGraw-Hill, New York).

A. Taylor and C. Halberg, *Calculus with Analytic Geometry: Functions of One Variable* (Prentice-Hall, Englewood Cliffs, N.J.).

G. Thomas, *Calculus and Analytic Geometry* (Addison-Wesley, Reading, Mass.).

O. Toeplitz, *The Calculus, A Genetic Approach* (Univ. of Chicago Press, Chicago).

MISCELLANEOUS

R. Courant and H. Robbins, *What is Mathematics?* (Oxford Univ. Press, London and New York).

R. Fisher and A. Ziebur, *Integrated Algebra and Trigonometry with Analytic Geometry* (Prentice-Hall, Englewood Cliffs, N.J.).

B. Gelbaum and J. Olmsted, *Counterexamples in Analysis* (Holden-Day, San Francisco, Calif.).

M. Marcus and H. Minc, *Elementary Functions and Coordinate Geometry* (Houghton Mifflin, Boston).

G. Polya, *How to Solve It* (Doubleday, New York).

G. Polya, *Mathematical Discovery*, Vols. 1 and 2 (Wiley, New York).

F. Sparks and P. Rees, *Plane Trigonometry* (Prentice-Hall, Englewood Cliffs, N.J.).

BOOKS OF TABLES

R. Burington, *Handbook of Mathematical Tables and Formulas* (Handbook Publ., Inc., Sandusky, Ohio).

H. Larsen, *Rinehart Mathematical Tables, Formulas and Curves* (Holt, Rinehart, and Winston, New York).

ANSWERS

SECTION 1

1.1 $(5, 4)$ and $(9, 2)$; yes, take the circle with diameter equal to the given diagonal. Then any point on the circle is a vertex of a rectangle satisfying the given condition.

1.2 $(9, 6)$ and $(6, 2)$; yes, take any line L passing through $(1, 2)$ but not $(4, 6)$; construct a second line L' through $(4, 6)$ parallel to L; measure off segments equal to the given side on L and L'.

1.3 (a) $(5, 0)$. (b) $(7, 6)$. (c) $(-1, 6)$.

SECTION 2

2.1 (a) right, area 10. (b) isosceles, area 28. (c) equilateral, area $4\sqrt{3}$. (d) isosceles, right, area 5/2. (e) none, area 19.

2.3 (a) $(4, 1)$. (b) $(3, \frac{1}{2})$.

2.4 (a) $(5, 1)$, $(6, 3)$. (b) $(11/3, -2)$, $(-\frac{2}{3}, 0)$.

2.5 (a) $(-31/6, 1)$. (b) $(-\frac{3}{4}, 19/4)$. (c) $(5, 8)$.

(d) $\left(x_1 + \dfrac{r}{r+s}(y_1 - x_1),\ x_2 + \dfrac{r}{r+s}(y_2 - x_2) \right)$.

SECTION 3

3.1 (a) collinear. (b) not collinear. (c) not collinear (d) collinear.

3.2 (a) $y = -\frac{3}{5}x + \frac{1}{5}$. (b) $(x/-2) + (y/4) = 1$. (c) $(x/a) + (y/b) = 1$. (d) $y = 3x - 5$. (e) $y = -5x - 3$. (f) $y = \frac{1}{2}x - 2$. (g) $y = -3x + 6$. (h) $y = -2x + 4$.

3.3 (a) $m = \frac{1}{2}, (0, \frac{3}{2}), (-3, 0)$. (b) $m = -2, (0, -6), (-3, 0)$. (c) $m = -5$, $(0, 2), (\frac{2}{5}, 0)$. (d) $m = -5, (0, 0)$.

SECTION 4

4.1 (a) $(-2, -1)$. (b) none. (c) $(1, 8)$. (d) $(27/17, -2/17)$. (e) $\dfrac{ce - bf}{ae - bd}$, $\dfrac{cd - af}{bd - ae}$.

4.2 (a) angle is $90°$; tangent is undefined. (b) 0. (c) 9/11. (d) $-17/7$.
(e) $\dfrac{ae - bd}{be + ad}$.

4.4 $(2, 1), (6, -4), (4, 4)$.

SECTION 5

5.1 (a) $\sqrt{5}/5$. (b) 0. (c) $7\sqrt{10}/5$.

5.2 (a) $3\sqrt{5}/5$. (b) $8\sqrt{5}/5$. (c) $\sqrt{10}/2$.

5.3 (a) $9\sqrt{10}/10$. (b) $3\sqrt{2}/2$. (c) $8\sqrt{37}/37$. (d) $4\sqrt{5}/5$. (e) $2\sqrt{13}/13$.

5.4 (a) $4\sqrt{85}/17, \sqrt{5}, 4\sqrt{5}$. (b) $\sqrt{5}, \sqrt{5}, \sqrt{10}/2$.

5.5 $(2\sqrt{2}/3, (3 - 4\sqrt{2})/3)$ or $(-2\sqrt{2}/3, (3 + 4\sqrt{2})/3)$.

SECTION 7

7.1 (a) $(x + 1)^2 + (y - 3)^2 = 25$. (b) $(x - 2)^2 + (y + 3)^2 = 4$.
(c) $(x - 1)^2 + (y + 1)^2 = 40$. (d) $(x - 2)^2 + (y - 3)^2 = 25$.
(e) $(x - 5/2)^2 + (y - 5/2)^2 = 45/2$. (f) $(x - 2)^2 + (y + 2)^2 = 41$.
(g) $(x - 11/2)^2 + (y - 5)^2 = 85/4$. (h) two answers, $(x - 1)^2 + (y - 6)^2 = 25$ and $(x - 2)^2 + (y + 1)^2 = 25$.

7.2 The graphs are circles with center C and radius r as follows.
(a) $C = (0, 0), r = \sqrt{5}$. (b) $C = (1, -2), r = 2$. (c) $C = (2, 3), r = 2\sqrt{3}$.
(d) $C = (-5, 8), r = 10$. (e) $C = (2, 0)$ $r = 4$.
(f) $C = (-C/2, -D/2), r = \frac{1}{2}(C^2 + D^2 - 4E)^{1/2}$ provided $C^2 + D^2 - 4E > 0$.

7.3 (a) $y = 3x - 5$. (b) $y = \frac{1}{2}x - 4$. (c) $y = -\frac{3}{4}x + (25/4)$.
(d) $y = (-a/b)x + (a^2 + b^2)/b$.

7.5 (a) $(x - 4)^2 + (y + 3)^2 = 25$. (b) $(x - 3)^2 + (y + 2)^2 = 4$.
(c) $(x - 8)^2 + (y - 9)^2 = 40$. (d) $(x - 1)^2 + (y + 1)^2 = 16/5$.
(e) $(x - 3)^2 + (y + 5)^2 = 25$.

SECTION 8

8.1 (4, 3) and (4, −3).

8.2 (3, 2) and (−3, 2).

8.3 (2, 2) and (−2, −2).

8.4 none.

8.5 none.

8.6 (4, 2).

8.7 $(1, \sqrt{3})$ and $(1, -\sqrt{3})$.

8.8 none.

8.9 (2, 0).

8.10 none.

8.11 $\left(\dfrac{3+\sqrt{7}}{2}, \dfrac{\sqrt{7}-1}{2}\right)$ and $\left(\dfrac{3-\sqrt{7}}{2}, \dfrac{-(1+\sqrt{7})}{2}\right)$.

8.12 (0, 1) and (−2, −3).

8.13 $\left(\dfrac{-15+2\sqrt{5}}{5}, \dfrac{10+4\sqrt{5}}{5}\right)$ and $\left(\dfrac{-15-2\sqrt{5}}{5}, \dfrac{10-4\sqrt{5}}{5}\right)$.

8.14 (−3, 4).

8.15 $E_3 = -2xh - 2yk + h^2 + k^2 - r^2 + s^2 = 0$. (a) if $h = k = 0$. (c) $r = s$

(d) no (f) $\begin{vmatrix} h^2 + k^2 + s^2 - r^2 \\ -h^2 - k^2 + s^2 - r^2 \end{vmatrix}$, ratio is 1 when $r = s$.

8.16 (a) $y = -x + 2$ and $y = \frac{1}{7}x + \frac{4}{7}$. (b) $y = x + 2$ and $y = -7x + 10$.

SECTION 9

9.1 $(y - 2)^2 = 12(x - 5)$.

9.2 (a) $x^2 = 32y$. (b) $x^2 = -16y$. (c) $(y - 2)^2 = -24(x - 3)$.
(d) $(y - 3)^2 = 16x$. (e) $(y - 3)^2 = 8(x + 2)$.

9.3 (a) $V = (0, 0), F = (0, \frac{1}{16})$. (b) $V = (0, 0), F = (-\frac{1}{32}, 0)$.
(c) $V = (-2, 1), F = (-2, 5/4)$. (d) $V = (-25/8, -\frac{1}{4}), F = (-3, -\frac{1}{4})$.
(e) $V = (-\frac{1}{4}, 25/8), F = (-\frac{1}{4}, 3)$. (f) $V = (-3, -2), F = (-3, -31/16)$.
(g) $V = (-1, 3), F = (-9/8, 3)$.

9.4 $(x - 4)^2 = -20y$. There are infinitely many solutions. Take any negative real number a, let $b = (16a^2 + 21)/16a$; then the parabola with $V = (4, a)$ and $F = (4, b)$ is tangent to the line $y = x + 1$.

9.5 $(x - 2)^2 = 2y$.

SECTION 10

10.1 $[(x-2)^2/16] + [(y-3)^2/12] = 1.$

10.2   (a) $y^2/16 + x^2/9 = 1.$ (b) $x^2/15 + y^2/21 = 1.$ (c) $x^2/36 + (y-3)^2/20 = 1.$
(d) $(y-1)^2/9 + (x-5)^2/4 = 1.$

10.3   We give the 4 vertices in each case.
(a) $(0, 5), (0, -5), (2, 0), (-2, 0).$
(b) $(0, 3), (0, -3), (2, 0), (-2, 0).$
(c) $(-2, 3), (-2, -1), (-2+\sqrt{3}, 1), (-2-\sqrt{3}, 1).$
(d) $(6, -1), (-2, -1), (2, 2), (2, -4).$
(e) $(1, 0), (-1, 0), (0, \sqrt{2}/4), (0, -\sqrt{2}/4).$
(f) $(2, 2), (-4, 2), (-1, 0), (-1, 4).$
(g) $(-1, -2), (-1, 6), (-3, 2), (1, 2).$

10.4 $[(x-4)^2/16] + (y^2/4) = 1.$

SECTION 11

11.1 $x^2/64 - y^2/36 = 1.$

11.2   (a) $x^2/16 - y^2/20 = 1.$ (b) $[(y+2)^2/4] - (x^2/5) = 1.$
(c) $x^2/36 - y^2/64 = 1.$ (d) $(x-2)(y-4) = 25.$ (e) $xy = -2.$

11.3   We give the vertices and the asymptotes in each case.
(a) $(2, 0), (-2, 0), y = 2x, y = -2x.$ (b) $(0, \frac{1}{4}), (0, -\frac{1}{4}), y = \frac{1}{4}x, y = -\frac{1}{4}x.$
(c) $(1, 2), (-3, 2), y = (\sqrt{2}/2)x + [2 + (\sqrt{2}/2)], y = (-\sqrt{2}/2)x + [2 - (\sqrt{2}/2)].$
(d) $(1, -1) (-1, -1), y = x - 1, y = -x - 1.$
(e) $(8, 8), (-8, -8), x = 0, y = 0.$
(f) $(-3, 3), (3, -3), x = 0, y = 0.$
(g) $(-1+\sqrt{2}, \sqrt{2}), (-1-\sqrt{2}, -\sqrt{2}), x = -1, y = 0.$
(h) $(2, 0), (0, -2), x = 1, y = -1.$
(i) $(\sqrt{3}, 1+\sqrt{3}), (-\sqrt{3}, 1-\sqrt{3}), x = 0, y = 1.$
(j) $(-3, 2), (-1, 0), x = -2, y = 1.$

11.4 $\dfrac{x^2}{9} - \dfrac{(y-3)^2}{(81/16)} = 1.$

SECTION 12

12.1   (a) $\left(\dfrac{1+\sqrt{65}}{8}, \dfrac{33+\sqrt{65}}{16}\right), \left(\dfrac{1-\sqrt{65}}{8}, \dfrac{33-\sqrt{65}}{16}\right).$

(b) $(1, 2), (-2, -1).$ (c) $(\frac{1}{2}, \frac{1}{2}), (-\frac{1}{2}, -\frac{1}{2}).$
(d) $(1, 1), (-1, 1).$ (e) $(\sqrt{2}/2, 1), (-\sqrt{2}/2, 1).$ (f) none.

(g) $([(1+\sqrt{17})/2]^{1/2}, 2[(1+\sqrt{17})/2]^{-1/2}), (-[(1+\sqrt{17})/2]^{1/2},$
$-2[(1+\sqrt{17})/2]^{-1/2})$

(h) $(\sqrt{3}/3, \sqrt{3}/3), (\sqrt{3}/3, -\sqrt{3}/3), (-\sqrt{3}/3, \sqrt{3}/3), (-\sqrt{3}/3, -\sqrt{3}/3).$

(i) $(3, 4), (4, 3).$ (j) $(3, 2), (21/4, -19/4).$

(k) $\left(\left(\dfrac{63+\sqrt{897}}{32}\right)^{1/2}, \dfrac{-1+\sqrt{897}}{32}\right), \left(-\left(\dfrac{63+\sqrt{897}}{32}\right)^{1/2}, \dfrac{-1+\sqrt{897}}{32}\right),$
$\left(\left(\dfrac{63-\sqrt{897}}{32}\right)^{1/2}, \dfrac{-1-\sqrt{897}}{32}\right), \left(-\left(\dfrac{63-\sqrt{897}}{32}\right)^{1/2}, \dfrac{-1-\sqrt{897}}{32}\right).$

(l) $\left(\dfrac{1+3\sqrt{5}}{2}, \left(\dfrac{21+3\sqrt{5}}{2}\right)^{1/2}\right), \left(\dfrac{1+3\sqrt{5}}{2}, -\left(\dfrac{21+3\sqrt{5}}{2}\right)^{1/2}\right),$
$\left(\dfrac{1-3\sqrt{5}}{2}, \left(\dfrac{21-3\sqrt{5}}{2}\right)^{1/2}\right), \left(\dfrac{1-3\sqrt{5}}{2}, -\left(\dfrac{21-3\sqrt{5}}{2}\right)^{1/2}\right).$

(m) $(4+\sqrt{10}, 2+\sqrt{3}/2), (4+\sqrt{10}, 2-\sqrt{3}/2),$
$(4-\sqrt{10}, 2+\sqrt{3}/2), (4-\sqrt{10}, 2-\sqrt{3}/2).$

12.2 $(0, 0), (1, 1), \left(\dfrac{1+\sqrt{29}}{2}, \dfrac{15+\sqrt{29}}{2}\right), \left(\dfrac{1-\sqrt{29}}{2}, \dfrac{15-\sqrt{29}}{2}\right).$

12.3 $(0, 0), (0, 2\sqrt{3}), (-3, \sqrt{3}), (3, \sqrt{3}), (-6, 0), (6, 0), (12, 6\sqrt{3}), (-12, 6\sqrt{3}).$

SECTION 13

13.25 $y = (4 - x^2)^{1/2}.$

13.26 $y = -(4 - x^2)^{1/2}.$

13.27 $y = \sqrt{x}.$

13.28 $y = -(1 - 4x^2)^{1/2}.$

13.29 $y = (1 + x^2)^{1/2}.$

SECTION 14

14.1   (a) $y = 10x - 5.$ (b) $y = -x.$ (c) $y = -\frac{1}{2}x + 2.$ (d) $y = 4x - 4.$
(e) $y = -2x + 3.$ (f) $x = 1.$ (g) $y = -1.$

14.2   at $(0, -1), \tan \beta = 0$;   at $(\sqrt{3}, 2), \tan \beta = 3\sqrt{3}/8$;   at $(-\sqrt{3}, 2), \tan \beta = 3\sqrt{3}/8.$

14.3   (a) $(\frac{1}{3}, \frac{1}{3}).$ (b) $(\frac{1}{2}, 4)$ and $(-\frac{1}{2}, -4).$ (c) $(2\sqrt{3}/3, \sqrt{3}/3)$ and $(-2\sqrt{3}/3, -\sqrt{3}/3).$

14.4 (a) at $(1, 1)$, $y = x$; at $(0, 0)$, no tangent. (b) at $(0, 1)$, $y = 1$; at $(1, 0)$ and $(-1, 0)$, no tangent.

14.5 (a) no, tangent line is $y = -x + 2$. (b) no, consider $x = 1$ and $y = 1$ in part (a).

SECTION 15

15.1 $f(x) = 2\pi x$, $D = R =$ all reals ≥ 0.

15.2 $f(x) = \pi x^2$, $D = R =$ all reals ≥ 0.

15.3 $f(x) = (4/3)\pi x^3$, $D = R =$ all reals ≥ 0.

15.4 (a) $R = \{1, 2, 3, 4, 5, 6\}$. (b) $R = \{0, 1, 2, 3, 4, 5, 6, 7, 8\}$.

15.5 (a) $R = \{2, 3, 4, 5, 6\}$. (b) $R = \{4, 9, 16, 25, 36\}$.

15.6 $1 - 1$, $D =$ all reals ≥ 1, $R =$ all reals ≥ 0.

15.7 $1 - 1$, $D = R =$ all reals.

15.8 $1 - 1$, $D = R =$ all reals.

15.9 $1 - 1$, $D =$ all reals $\neq 2$, $R =$ all reals $\neq 0$.

15.10 not $1 - 1$, $D = R =$ all reals $\neq 0$.

15.11 not $1 - 1$, $D =$ all reals, $R =$ all reals ≥ 0.

15.12 not $1 - 1$, $D =$ all reals $\neq -1$, $R =$ all reals > 0.

15.13 not $1 - 1$, $D =$ all reals in the interval $-2 \leq x \leq 2$, $R =$ all reals in $0 \leq y \leq 2$

15.14 not $1 - 1$, $D =$ all reals in the interval $-2 \leq x \leq 2$, $R =$ all reals in $-2 \leq y \leq 0$.

15.15 not $1 - 1$, $D =$ all reals $\neq 0$, $R = \{-1, 1\}$.

15.16 not $1 - 1$, $D =$ all reals, $R =$ all reals ≥ 0.

15.17 $1 - 1$, $D =$ all reals, $R =$ all reals ≥ 1.

15.18 not $1 - 1$, $D =$ all reals, $R =$ all reals in the interval $-1 \leq y \leq 1$.

15.19 not $1 - 1$, $D =$ all reals $\neq (2n + 1)(\pi/2)$, where n is any integer, $R =$ all reals.

SECTION 16

16.20 (a) $f(x) = 2(1 - x^2)^{1/2}$, $g(x) = -2(1 - x^2)^{1/2}$. (b) $f(x) = x^{2/3}$.
(c) $f(x) = -x$. (d) $f(x) = \frac{1}{2}[-x + (x^2 + 4)^{1/2}]$, $g(x) = \frac{1}{2}[-x - (x^2 + 4)^{1/2}]$.

SECTION 17

17.1 $x^2 + \sqrt{x} + 1$, $D = $ all reals ≥ 0, $R = $ all reals ≥ 1.

17.2 $x^2 - \sqrt{x} + 1$, $D = $ all reals ≥ 0, $R = $ all reals $\geq 1 - \frac{1}{2}\sqrt[3]{\frac{1}{4}}$

17.3 $x^2\sqrt{x} + \sqrt{x}$, $D = $ all reals ≥ 0, $R = $ all reals ≥ 0.

17.4 $x + 1/x$, $D = $ all reals $\neq 0$, $R = $ all reals except those in the interval $-2 < y < 2$.

17.5 $x^{-3/2}$, $D = R = $ all reals > 0.

17.6 $\sqrt{x}/(x^2 + 1)$, $D = R = $ all reals ≥ 0.

17.7 $x + 1$, $D = R = $ all reals.

17.8 $(x^2 + 1)^{1/2}$, $D = $ all reals, $R = $ all reals ≥ 1.

17.9 $(1/x^2) + 1$, $D = $ all reals $\neq 0$, $R = $ all reals > 1.

17.10 $(x^2 + 1)^{-1}$, $D = $ all reals, $R = $ all reals in the interval $0 < y \leq 1$.

17.11 $\left(x^2 + \dfrac{1}{x} + 1\right)^{1/2}$, $D = $ all reals $\neq 0$.
  $R = $ all reals $\geq (2^{-1/3} + 2^{1/3} + 1)^{1/2}$.

17.12 $\sqrt{x} + 1/\sqrt{x}$, $D = $ all reals > 0, $R = $ all reals ≥ 2.

17.13 $x^2\sqrt{x} + \sqrt{x} + 1/x + x$, $D = $ all reals > 0, $R = $ all reals ≥ 4.

17.14 $(1/x) + 1$, $D = $ all reals $\neq 0$, $R = $ all reals $\neq -1$.

17.17 (a) $x^{1/5}$. (b) $\frac{1}{6}(x + 3)$. (c) none. (d) none. (e) none. (f) $(x - 1)^{1/2}$.

SECTION 18

18.1 $\frac{1}{2}, \frac{1}{4}, \frac{1}{8}, \frac{1}{16}, \frac{1}{32}, \frac{1}{64}$; limit 0.

18.2 $2, \frac{3}{2}, \frac{4}{3}, \frac{5}{4}, \frac{6}{5}, \frac{7}{6}$; limit 1.

18.3 $1, 0, -1, 1, 0, -1$; no limit.

18.4 $\frac{1}{3}, -1/9, 1/27, -1/81, 1/243, -1/729$; limit 0.

18.5 $0.6, 0.66, 0.666, 0.6666, 0.66666, 0.666666$; limit 2/3.

18.6 $0.5, 0.55, 0.555, 0.5555, 0.55555, 0.555555$; limit 5/9.

18.7 $1, \frac{3}{2}, 9/5, 2, 15/7, 9/4$; limit 3.

18.8 $0, \frac{3}{5}, \frac{4}{5}, 15/17, 12/13, 35/37$; limit 1.

18.9 $2, 5/3, 2, 17/7, 26/9, 37/11$; limit ∞.

18.10   2, 7/5, 1, 13/17, 8/13, 19/37; limit 0.

18.11 $\frac{3}{4}$, 21/24, 91/108, 273/320, 651/750, 1333/1512; limit 1.

18.12 $\frac{3}{4}$, 7/3, 91/28, 273/65, 651/126, 1333/217; limit ∞.

18.13 $\frac{3}{4}$, 21/33, 91/244, 273/1025, 651/3126, 1333/7777; limit 0.

SECTION 19

19.1 x_n: $\frac{1}{4}$, $-\frac{1}{7}$, $-3/10$, $5/13$, $-7/16$; limit $-\frac{2}{3}$.
y_n: 5, 7/2, 3, 11/4, 13/5; limit 2.
$x_n + y_n$: 21/4, 47/14, 243/90, 492/208, 865/400; limit 4/3.
$x_n - y_n$: $-19/4$, $-102/28$, $-297/90$, $-652/208$, $-1215/400$;
limit $-8/3$. $x_n y_n$: 5/4, $-\frac{1}{2}$, $-9/10$, $-220/208$, $-455/400$;
limit $-4/3$. x_n/y_n: 1/20, $-4/98$, $-1/10$, $-80/572$, $-175/1040$;
limit $-1/3$.

19.2 x_n: 1, -2, 3, -4, 5; no limit.
y_n: 1, $-\frac{1}{2}$, $\frac{1}{3}$, $-\frac{1}{4}$, $\frac{1}{5}$; limit 0.
$x_n + y_n$: 2, $-5/2$, 10/3, $-17/4$, 26/5; no limit.
$x_n - y_n$: 0, $-\frac{3}{2}$, 8/3, $-15/4$, 24/5; no limit.
$x_n y_n$: 1, 1, 1, 1, 1; limit 1.
x_n/y_n: 1, 4, 9, 16, 25; limit ∞.

19.3 x_n: 2, 7/5, 1, 13/17, 16/26; limit 0.
y_n: 2, 5/3, 2, 17/7, 26/9; limit ∞.
$x_n + y_n$: 4, 46/15, 3, 380/119, 820/234; limit ∞.
$x_n - y_n$: 0, $-4/15$, -1, $-198/119$, $-532/234$; limit $-\infty$.
$x_n y_n$: 4, 7/3, 2, 13/7, 16/9; limit $\frac{3}{2}$.
x_n/y_n: 1, 21/25, $\frac{1}{2}$, 91/289, 144/676; limit 0.

19.4 x_n: 1, $\frac{3}{2}$, 9/5, 2, 15/7; limit 3.
y_n: $\frac{1}{2}$, 7/5, 17/10, 31/17, 49/26; limit 2.
$x_n + y_n$: $\frac{3}{2}$, 58/20, 175/50, 390/102, 733/182; limit 5.
$x_n - y_n$: $\frac{1}{2}$, $\frac{1}{10}$, $\frac{1}{10}$, 18/102, 47/182; limit 1.
$x_n y_n$: $\frac{1}{2}$, 42/20, 153/50, 372/102, 735/182; limit 6.
x_n/y_n: 2, 30/28, 18/17, 204/186, 390/343; limit $\frac{3}{2}$.

19.5 x_n: 1, -1, 1, -1, 1; no limit.
y_n: 1, $-\frac{1}{2}$, $\frac{1}{3}$, $-\frac{1}{4}$, $\frac{1}{5}$; limit 0.
$x_n + y_n$: 2, $-\frac{3}{2}$, 4/3, $-5/4$, 6/5; no limit.
$x_n - y_n$: 0, $-\frac{1}{2}$, $\frac{2}{3}$, $-\frac{3}{4}$, $\frac{4}{5}$; no limit.
$x_n y_n$: 1, $\frac{1}{2}$, $\frac{1}{3}$, $\frac{1}{4}$, $\frac{1}{5}$; limit 0.
x_n/y_n: 1, 2, 3, 4, 5; limit ∞.

19.6 $x_n : 0, \frac{1}{2}, \frac{2}{3}, \frac{3}{4}, \frac{4}{5}$; limit 1.

$y_n : 1, 2, 3, 4, 5$; limit ∞.

$x_n + y_n : 1, 5/2, 11/3, 19/4, 29/5$; limit ∞.

$x_n - y_n : -1, -\frac{3}{2}, -7/3, -13/4, -21/5$; limit $-\infty$.

$x_n y_n : 0, 1, 2, 3, 4$; limit ∞.

$x_n/y_n : 0, \frac{1}{4}, 2/9, 3/16, 4/25$; limit 0.

SECTION 20

20.1 (a) $2, 4, -4, 8$. (b) $1, 2, 0, 33$. (c) $1, 0, 2, 1$. (d) $1, \sqrt{2}, 0, \sqrt{3}$.
(e) $1, 2, \frac{1}{2}, 4$. (f) $\infty, 2, \frac{1}{2}, \sqrt{2}$. (g) $\frac{1}{2}, \frac{2}{3}, 2, \frac{1}{2}$.

20.2 (a) 0. (b) 1. (c) $\frac{2}{3}$.

20.4 (a) left limit -1, right limit 1.
(b) continuous for all real $x \neq 0$.

20.7 (a) not continuous for any x. (b) continuous for all irrational x and for $x = 0$.

SECTION 21

21.1 (a) $2, 10, -10, 56$. (b) $1, -2, 12, 39$. (c) $-1, 0, -2, 31$.
(d) $1, -1, 25, -17$. (e) $7, 2, 8, 37$. (f) $2, 4, 4, 32$. (g) $\frac{1}{2}$, no limit, $\sqrt{2}/2, 2$.
(h) $1, 32, 0, (153)^5$. (i) $\frac{1}{9}, \frac{1}{12}, \frac{3}{4}, 0$. (j) $-1, 1, \frac{3}{5}, 9/49$. (k) $-2, -\frac{2}{3}, 6/5, 0$.
(l) $2, 12, 0, 204$. (m) $1, 0, 0, 135$.

21.2 (b) $1, 3, \sqrt{3}$.

21.5 (b) $\pi\sqrt{2}/8, \pi/2, 0$. (c) $2\sqrt{2}/\pi, 0, -1/\pi$. (d) $1, 0, -1, 1$. (e) $1, 0, -1, 1$.
(f) $\sin 1, -\sin 1$.

21.7 (a) continuous for all real x. (b) continuous for all real x. (c) continuous for al real $x \neq (2n + 1)(\pi/2)$, where n is any integer. (d) continuous for all real $x \neq 0$. (e) continuous for all real x.

SECTION 22

22.1 (a) $-2, 6, 8$. (b) $3, 3, 12$. (c) $-2, 3, -1$.

22.2 (a) $1, 3$. (b) $96, 5$.

22.3 (a) $3\pi/2, 5\pi/2$. (b) π.

22.4 (a) $112\pi/3, 76\pi/3$. (b) 16π.

22.5 (a) m. (b) $2ax_0 + b$. (c) tangent: $y = (2ax_0 + b)x - ax_0^2 + c$, normal:

$$y = \left(\frac{-1}{2ax_0 + b}\right)x + \frac{x_0}{2ax_0 + b} + ax_0^2 + bx_0 + c.$$

22.8 $f'(-1) = f'(1) = 0; f'(0)$ does not exist.

22.9   (c) left-hand derivative does not exist; right-hand derivative is 0.

SECTION 23

23.1   (a) $-7x^{-8}$.   (b) $(7/2)x^{5/2}$.   (c) $-\frac{3}{2}x^{-5/2}$.   (d) $7x^6 - 15x^4 + 6x^2 - 7$.
  (e) $11x^{10} - 72x^8 + 70x^6 - 30x^4 - 3x^2 + 3$.
  (f) $\dfrac{(x^6 + x^4 + x^2 + 1)(5x^4 - 3x^2 + 1) - (x^5 - x^3 + x)(6x^5 + 4x^3 + 2x)}{(x^6 + x^4 + x^2 + 1)^2}$.

  (g) $\dfrac{2}{3}\left(\dfrac{1}{x^2} - x^3\right)^{-1/3}\left(-\dfrac{2}{x^3} - 3x^2\right)$.

  (h) $\frac{3}{2}x(x^2 + 1)^{1/2}(2x) = (x^2 + 1)^{3/2}$.

  (i) $\dfrac{(\sqrt{x} + 1)(1/2\sqrt{x}) - \frac{1}{2}}{(\sqrt{x} + 1)^2}$.   (j) $\dfrac{-(2x + 2)}{(x^2 + 2x + 7)^2}$.

23.2   (a) $0, 4$.   (b) $7\sqrt{2}/4, 7$.   (c) $-8, 2^9$.   (d) $\frac{1}{3}$, undefined, $1/12$.
  (e) $\sqrt{2}/3$, undefined.

23.3 $5, 36$.

23.4   (b) $\frac{3}{2}x^{1/2}(x^3 + 1)^{1/2}(x^2 - 1)^8 + x^{3/2}(x^2 - 1)^8 \dfrac{(3x^2)}{2(x^3 + 1)^{1/2}}$
  $+ x^{3/2}(x^3 + 1)^{1/2}8(x^2 - 1)^7(2x)$.

SECTION 24

24.1   (a) $\frac{1}{2}x^{-1/2}, -\frac{1}{4}x^{-3/2}, \frac{3}{8}x^{-5/2}$.
  (b) $\frac{1}{2}(\frac{1}{2} - 1)(\frac{1}{2} - 2) \cdots (\frac{1}{2} - [n - 1])x^{1/2 - n}$.

24.2 $(-1)^n n! x^{-1-n}$.

24.3   (a) $7x^6 + 6x^5 + 5x^4 + 4x^3 + 3x^2 + 2x + 1$,
      $42x^5 + 30x^4 + 20x^3 + 12x^2 + 6x + 2$,
      $210x^4 + 120x^3 + 60x^2 + 24x + 6$,
      $840x^3 + 360x^2 + 120x + 24$,
      $2520x^2 + 720x + 120$.

  (b) $\dfrac{(-x^4 - 3x^2 + 2x)}{(x^3 + 1)^2}$,

      $\dfrac{(x^3 + 1)^2(-4x^3 - 6x + 2) - (-x^4 - 3x^2 + 2x)6x^2(x^3 + 1)}{(x^3 + 1)^4}$.

  (c) $5(x^3 + 2x)^4(3x^2 + 2)$,
  $20(x^3 + 2x)^3(3x^2 + 2)^2 + 5(x^3 + 2x)^4(6x)$,
  $60(x^3 + 2x)^2(3x^2 + 2)^3 + 60(x^3 + 2x)^3(3x^2 + 2)(6x) + 30(x^3 + 2x)^4$.

24.4   (a) $100!$ (b) 0.

24.5 (a) $(1-x)^{-2}, 2(1-x)^{-3}, 6(1-x)^{-4}, 24(1-x)^{-5}, 120(1-x)^{-6}$.
(b) $n!(1-x)^{-1-n}$.

24.7 (c) $1-2x+3x^2-4x^3$, infinitely many terms.
(d) $1+\frac{1}{2}x-\frac{1}{8}x^2+\frac{1}{16}x^3$, infinitely many terms.

24.8 (a) $-\frac{4}{5}$, (b) $-(65/68)$. (c) $13/8$. (d) $-(3/14)$.

25.9 (a) $y-2=(5/4)(x-1)$.
(b) $y-\frac{1}{2}=(68/65)(x-2)$.
(c) $y+1=-(8/13)(x-2)$.
(d) $y-1=(14/3)(x-1)$.

24.10 (a) $f': \dfrac{-(2x+y)}{2y+x}, f'': \dfrac{-6(y^2+x^2+xy)}{(2y+x)^3}=\dfrac{-42}{(2y+x)^3}$.

(b) $f': \dfrac{-(3x^2+y)}{(x-3y^2)}, f'': \dfrac{54xy(-y^3+xy+x^3)+2xy}{(x-3y^2)^3}=\dfrac{-592xy}{(x-3y^2)^3}$.

24.11 $\sqrt{3}/6, -\sqrt{3}/6$

24.12 $f': \dfrac{y-x^2}{y^2-x}, f'\left(\dfrac{-1}{\sqrt[3]{2}}\right)=\dfrac{2-\sqrt[3]{2}-\sqrt[3]{4}}{2+\sqrt[3]{2}-\sqrt[3]{4}}$.

24.13 (a) $2, -2, -2, 2$. (b) $-\frac{1}{2}, \frac{1}{2}, \frac{1}{2}, -\frac{1}{2}$. (c) $-(5/4), -(5/4)$.

SECTION 25

25.1 (a) rel min $(-\frac{1}{3}, 23/27)$, rel max $(-1, 1)$, infl pt $(-\frac{2}{3}, 25/27)$. (b) rel min $(\frac{1}{3}, -113/27)$, rel max $(-1, -3)$, infl pt $(-\frac{1}{3}, -97/27)$, abs min $(-2, -6)$, abs max $(2, 6)$. (c) rel min $(2, -3)$, rel max $(0, 1)$, infl pt $(1, -1)$. (d) rel min $(3, -26)$, rel max $(-1, 6)$ infl pt $(1, -10)$, abs min $(-4, -75)$, abs max $(6, 55)$. (e) rel min $(1, -1)$ and $(-1, -1)$, rel max $(0, 0)$, infl pt $(\sqrt{3}/3, -5/9)$ and $(-\sqrt{3}/3, -5/9)$. (f) rel min $(\frac{3}{4}, -27/256)$, horiz infl pt $(0,0)$, infl pt $(1/2, -1/16)$, abs max $(2, 8)$. (g) rel min $(3, -165)$ and $(-2, -40]$, rel max $(0, 24)$, infl pts when $x=(1\pm\sqrt{19})/3$. (h) rel min $(1, -4)$, rel max $(-1, 4)$, infl pt $(0, 0)$.

25.3 (a) rel min $(1, 2)$, rel max $(-1, -2)$.
(b) rel min $(-1, -\frac{1}{2})$, rel max $(1, \frac{1}{2})$ infl pts $(0, 0), (\sqrt{3}, \sqrt{3}/4)$ and $(-\sqrt{3}, -\sqrt{3}/4$.
(c) rel min $(0, 0)$, rel max $(-4/5, 3456/3125)$, horiz infl pt $(-2, 0)$, infl pts when $x=(-4\pm\sqrt{6})/5$.
(d) abs min $(-1, 0)$ (derivative does not exist).
(e) abs min $(2, 0)$ and $(0, 0)$ (derivative does not exist), rel max $(1,1)$.

25.4 (a) vert infl pt $(0, 0)$. (b) vert tangent $(0, 0)$ (not an infl pt).
(c) abs min $(0, 0)$. (d) rel min $(-\frac{2}{3}, -\frac{2}{3}\sqrt{3})$. (e) rel max when $x=-\frac{2}{5}$, vert infl pt $(0, 0)$, infl pt when $x=\frac{1}{5}$.
(f) critical value for each nonnegative integer n (not cont or diff for any nonnegative integer n).
(g) same as (f).

25.5   (a) $k = 9$.   (b) any $k < 0$.

25.6 $x = \frac{1}{3}(a + b + c)$.

25.7   (a) $y - 2 = k(x^3 - 3x)$ for any $k > 0$.
   (b) $y - 5 = k(-x^3 + 12x)$ for any $k > 0$.

25.10   (a) 1, 3.   (b) $\sqrt{3}$.

25.12   (a) 4/9.   (b) $2\sqrt{3}/3$.   (c) $\sqrt{3}/3, -\sqrt{3}/3$.   (d) $2 - \frac{2}{3}\sqrt{3}$.
   (e) $2 + \frac{1}{3}\sqrt{3}, 2 - \frac{1}{3}\sqrt{3}$.   (f) $\sqrt{2}$.

SECTION 26

26.1   (a) (1, 2).   (b) $(\frac{1}{3}, 11/27)$.   (c) $(-\frac{1}{2}, -2)$.

26.2 $\frac{3}{4}$.

26.5   (a) min when perimeter of square is $48/(4 + \pi)$; max when perimeter of square
   is 0.   (b) min when perimeter of square is $4\sqrt{3}/3$; max when perimeter of square
   is 12.   (c) min when perimeter of equilateral triangle is $216/[\sqrt{3}(2 + \sqrt{2})^2 + 18]$;
   max when perimeter of equilateral triangle is 12.   (d) min when perimeter of
   isosceles triangle is $[12(2 + \sqrt{2})^2]/[2\pi + (2 + \sqrt{2})^2]$; max when perimeter of
   isosceles triangle is 0.

26.6   min when perimeter of square is 6 and of triangle is $108/(9 + \pi\sqrt{3})$; max
   when perimeter of square is 6 and of triangle is 0.

26.8   113

26.9   (a) $33/2, -63, 384$.   (b) $t = 2, s = 3$.
   (c) slowing down over $\frac{1}{4} < t < 2$ and speeding up over $2 < t < \infty$.   (d) 1.   (e) 0.

26.10   49 ft, 3 sec, 56 ft/sec.

26.11   (a) $1600\sqrt{3}$.   (b) 5.   (c) 400.

26.12 $d(t) = \frac{1}{2}(25t^2 - 72t + 144)^{1/2}, d'(t) = \dfrac{\frac{1}{4}(50t - 72)}{(25t^2 - 72t + 144)^{1/2}}, t = 36/25$.

26.13   600 cubic in./sec.

26.14   (a) $(\pi t^2)^{-1/3}, \frac{1}{4}\pi^{-1/3}$ (b) $-\frac{1}{4}\pi$.

26.15 $(1 + \sqrt[3]{9})^{3/2}$.

26.16   (a) $\sqrt{2}/2$.   (b) $\sqrt{3}$.

26.17 $\sqrt{700/3}$.

SECTION 27

27.1 (a) 0. (b) $\frac{1}{2}$. (c) $2/(2 + \pi)$.

27.2 (a) $\dfrac{3x[\sin(x^2 + 1)^{1/2}]^2 \cos(x^2 + 1)^{1/2}}{(x^2 + 1)^{1/2}}$.

(b) $\dfrac{-(x + 1) \sin(x^2 + 2x)}{[\cos(x^2 + 2x)]^{1/2}}$.

27.3 (a) f': $\sin x + x \cos x$, f'': $2 \cos x - x \sin x$.

(b) f': $\dfrac{-x \sin x - \cos x}{x^2}$, f'': $\dfrac{-x^2 \cos x + 2x \sin x + 2 \cos x}{x^3}$.

(c) f': $2 \sin x \cos x$, f'': $2 \cos^2 x - 2 \sin^2 x$.

(d) f': $2x \cos(x^2)$, f'': $-4x^2 \sin(x^2) + 2 \cos(x^2)$.

(e) $f'(x) = \cos x$ for $0 < x < \pi$;
$f'(x) = -\cos x$ for $\pi < x < 2\pi$;
$f''(x) = -|\sin x|$ for all $x \ne \pi$.

(f) $f'(x) = \sec^2 x$ for $0 < x < \pi/2$.;
$f'(x) = -\sec^2 x$ for $-\pi/2 < x < 0$;
$f''(x) = |2 \tan x \sec^2 x|$ for all $x \ne 0$.

(g) f': $1 - 2 \cos x$, f'': $2 \sin x$.

(h) f': $1 - \sin x$, f'': $-\cos x$.

(i) f': $-4(\sin 4x + \sin 2x)$; f'': $-8(2 \cos 4x + \cos 2x)$.

27.4 (a) rel min $(0, 0)$, rel max when $-x = \tan x$.

(b) rel min when $-x = \cot x$.

(c) rel min when $x = k\pi$ for any integer k, rel max when $x = (2k + 1)(\pi/2)$ for any integer k.

(d) rel min $(0, 0)$, rel max when $x = \pm(\pi/2)^{1/2}$.

(e) rel max when $x = \pi/2$, $3\pi/2$; abs min when $x = \pi$ (derivative does not exist).

(f) abs min $(0, 0)$ (derivative does not exist).

(g) abs min when $x = -\pi$, abs min when $x = \pi$, rel min when $x = \pi/3$, rel min when $x = -\pi/3$.

(h) horizontal tangents when $x = (\pi/2) + 2\pi k$ for any integer k.

(i) rel max $(0, 3)$, rel min when $x = \pi/3$, $-\pi/3$.

27.5 (a) 1. (b) $2/\pi$. (c) slope does not exist.

27.6 (a) cont and diff at $x = 1/\pi$, not cont or diff at $x = 0$.

(b) cont and diff at $x = 1/\pi$, cont but not diff at $x = 0$.

(c) cont and diff at $x = 1/\pi$ and at $x = 0$.

(d) cont and diff at $x = 1/\pi$ and at $x = 0$.

(e) cont and diff at $x = 1/\pi$ and at $x = 0$.

27.7 (a) f' is cont and diff at $x = 1/\pi$; f' is not cont or diff at $x = 0$.

(b) f' is cont and diff at $x = 1/\pi$; f' is not cont or diff at $x = 0$.

(c) f' is cont and diff at $x = 1/\pi$;
f' is not cont or diff at $x = 0$.

(d) f' is cont and diff at $x = 1/\pi$;
f' is cont but not diff at $x = 0$.
(e) f' is cont and diff at $x = 1/\pi$;
f' is cont and diff at $x = 0$.

27.8 (a) $-2/\sqrt{3}, 2/\sqrt{3}, 0$. (b) $\dfrac{\cos \sqrt{\pi/4}}{\sqrt{\pi}}$.

27.10 (a) $f'(x) = \sec^2 x$. (b) 2.

27.12 (b) $\sec x \tan x$.
(c) domain $1 \le x < \infty$, range $0 \le y < \pi/2$.
(d) $1/x(x^2 - 1)^{1/2}$ for $1 \le x < \infty$.

27.13 (b) $\sec x \tan x$. (c) domain $-\infty < x \le -1$, range $\pi < y \le 3\pi/2$.
(d) $1/x(x^2 + 1)^{1/2}$ for $-\infty < x \le -1$.

27.14 (a) $\arccos(2/\pi)$. (b) $\arcsin(2/\pi)$. (c) $(1/\pi)(\pi^2 - 4)^{1/2}$. (d) $[(4 - \pi)/\pi]^{1/2}$

SECTION 28

28.1 (a) rel max $(1, 1/e)$.
(b) rel max $(\sqrt{2}/2, \sqrt{2}/2\sqrt{e})$, rel min $(-\sqrt{2}/2, -\sqrt{2}/2\sqrt{e})$.
(c) rel max $(2, 4/e^2)$, rel min $(0, 0)$.
(d) rel max $(-1, -e)$. (e) rel min $(0, 1)$.
(f) rel max $(0, 1)$. (g) rel max $(0, -1)$.

28.2 (b) $\sqrt{6}/3, -\sqrt{6}/3$.

28.4 rel min $(0, 0)$, rel max for $x = 2/\log a$.

28.8 (a) rel min $(1, 1)$.
(b) critical points for all x such that $x = \log(\cos x)$, in particular, rel min $(0, 1)$.
(c) critical points for all x such that $\tan x = -1$, in particular,

$$\text{rel max}\left(-\pi/4, \frac{\sqrt{2}}{2} e^{\pi/4}\right) \text{ and rel min}\left(3\pi/4, \frac{-\sqrt{2}}{2} e^{-(3\pi/4)}\right).$$

(d) critical points for all x such that $x = \log(\sin x)$.

28.9 (a) -1. (b) -1. (c) -1. (d) $-\pi$.

28.10 (a) $\dfrac{(1/x)\sin x - (\log x)(\cos x)}{\sin^2 x}$.

(b) $2x \sin e^{2x} + 2x^2 e^{2x} \cos e^{2x}$.
(c) $e^x \sin(5e^x) + 5e^{2x} \cos(5e^x)$. (d) $-\tan x$.
(e) $-e^x \tan e^x$. (f) $e^x(\cos x - \sin x)$. (g) $-(1 - x^2)^{-1/2}$.
(h) $\dfrac{(\cot(\arctan e^{x^2})^{1/2})(xe^{x^2})}{(1 + e^{2x^2})(\arctan e^{x^2})^{1/2}}$.

(i) $\dfrac{(\cos (\arctan x^2)^{1/2})(x)}{(1 + x^4)(\arctan x^2)^{1/2}}$.

(j) $(\cos x - x \sin x)\, e^{x \cos x}$.

(k) $(1/x)\cos(\log x)e^{\sin(\log x)}$.

(l) $[\cos x \cot x - \sin x \log(\sin x)](\sin x)^{\cos x}$.

(m) $[e^x \log(\sin x) + e^x \cot x](\sin x)^{e^x}$.

(n) $\left[2x \log(x^2 + x) + \dfrac{x(2x + 1)}{x + 1} \right](x^2 + x)^{x^2}$.

(o) $x^{x^2}(x + 2x \log x)$.

(p) $x^x x^{x^x}(\log^2 x + \log x + 1/x)$.

(q) $x^{x^2}x^{x^{x^2}}(2x \log^2 x + x \log x + 1/x)$.

28.11   (a) $1, 32(1 + 2 \log 2)$.   (b) 0.   (c) $2/\pi$.

28.12   (a) $f': -y/(xy + 2x)$.   (b) $-e/3$.

28.13   (a) $\log(e - 1)$.   (b) $e - 1$.   (c) $1 - \log_{10}(\log_e 10)$.

28.14   (a) $(0, 0)$ (derivative does not exist).
(b) cont, not diff (left-hand derivative is 1, right-hand derivative is 0).

28.15 $a = e^{1/e}$, point of tangency is (e, e).

SECTION 29

29.1   (a) 2.   (b) 0.   (c) ∞.   (d) no limit.   (e) 1.   (f) 1.   (g) 1.   (h) 0.

29.2   (a) ∞.   (b) ∞.   (c) 0.   (d) 1.   (e) ∞.   (f) 0.   (g) ∞.   (h) 1.   (i) 1.

29.3   (a) -1.   (b) -2.   (c) 1.   (d) -4.   (e) 0.   (f) ∞.

29.7   (a) not cont or diff.   (b) cont, not diff.   (c) cont and diff.   (d) cont and diff.

29.8   (a) not cont or diff (not defined).   (b) not cont or diff.   (c) cont, not diff.
(d) cont and diff.

SECTION 30

Note that in the following, where antiderivatives are given as answers, the constant C has been excluded for the sake of brevity.   You should, however, include it as part of your answer.

30.1   (a) antiderivative of $e^{f(x)}f'(x) = e^{f(x)}$.
(b) $e^{5x}, -e^{-x}, \tfrac{1}{2}e^{x^2}, e^{\sin x}$.

30.2   (a) antiderivative of $(\sin f(x))f'(x) = - \cos f(x)$.
(b) $-\tfrac{1}{3} \cos x^3, -\tfrac{1}{5} \cos 5x, -\cos (\log x)$.

30.3   (a) antiderivative of $f'(x)/(1 + f^2(x)) = \arctan f(x)$.
  (b) $\frac{1}{2}\arctan 2x$, $\arctan x^2$, $\frac{1}{3}\arctan(x/3)$.

30.4   (a) $\frac{1}{5}x^5 - x^3 + x^2 + x$.   (b) $\frac{1}{3}(x^2 + 1)^{3/2}$.
  (c) $\frac{1}{18}(x^3 + 1)^6$.   (d) $-\frac{1}{5}\cos^5 x$.
  (e) $\log(x^3 + x^2 + 1)$.   (f) $-(x^3 + x^2 + 1)^{-1}$.
  (g) $-\frac{1}{2}(x^3 + x^2 + 1)^{-2}$ (h) $-\log(\cos x)$.   (i) $\tan x$.   (j) $\frac{1}{4}\sin(2x^2)$.
  (k) $\frac{1}{2}\arcsin 2x$.   (l) $\arcsin(x/2)$.   (m) $\frac{1}{2}\log^2 x$.   (n) $\frac{1}{3}\arctan^2 x$.   (o) $e^{\sqrt{x}}$.
  (p) $(-2/9)(1 - 3x^3)^{1/2}$.   (q) $(2/7)x^{7/2} - (10/3)x^{3/2} + 2x^{1/2}$.

30.5   (a) $f(x) = x^3 + x^2 + x + 3$.
  (b) $f(x) = \frac{1}{3}x^3 + \frac{1}{2}x^2 + 2x + (11/6)$.

30.6   (a) $f(x) = \frac{1}{3}ax^3 - ax + 2$ for any real number $a < 0$.
  (b) $f(x) = \frac{1}{5}ax^5 - \frac{1}{3}ax^3 + 2$ for any real number $a < 0$.
  (c) $f(x) = \frac{1}{3}a(x^2 + 2)^3 + 5 - (8/3)a$ for any real number $a \neq 0$.
  (d) $f(x) = \frac{1}{4}ax^4 - ax^3 + 1$ for any real number $a < 0$.

30.7   (a) $s(t) = \frac{1}{4}t^4 - \frac{1}{3}t^3 - \frac{1}{2}t^2$.
  (b) $s(t) = \frac{1}{4}t^4 - \frac{1}{3}t^3 - \frac{1}{2}t^2 + t$.
  (c) in part (a), changes direction when $t = \frac{1}{2}[1 + \sqrt{5}]$. At the point
    $s(\frac{1}{2}[1 + \sqrt{5}])$; in part (b), no change of direction occurs.

30.8   (a) $h(t) = -16t^2 + 60t + 100$.
  (b) $h(t) = -16t^2 - 60t + 100$.
  (c) In part (a), $t = 5$, speed 100; in part (b), $t = 5/4$, speed 100.

30.9 $t = 5$, $K = 8$.

SECTION 31

31.1   (a) 15.   (b) $b^2 - a^2$.   (c) 2.   (d) 1.

31.3   (a) $5(b^2 - a^2)$.   (b) 36.   (c) 3.

SECTION 32

32.2   (a) 44/3.   (b) 5/64.   (c) $\frac{1}{3}\log 2$.   (d) $\frac{1}{4}$.   (e) $\frac{1}{2}(e^2 - 1)$.   (f) $\frac{1}{2}(e - 1)$.
  (g) 2.   (h) $\frac{1}{4}\sin(2\pi^2)$.   (i) 1.   (j) $\frac{1}{3}(2\sqrt{2} - 1)$.   (k) 13/24.

32.3   (a) 1.   (b) 2.

32.4   (a) $(5/4)e$.   (b) e.   (c) $\frac{1}{4}\pi^3 \log \pi$.

32.5   (a) $\int_0^z \sin x \, dx$.   (b) $\int_0^z (x + \cos x) \, dx$.   (c) $\int_0^z \sec^2 x \, dx$.

SECTION 33

33.1 $n/(n + 1)$, area approaches 1.

33.2   (a) $\log n$, area approaches ∞.
  (b) $(n - 1)/n$, area approaches 1.

33.3 (a) rel max (0, 1), infl pts ($\sqrt{3}/3$, 3/4) and ($-\sqrt{3}/3$, 3/4).

(b) 2 arctan(n), area approaches π.

33.4 (a) 1. (b) 4. (c) 1.

33.5 (a) $(5/4)e$. (b) e.

33.6 (a) 15/4. (b) 17/4.

33.7 (a) 0. (b) 4.

33.8 $\dfrac{1}{\log 2} - \dfrac{1}{\log 4}$.

33.9 (b) 1, $(4\sqrt{2}/3) - 1$, $(2/3)2^{3/4} + 1 - (4\sqrt{2}/3)$.

33.10 (a) 8/3. (b) 16/3.

33.11 (b) $\sqrt{2} - 1, 2\sqrt{2}, \sqrt{2} + 1$.

33.12 (b) 4.

33.13 (b) 4.

33.14 (a) 0. (c) $(1 - e^{-n^2})$. (d) the area approaches 1.

33.15 (a) $2\sqrt{3}/3$ (b) $e - 1$. (c) $\pi/4$ or $3\pi/4$.

SECTION 34

34.1 (a) $129\pi/7$. (b) $66\pi/5$.

34.2 (a) $8\pi/3, 4\pi/3$. (b) $4\pi/3, 2\pi\sqrt{3}$.

34.3 (a) $2\pi/3$. (b) $92\pi/15$.

34.4 (a) $(25/16)e^2\pi$. (b) πe^2.

34.5 (a) $\pi(1 - 1/b)$. (b) volume approaches π.

(c) $(\pi/3)(1 - 1/b^3)$, volume approaches $\dfrac{\pi}{3}$.

34.6 (a) $1 - e^{-b}$. (b) area approaches 1. (c) $(\pi/2)(1 - e^{-2b})$.
(d) volume approaches $\pi/2$.

34.7 (a) $(\pi/6)(e^2 - 1)$. (b) $(\pi/10)(e^2 - 1)$.

34.8 $\frac{1}{2}\pi^2$.

34.9 $(\pi/2)(1 - e^{-2})$.

34.10 (a) $2\pi - \frac{1}{4}\pi^2$. (b) $\pi \int_0^1 \arccos^2 y \, dy$.

34.11   (a) $(\pi/2)(e^2 - 4e + 5)$.
  (b) $\pi(e - 1) - \pi \int_1^e \log^2 y \, dy$.

34.12   (a) $41\pi/105, 1849\pi/105, 71\pi/105$.
  (b) $13\pi/30, 67\pi/30$.   (c) $31\pi/70, 81\pi/70$.
  (d) $22\pi/5, 53\pi/30$.

34.16   (a) $8\pi/3, 16\pi/3$.   (b) $\frac{2}{3}\pi(1 - e^{-1})$.   (c) $2\pi^2$.   (d) $4\pi^2$.

34.17   (a) $66\pi/5$.   (b) $2\pi \int_0^1 x(e^x - 1) \, dx = \pi$.   (c) $13\pi/30, 67\pi/30$.
  (d) $22\pi/5, 53\pi/30$.

SECTION 35

Note that in most of the following, the answers represent the number of foot-pounds required to do the given work.

35.1   (a) 4500π.   (b) 7125π.

35.2   (a) 750π.   (b) $(4375)/8)\pi$.

35.3   (a) 6250π.   (b) $(1130/3)\pi$.   (c) $(5120/3)\pi$.

35.4   (a) $(10/\pi)^{1/2}$ft.   (b) $(500/3\pi)^{1/2}$ ft.

35.5   (a) 1500π.   (b) $(8000/9)\pi$.

35.6   (a) $22,000\pi$.   (b) $40,000\pi$.

35.7   (a) $(62.5)\pi r^4/4$.   (b) $(62.5)s^2h^2/4$.   (c) $(62.5)(7/12)\pi s^2 h^2$.

35.8   (a) $(245/3)$in.-lb.   (b) 25 in.-lb.

35.9 $92,250$.

35.10   (a) $\frac{1}{4}kAB$.   (b) $kAB(1 - 1/d)$.   (c) kAB.

35.12 $(7,812,500/3)$lb.

35.13 $(12,500,000/3)$lb.

35.14 $312,500$ lb.

35.15 $62.5 \int_{-1}^1 (6 - y)(1 - y^2)^{1/2} \, dy$ lb. or $62.5 \int_0^2 (6 - y)(2y - y^2)^{1/2} \, dy$ lb.

SECTION 36

Note that in Problems 36.1 through 36.42, add a constant C to the given answer.

36.1 $x + \sin^2 x$.

36.2 $\frac{1}{2} \sin x^2$.

36.3 $-\frac{1}{2} \cos x^2$.

36.4 $\tan(x/2)$ or $\csc x - \cot x$.

36.5 $-2(1 - \sin x)^{1/2}$.

36.6 $\frac{1}{2}e^{x^2}$.

36.7 $\frac{1}{3}e^{x^3}$.

36.8 $e^x \sin x$.

36.9 $\log(e^x/[1 + e^x])$.

36.10 $\log(e^x + e^{-x})$.

36.11 e^{e^x}.

36.12 $(1/\log 2)2^x$.

36.13 $(1/\log a)a^x$.

36.14 $e^x/(1 + x)$.

36.15 $(x^2/2)(\log x - \frac{1}{2})$.

36.16 $(x^2/2)(\log^2 x - \log x + \frac{1}{2})$.

36.17 $(x^2/2)(\log x^2 - 1)$.

36.18 $(x^3/3)(\log x^2 - \frac{2}{3})$.

36.19 $-(1/x)(\log x + 1)$.

36.20 $(x/2)[\sin(\log x) - \cos(\log x)]$.

36.21 $\frac{1}{2}(x^2 \arctan x + \arctan x - x)$.

36.22 $\frac{1}{2}(x^2 \arctan x^2 - \frac{1}{2}\log(1 + x^4))$.

36.23 $\frac{1}{4}((2x^2 - 1)\arcsin x + x(1 - x^2)^{1/2})$.

36.24 $e^x \arcsin e^x + (1 - e^{2x})^{1/2}$.

36.25 $\frac{2}{5} e^x(\cos^2 x + 2 \sin x \cos x + 2)$.

36.26 $\frac{1}{2}\log(x^2 + 1) - \log(x + 1)$.

36.27 $-\log(x^2 + 1)\arctan + x + 2\log(x - 1)$.

36.28 $\frac{2}{3}(x + 2)^{3/2} - 4(x + 2)^{1/2}$.

36.29 $\frac{1}{2}\{x[x^2 - 1]^{1/2} - \log(x + [x^2 - 1]^{1/2})\}$.

36.30 $\frac{1}{2}[(x - 5)(-x^2 + 10x - 24)^{1/2} + \arcsin(x - 5)]$.

36.31 $\arctan(x + 2)$.

36.32 $\arcsin(x - 1)$.

36.33 $-(1 + x^2)^{1/2} + \frac{1}{3}(1 + x^2)(1 + x^2)^{1/2}$.

36.34 $-\frac{1}{3}(1 - x^2)^{3/2} + \frac{1}{5}(1 - x^2)^{5/2}$.

36.35 $\frac{1}{2}\{x[1+x^2]^{1/2}-\log(x+[1+x^2]^{1/2})\}.$

36.36 $\frac{1}{2}\left(\frac{x}{1+x^2}+\arctan x\right).$

36.37 $\frac{1}{2}\{x[x^2-4]^{1/2}-4\log(x+[x^2-4]^{1/2})\}.$

36.38 $\frac{1}{2}\{x[x^2+4]^{1/2}+4\log(x+[x^2+4]^{1/2})\}.$

36.39 $\arcsin(x/5).$

36.40 $\frac{1}{2}\arcsin x^2.$

36.41 $-(1-x^2)^{1/2}+\arcsin x.$

36.42 $(1-x^2)^{1/2}+2\arcsin\left(\frac{1+x}{2}\right)^{1/2}.$

36.43 $\log\sqrt{2}.$

36.44 $9/\log 10.$

36.45 $2(1-\log 2).$

36.46 $15-2\log 4.$

36.47 $16\sqrt{2}/3.$

36.48 $\pi/6.$

36.49 $25\pi/4.$

36.50 $\frac{1}{3}\arctan 3.$

36.51   If we use the left endpoint of each interval, the following approximations correspond to the given values of n. $n=2, 2.414; n=3, 2.66; n=4, 2.78; n=5, 2.87; n=6, 3.03.$ If we use the right endpoint of each interval, the approximations are $n=2, 4.414; n=3, 4.00; n=4, 3.78; n=5, 3.67; n=6, 3.59.$

36.52   1.61 (correct to two decimal places).

36.53   (a) $-x(1+x^2)^{-1/2}+\log(x+[1+x^2]^{1/2}).$

(b) $-\frac{1}{2}\log\left(\frac{1+(1+x^2)^{1/2}}{-1+(1+x^2)^{1/2}}\right).$

36.54   (a) $\frac{3\pi}{2}-\log(\sqrt{2}+1)-3\arcsin(\frac{2}{3})^{1/2}.$

(b) $\pi(\frac{2}{3}+2\sqrt{3}-8\sqrt{2}/3).$ (c) $8\pi/3.$

36.55   (a) $\frac{1}{2}(2\sqrt{3}+\log(\sqrt{3}+2)).$ (b) $19\pi/6.$

SECTION 37

37.1   3.28.

37.2   0.48.

37.7   0.89.

37.4   2.27.

37.5   0.75.

37.6   2.98.

37.7 0.67π.

37.8 0.67π.

SECTION 38

38.2   (a) $2(1 - e^{-1})$.   (b) 1.

38.3   (a) $3, 3\pi/5$.   (b) ∞, ∞.   (c) $\infty, 3\pi$.   (d) $3, \infty$.   (e) $4/9, \pi/4$.   (f) $1, \pi/4$.

38.4   (a) 1.   (b) $\pi/2$.   (c) ∞.   (d) e^{-1}.   (e) -1.   (f) $-\frac{1}{4}$.   (g) $-\infty$.
(h) 6.   (i) ∞.   (j) ∞.   (k) π.   (l) does not exist.   (m) ∞.   (n) $2(1 - \cos 1)$.
(o). ∞.   (p) $\frac{1}{2}$.

38.5   (a) exists (since $|(\sin x)/x^2| \le 1/x^2$ for all $x \ge 1$).
(b) does not exist (since $e^x/x \ge 1/x$ for all $x \ge 1$).
(c) exists (since $e^{-x^2} \le e^{-x}$ ror all $x \ge 0$).

(d) exists $\left(\text{since } \dfrac{x}{x^3 + 1} = \dfrac{1}{x^2 + \dfrac{1}{x}} < \dfrac{1}{x^2} \text{ for all } x \ge 1 \right)$.

SECTION 39

39.1   (a) $(13\sqrt{13} - 8)/27, (8/27)(10\sqrt{10} - 1)$.
(b) $(\sqrt{5}/2) + \frac{1}{4}\log(\sqrt{5} + 2)$.
(c) $\int_0^1 (4x^4 + 8x^2 + 1)^{1/2}\, dx$.   (d) 2π.
(e) $\int_0^{1/2} [1 + 16x^2(1 - 4x^2)^{-1}]^{1/2}\, dx$.
(f) $\sqrt{26} + \log(\frac{1}{5}[\sqrt{26} - 1]) - \sqrt{2} - \log(\sqrt{2} - 1)$.
(g) $\log\sqrt{3}$.   (h) $\log\sqrt{3}$.   (i) $(-3/4) + \log 7$.   (j) $e - e^{-1}$.

39.2   (a) $n = 4, 4.24; n = 8, 4.31$.   (b) $n = 2, 9.68; n = 4, 8.88$.
(c) $n = 4, 4.25; n = 8, 4.17$.   (d) $n = 2, 2.27; n = 4, 2.24$.
(e) $n = 2, 1.10; n = 4, 1.10$.

39.3 (a) $8\sqrt{65}\,\pi$.   (b) $(\pi/9)(2\sqrt{2}-1)$.
(c) $\pi(2\sqrt{2}+\log(1+\sqrt{2})-\log(-1+\sqrt{2}))$.
(d) $\pi(-2\sqrt{2}+\log(-1+\sqrt{2})-\log(1+\sqrt{2}))$.

(e) $2\pi \int_1^{2\sqrt{2}} (x^2+1)^{1/2}\left(\dfrac{\log x}{x}\right) dx$.

(f) $\pi\left(\sqrt{2}-\dfrac{\sqrt{5}}{4}+\log(\sqrt{2}+1)-\log\left(\dfrac{\sqrt{5}}{2}+\dfrac{1}{2}\right)\right)$.

(g) $\dfrac{\pi}{2} \int_0^1 (e^x+e^{-x})(e^{2x}+e^{-2x}+2)^{1/2}\,dx$.

39.4 (a) $\pi\{\sqrt{2}-(1/e)(1+e^{-2})^{1/2}+\log(1+\sqrt{2})-\log([1/e]+[1+e^{-2})^{1/2}])\}$.

(b) $2\pi \int_0^1 x\left(\dfrac{2-x^2}{1-x^2}\right)^{1/2} dx$.   (c) 2π.

39.5 4π. (b) $2\pi\left[1+\dfrac{\sqrt{3}}{12}\log\left(\dfrac{2+\sqrt{3}}{2-\sqrt{3}}\right)\right]$.

39.6 (a) $64\pi(1+(2/3)(2\sqrt{2}-1))$.
(b) $\pi(16+24\sqrt{2}-4\log(3+2\sqrt{2}))$.

39.7 (a) $n=2,\ 1.72\pi;\ n=4,\ 1.76\pi$.
(b) $n=2,\ 2.13\pi;\ n=4,\ 1.39\pi$.

SECTION 40

40.1 (a) $y=2x-4$.   (b) $y=x^2-2x+1$.
(c) $y=x^4+(1/x^2),\ 0<x\le 2$.
(d) $y=-x(x+1)^{1/2},\ -1\le x\le 0$.
(e) $y=x(x+1)^{1/2},\ -1\le x\le 0$. (f) not possible.
(g) $y=\pm(x^2-1)^{1/2},\ 1\le x<\infty$. (h) not possible.

40.2 (a) at $(1,0)$, $y=0$; at $(2,1)$, $y=2x-3$; at $(0,1)$, $y=-2x+1$.
(b) at $(0,0)$, $y=x$ and $y=-x$; at $(-1,0)$, $x=-1$; at $(3,6)$, $y=(11/4)x-(9/4)$;
at $(-2/3,\ 2\sqrt{3}/9)$, $y=(2\sqrt{3})/9$. (c) at $(4,0)$, $x=4$; at $(0,3)$, $y=3$;
at $(2,3\sqrt{3}/2)$, $y=(-\sqrt{3}/4)x+2\sqrt{3}$; at $(2,-3\sqrt{3}/2)$, $y=(\sqrt{3}/4)x-2\sqrt{3}$.
(d) at $(1,0)$, $y=x-1$; at $(-e^\pi,0)$, $y=x+e^\pi$; at $(0,e^{\pi/2})$, $y=-x+e^{\pi/2}$.

40.3 (a) $(-2/3,2\sqrt{3}/9)$ and $(-2/3,-2\sqrt{3}/9)$. (b) $(-3,1)$. (c) $(-e^{3\pi/4}(\sqrt{2}/2),$
$e^{3\pi/4}(\sqrt{2}/2))$ and $(e^{7\pi/4}(\sqrt{2}/2),\ -e^{7\pi/4}(\sqrt{2}/2))$.

40.4 (a) $\sqrt{17}+\frac{1}{4}\log(4+\sqrt{17})$.
(b) $(8/27)(10^{3/2}-(13/4)^{3/2})$.
(c) same as part (b).   (d) π.   (e) $3\sqrt{2}/4$.

(f) $\sqrt{2}-\dfrac{\sqrt{5}}{2}+\log\left(\dfrac{2+\sqrt{5}}{1+\sqrt{2}}\right)$.

40.5 (b) 116/15.

40.6 (a) 4. (b) 8/3. (c) 4/15. (d) $\pi/2$. (e) $\frac{1}{2}$. (f) $\pi ab/4$.

40.8 (a) $56\pi/15$. (b) $117\pi/4$. (c) $\pi/12$. (d) 8π.

40.10 (a) $(\pi/2)(7\sqrt{2} + 3\log(1 + \sqrt{2}))$.
 (b) $2\pi \int_1^2 (t^3 - t)(9t^4 - 2t^2 + 1)^{1/2}\, dt$.
 (c) $(20\sqrt{5}\,\pi/3)(8\sqrt{2} - 1)$. (d) 4π. (e) $\sqrt{2}\pi$. (f) $(2\sqrt{2}\pi/5)(2e^\pi + 1)$.

SECTION 41

41.2 (a) $(\frac{3}{2}, 3\sqrt{3}/2)$. (b) $(3\sqrt{3}, -3)$. (c) $(-2, 0)$. (d) $(-\frac{3}{2}, -3\sqrt{3}/2)$.
 (e) $(0, 4)$. (f) $(-\sqrt{3}/2, \frac{1}{2})$. (g) $(-\sqrt{2}, -\sqrt{2})$. (h) $(-\sqrt{2}, -\sqrt{2})$.
 (i) $(-\sqrt{2}, 2)$.

41.3 (a) $(3, \pi/2)$. (b) $(-1, 0)$. (c) $(5, \arccos(-\frac{4}{5}))$. (d) $(13, \arcsin 12/13)$.
 (e) $(0, \theta)$, where θ is any real number. (f) $(2, \pi/4)$.

41.5 (a) $0, \sqrt{3}$, slope does not exist (vertical tangent).

 (b) $0, \dfrac{3\sqrt{3} + \pi}{3 - \sqrt{3}\pi}, \pi$. (c) $0, -1$.

 (d) $-\sqrt{3}, 0, \sqrt{3}, -\sqrt{3}, 0$.

 (e) $1, \dfrac{\sin(\log 2) + \cos(\log 2)}{\cos(\log 2) - \sin(\log 2)}, 1$

41.7 (a) $\sqrt{3}/3, 1$. (b) $\pi/3, \pi, \theta$. (c) $-\sqrt{3}, 1$.
 (d) at $(1, \pi/6)$, $\tan\beta$ does not exist ($\beta = \pi/2$); at $(\sqrt{2}/2, \pi/4)$, $\tan\beta = -\frac{1}{3}$.
 (e) $\tan\beta = 1$ at every point on the graph.

41.8 (a) $\pi/2$.
 (b) $\frac{3}{2}[2\pi(1 + 4\pi^2)^{1/2} + \log(2\pi + [1 + 4\pi^2]^{1/2})]$
 (c) $\sqrt{2}$. (d) $\sqrt{2}/2$.
 (e) $\int_0^{\pi/4} (1 + 3\cos^2 2\theta)^{1/2}\, d\theta$.

41.9 (a) $(\pi/24) + (\sqrt{3}/16)$.
 (b) $\pi^3/96, 13\pi^3/6$. (c) $\pi/12$. (d) $\pi/12$. (e) $\pi/8$.
 (f) $(3\pi - 8)/8$. (g) $(3\pi + 8)/8$. (h) $\frac{3}{4}$. (i) $\sqrt{3}/2$. (j) $3\pi/32$.

41.11 (a) $\pi^2/8 + 5\pi/12$. (b) $\pi/6$.
 (c) $8\pi \int_\pi^{\pi/2} \theta^2 \sin^2 \theta(\cos\theta - \theta\sin\theta)\, d\theta$. (d) $5\pi/2$.

41.13 (a) $\pi + (\pi^2/2)$. (b) π. (c) $8\pi \int_{\pi/2}^\pi \theta(1 + \theta^2)^{1/2} \sin\theta\, d\theta$.
 (d) $[4\pi(8 - \sqrt{2})]/5$. (e) $[2\sqrt{2}\pi e^\pi(e^\pi - 2)]/5$.

INDEX